AGGREGATES:

Sand, Gravel and Crushed Rock Aggregates for Construction Purposes

edited by

L. Collis† & R. A. Fox*
†Messrs Sandberg, 40 Grosvenor Gardens, SW1W OLB
* RMC (UK) Ltd, RMC House, 53–55 High Street, Feltham, Middx TW13 4HA

1985
Published by
The Geological Society

Published by

The Geological Society
Burlington House,
Piccadilly,
London W1V 0JU

First published 1985

British Library Cataloguing in Publication Data

Aggregates: sand, gravel and crushed rock
aggregates for construction purposes.—
(Engineering geology special publications,
ISSN 0267-9914; 1)
1. Aggregates (Building materials)
I. Collis, L. II. Fox, R. A. III. Geological
Society of London
624.1′891 TA455.A/

ISBN 0-903317-28-1
ISBN 0-903317-29-X (paperback)

Printed in The United Kingdom
by The Universities Press, (Belfast) LTD.

Foreword

This book is the Report of a Working Party of the Engineering Group of the Geological Society of London. It is preceded by four earlier Working Party Reports, all published in the Quarterly Journal of Engineering Geology, one of the journals of the Geological Society. The earlier reports on, Logging of Rock Cores for Engineering Purposes, 1970, the Preparation of Maps and Plans in Terms of Engineering Geology, 1972 and the Description of Rock Masses for Engineering Purposes in 1977, were well received at the time of their publication. In the fullness of time they have all become milestones which to a significant extent have changed the courses of engineering geology, soil and rock mechanics, especially in the qualitative aspects of these disciplines, in UK and overseas. At least one British Standard, notably BS 2001 (1981) on Site Investigation, has drawn considerably from these works. The most recent report before this one, Land Surface Evaluation for Engineering Purposes, 1982, is very informative but seen as somewhat parochial by some overseas readers.

This Working Party Report is somewhat different to its distinguished predecessors. For a start, it is considerably longer and had to be published in book form, but so as to make it freely available to the Engineering Group members, the great majority of whom take the Quarterly Journal of Engineering Geology, and to maintain the tradition of venue of publication, it is also published as an extended summary in that journal.

The principal difference lies in its approach of the Report. Hitherto there was not readily available, in one source, a reasonably comprehensive book or document encompassing all the topics logically through from aggregates in their green field condition, to their working life in a finished structure. Such an ambitious concept has meant of course that no particularly in-depth coverage could be attempted, so many references for further study have been given. Inevitably there are errors and omissions and it will not be possible to please all of the readers all the time. It is the wide, practical and comprehensive nature of the report which it is hoped will be of value, use and interest.

The Report has been some time in its preparation, and with the rapid changes occurring in some aspects it is already in places a little dated. This lateness has stemmed, not from any intrinsic difficulty in the subjects or access to literature, but simply because it has been a committee authorship. The committee in common with most in British Scientific and Engineering Institutions, has been composed of hard pressed, press ganged volunteers, who like their many predecessors and no doubt many successors, gave of their time, or their companies' time, as and when they could.

The work has been long and for the most part very enjoyable. All the committee, and indeed everybody who has helped in so many ways, including understanding spouses and neglected families, deserve thanks and praise. The brunt of the early work, until his untimely death, was borne by the first secretary to the committee, Dr. A. Hartley, and of the latter work by the two editors, Richard Fox from a Producer, and Laurence Collis, from a Materials Consulting Practice to whom special thanks. I hope you agree it has been worthwhile.

P. G. Fookes
Chairman of the Working Party
Caterham, September 1985

Working Party Members

Professor P G Fookes (Chairman) (1978–1984)*	Consultant Engineering Geologist Caterham, Surrey
Dr A D Burnett (1978–1981)	Senior Geotechnical Engineer Public Works Department, Hong Kong
L Collis (1978–1984)	Partner Messrs. Sandberg, Consulting, Inspecting and Testing Engineers, London
R A Fox (1978–1984)	Manager Land Search & Exploration Department, Ready Mixed Concrete (United Kingdom) Limited, Feltham, Middlesex
P M Harris (1978–1984)	Principal Scientific Officer British Geological Survey, Keyworth, Nottinghamshire
Dr A Hartley** (1978–1983)	Director of Civil Engineering Studies School of Construction Studies, Leeds Polytechnic, Leeds, Yorkshire
I E Higginbottom (1978–1984)	Chief Geologist Wimpey Laboratories Limited Southall, Middlesex
Dr G Lees (1978–1984)	Head of Department Department of Transportation and Highway Engineering, University of Birmingham, Birmingham
Professor D M Ramsay (1978–1984)	Head of Department Department of Geology, University of Dundee, Dundee
Dr I Sims (1981–1984)	Associate Messrs. Sandberg, Consulting Inspecting and Testing Engineers, London
Dr R G Thurrell (1978–1984)	Head of Programme Planning Group British Geological Survey, Keyworth, Nottinghamshire

* Dates in parenthesis give time served on Committee.

Special Acknowledgement

** Dr A Hartley was Secretary to the Working Party from the time of its formation in early 1978, until his untimely death in December 1983. The members of the Working Party are particularly indebted to him for his considerable enthusiasm, work and support during the preparation of this publication.

R A Fox was Secretary from December 1983 onward.

Acknowledgements

A large number of individuals and organisations were of particular assistance to the Authors, acting as corresponding members to the Working Party, both in the preparation of the various Chapters leading to the publication of the Draft Report for comment at the Engineering Group Regional Meeting at Bangor in September 1981 and afterwards.

The Working Party is particularly grateful to the following—R B Adamson, D J Ayres, W H Baxter Ltd., N E Beningfield, K M Brooke, W R Dearman, M S Eglinton, W J French, Holmpress Ltd., J R Hoskings, J P R Jordan, M F Maggs, J M Morrison, J Newbery, D J Pain, R D Pass, K R Peattie, D C Pike, A B Poole, A Roeder, G F Salt, A C E Sandberg, J M P Soper, F S Strongman, P Vaughan. If other supporting individuals have been omitted in error it is regretted, but our thanks to them all are recorded.

Grateful appreciation is also given for the clerical and other similar assistance provided by Leeds Polytechnic, Ready Mixed Concrete (UK) Limited, and Messrs. Sandberg. A special thanks is extended to the Directors of Wimpey Laboratories Limited for the free use of the Company's word processing equipment and for the devotion of its patient and most efficient Operator, Mrs Lina Burt.

After the completion of the Draft Report a number of Organisations agreed to comment on, and appraise, the document as appropriate, and the Working Party is most grateful for their contributions. They are—American Society for Testing & Materials, Asphalt & Coated Macadam Association, Association of Consulting Engineers, Association of Professional Geological Scientists (USA), British Quarrying & Slag Federation (now part of British Aggregate Construction Materials Industries), British Readymixed Concrete Association, Cement and Concrete Association, Aerated Concrete Producers Association, Concrete Society, County Surveyors Society, Department of Transport, Federation of Civil Engineering Contractors, Institution of Asphalt Technology, Institution of Civil Engineers, Institute of Concrete Technology, Institute of Geological Sciences (now British Geological Survey), Institution of Geologists, Institution of Highways Transportation, Institution of Mining Engineers, Institution of Mining and Metallurgy, Institution of Structural Engineers, Institute of Quarrying, National Crushed Stone Association, Refined Bitumen Association Limited, Society of Chemical Industry, Royal Institute of British Architects, Sand and Gravel Association Ltd., and Transport and Road Research Laboratories.

There were also a number of written contributions received from individuals who attended the Bangor Conference (1981) and the Working Party appreciate their assistance.

Units; Codes and Standards

The units adopted in this Report are those employed in the current versions of the various Standards or Codes of Practice. In general these are metric but Imperial units are still used in some overseas (e.g. North American) standards or in older British Standards. In such cases, Imperial units are quoted in this report, accompanied by the nearest metric equivalent or by the rationalised metric equivalent where this is well established.

Numerous revisions to British and other codes and specifications were under consideration, or were published at various times during the production of this report. The reader should therefore refer to the editions current at the time of reference.

Contents

Chapter 1. Introduction

1.1 Background

Natural sand, gravel and crushed rock aggregates represent a large proportion of the materials used in the construction industry and consideration of potential sources is commonly made during the pre-construction investigation for major projects. Aggregates are defined here as particles of rock, which when brought together in a bound or unbound condition, form part or whole of an engineering or building structure.

Production of aggregates for civil engineering and building construction in the United Kingdom is a major industry. It reached a peak in 1973, but the current (1984) delivered value of aggregate produced annually in Britain is probably in the order of £1000M: further production statistics are given in the Introduction to Chapter 4.

Despite the importance of geological materials in civil engineering design and construction, the discipline of engineering geology has traditionally been concerned more with interaction between the engineering structure and its geological environment than with the performance of rock aggregates in the structure itself. Recently, however, there are signs that this imbalance is being recognised and adjusted in the fields of teaching, practice and research, but an inhibiting factor may well be the profuse but highly dispersed nature of the literature on aggregate materials. There is a notable lack of reference manuals and textbooks, or even broadly based review publications, giving a comprehensive coverage of the subject. The available information is scattered through innumerable Standards, specifications, research reports, professional papers and articles in the commercial press. Consequently, those who are concerned with construction but are not materials specialists, often find difficulty in identifying the literature to guide them in defining and implementing specification requirements.

The Engineering Group of the Geological Society therefore established a Working Party whose aim was to create a reasonably comprehensive document to overcome this difficulty in the construction industry and related fields. The members of the Working Party were drawn from persons known by the committee of the Engineering Group of the Geological Society to have been particularly concerned with construction materials, either as producers, specifiers, users or researchers, to provide a broad range of expertise. Besides a wide general experience, each member has special knowledge reflected in the subject matter of the sections to which he has contributed and, in most cases, in his previous publications.

1.2 Objectives

At the first meeting of the Working Party in 1978 the objectives were defined. They remained little changed during the course of the work and are outlined below.

- to review the influences of the occurrence, mineral composition and geological history on the engineering properties of aggregate materials. These would include both coarse and fine natural or crushed rock aggregates, whether for use in a bound or unbound condition, the main fields of use being in concrete and mortar, road pavements, railway track ballast and biological or drainage filter media. No account would be taken of artificial aggregate materials, such as slag, expanded clay, calcined bauxite or sintered flyash, or of completely synthetic products such as expanded polystyrene beads or additives. Aggregates used for protective purposes (e.g. for rockfill breakwaters, rip-rap or gabions), or as general fill materials, would also be excluded, except for purposes of comparison or explanation. Soils bound by cement or bitumen would not be considered as aggregates for the purposes of this publication.
- to identify the main factors influencing aggregate behaviour in various operational and environmental conditions.
- to identify relevant Standard and other specifications and Codes of Practice and, where appropriate, to summarise and discuss some of the currently accepted limits.
- to explore some of the problems commonly encountered in applying Specifications and Codes. For example: limits for different properties might be incompatible with each other; properties might be specified which would tend to be mutually exclusive for geological or other reasons; or particular specifications might not be appropriate to the available materials or to the local environment. It would not be the intention to propose alternative specifications so much as to assist the reader in the interpretation and implementation of selected portions of the existing Standards and Codes of

Practice, and in the development of specifications where these are not already available.

- to identify areas in which present knowledge is deficient.
- to concentrate primarily on aggregates produced in the United Kingdom, but to refer, whenever possible to aggregates in some other countries where British specifications, design and construction practices are, or are likely to be, followed.
- to consider aggregates essentially in terms of their intrinsic properties as engineering and geological materials. To some extent their engineering performance may be a function of design, but the latter has been seen as a mainly extrinsic factor to which only limited consideration has been given.
- to review the different geological situations in which sources of aggregate materials can occur. However, no attempt would be made to quantify UK national resources, or to forecast trends in supply and demand, already considered at length elsewhere (Verney, 1976). It was decided that it would be beyond the scope of this publication, to deal in any detail with the many commercial and environmental factors which influence the practical usefulness of a source and contribute to a proper evaluation of its reserves.
- to achieve the maximum possible breadth and balance by inviting advice and criticism from a range of individual specialists, professional institutions, learned societies, industrial associations and research bodies.

Although the objectives focus on specific areas within a wide and ill-defined field, it is recognised that difficulties remain. For example, Standards and Codes of Practice are under more-or-less continuous revision throughout the world, with the rapid progress of experience and research in this subject making it virtually impossible for a review such as this to be up-to-date and wholly comprehensive in all aspects. An example is the Trade Group Classification in BS 812:1975 which was under active revision during the period that this report was being prepared. This classification has for decades been a source of much academic discussion and practical difficulty, which has led to a view that it is unsatisfactory in various respects. Chapter 6 reviews various classification schemes, including the current British Standard schemes, including the current British Standard scheme, together with the Working Party's proposals for an alternative system, designed to overcome some of the more misleading features of the current methods.

Moreover, it should be clear that the concepts and investigative techniques involved are drawn from a wide scientific spectrum, in which geologists, chemists, materials scientists and engineers, physicists, and statisticians, to name the more obvious disciplines, all collaborate with professional construction engineers and production staff. Any document which links such diverse fields to a common theme will inevitably be considered by persons of widely differing vocational backgrounds. Consequently, basic material is included which an expert in a particular field might consider over-simplified. However, it was believed that this approach was justified in order to serve a wide readership and, moreover, the Working Party hoped that many elementary facts might assume a new significance when presented in relation to the central theme.

The first complete draft of the Report was submitted for open discussion in September 1981 during the 17th Annual Region Meeting of the Engineering Group, held at The University College of North Wales, Bangor. The suggestions and criticisms made then and subsequently received from delegates have been considered in the preparation of this publication, together with those invited from particular individuals and organisations. However, the views expressed in this Report are those of the Working Party members.

1.3 Geographical factors

Although mainly concentrating on British practice, it is hoped that this publication will have a wider geographical application, subject to due consideration of the effects of local climate and geological history. Weathering can be of much greater importance in wet tropical or hot desert conditions, and some of the generalisations relating to the performance of certain rock types in Britain could be misleading if applied in very different climatic regimes. For example, igneous or metamorphic rocks may be impaired by deep weathering in wet tropical regions and, as a result, provide aggregates which could deteriorate rapidly in service; another example is when inherently weak limestones may be transmuted locally into strong calcretes under hot desert conditions. Local geological history will have an equally important influence, for instance the Mesozoic and Cainozoic sedimentary formations in Britain are almost devoid of rocks capable of yielding hard coarse aggregate, but strata of the same age in the Alps, which have recrystallised under conditions of high temperature and stress, are generally hard and durable. Geological age of itself therefore, will have only a local significance as an element in the description of an aggregate, and should always be considered in conjunction with the graphical location of the material.

Reference

VERNEY, Sir R. B. (Chairman) 1976. "Aggregates: the way ahead". Report of the Advisory Committee on Aggregates. (HMSO, for Dept. of the Environment, London)

Chapter 2. Occurrences and associations of sources

2.1 Introduction

This section is intended primarily as an introduction and guide for non-geologists engaged in the aggregates industry. It is intended to be useful in understanding the more straightforward day to day geological situations that arise in the development and management of workings, and it should provide a basis for recognising the more complex and intractable problems for which specialist geological advice is likely to be needed.

Although most hard rocks are potentially useful for coarse aggregate, since they satisfy a wide variety of specifications and uses, the substantial quantities of material used for road pavement and concrete aggregate are required to meet fairly stringent specifications, and may be termed 'good-quality'. Whilst the suitability of an aggregate for demanding applications can only be determined by empirical laboratory tests, there is nevertheless an approximate correlation with rock porosity; it may often be taken as a rough working rule that an otherwise suitable rock with a water absorption value of less than about 2% will usually produce a good-quality aggregate and that rocks with values exceeding about 4% may not.

High-quality aggregates therefore are commonly derived from the older, more compact sedimentary formations, particularly limestones and sandstones, from igneous and metamorphic rock masses of any age and from naturally-occurring particulate deposits (sand and gravel). The last-named are accumulations of the more durable rock fragments and mineral grains which have been abraded and sorted by the action of ice (glaciers) water (rivers and waves) and the wind; the finer grained, weaker and more porous material is effectively separated and removed by these natural agents of transport and erosion.

Much useful information in addition to that given below on the nature and occurrence of these aggregate-bearing deposits (amongst others) including guidance on their identification and description, will be found in the British Standard Code of Practice for Site Investigations, particularly Section 8 and Appendix G (BS 5930:1981).

2.2 The effects of alteration and weathering

In all parts of the world, particularly in the tropics and desert regions, local aggregate materials affected by widely different geological and climatic conditions might carry impurities, such as leached and evaporitic substances, which would not be expected in unacceptably large concentrations in temperate countries. The physical and chemical stability of rock materials in the ground (and in use) can vary widely: for example, joints and fractures in a source of aggregate in temperate conditions may not raise serious doubts about its potential suitability (indeed they may assist quarrying and processing) but in tropical conditions, they may greatly accelerate the effects of already rapid weathering processes and in a hot arid environment may allow deleterious substances to penetrate into the rock mass. Thus, the recognition of familiar rock types or geological formations is no guarantee that usable aggregates will be obtainable from them. Ideally, only fresh or faintly weathered rock material should be used for aggregate (Table 2.1).

2.2.1 Alteration. Destructive alteration of rocks, as distinct from weathering, arises from internal deep-seated and often wide-scale geochemical processes. Thus, at a late stage in the solidification of an igneous rock, release of accumulated volatile substances or mineralising fluids may extensively alter the rock as a whole, by affecting the chemical and hence physical stability of widely distributed minerals, or selectively by invading previously-formed discontinuities in the mass.

There may be a gradation, for example, from fresh granite to completely kaolinised granite, in which the feldspars have been altered to the clay-mineral kaolinite. Alteration by *in situ* recrystallisation of existing rocks due to thermal and regional metamorphism gives rise to a wide variety of rocks, many of which may yield higher-grade aggregate than the unaltered parent material. For example shales and slates may be recrystallised to become hornfels, and similarly sandstone may be converted to metamorphic quartzite (2.5.2).

2.2.2 Weathering. Weathering is brought about by the exposure of rock materials, commonly during long periods of geological time, to the varying climatic conditions at the earth's surface, and various definitions and discussions of the engineering effects of weathering have been published (Weinert 1964, Sanders and Fookes 1970, Dearman 1974, Ollier 1975).

There are two main types of weathering, one dominated by mechanical disintegration, the other by chemical decomposition including removal by solution. Generally, both effects act together, but,

TABLE 2.1. *Weathering & alteration grades*

Term	Grade*	Description	Materials characteristics†
Fresh	IA	No visible sign of rock material weathering.	Aggregate properties not influenced by weathering. Mineral constituents of rock are fresh and sound.
Faintly weathered	IB	Discolouration on major discontinuity (e.g. joint) surfaces.	Aggregate properties not significantly influenced by weathering. Mineral constituents sound.
Slightly weathered (this grade is capable of further sub-division)	II	Discolouration indicates weathering of rock material and discontinuity surfaces. All the rock material may be discoloured by weathering and may be somewhat weaker than in its fresh condition.	Aggregate properties may be significantly influenced by weathering. Strength and abrasion characteristics show some weakening. Some alteration of mineral constituents with micro-cracking.
Moderately weathered	III	Less than half of the rock material is decomposed and/or disintegrated to a soil. Fresh or discoloured rock is present either as a continuous framework or as corestones.	Aggregate properties will be significantly influenced by weathering. Soundness characteristics markedly affected. Alternation of mineral constituents common and much micro-cracking.
Highly weathered	IV	More than half of the rock material is decomposed and/or disintegrated to a soil. Fresh or discoloured rock is present either as a discontinuous framework or as corestones.	Not generally suitable for aggregate but may be suitable for lower parts of road pavement and hardcore.
Completely weathered	V	All rock material is decomposed and/or disintegrated to soil. The original mass structure is still largely intact.	Not suitable for aggregate, or pavment but may be suitable for select fill.
Residual soil	VI	All rock material is converted to soil. The mass structure and material fabric are destroyed. There is a large change in volume, but the soil still has not been significantly transported.	May be suitable for random fill.

* After Geological Society (1977). The description of rock masses for engineering purposes.
† For general engineering characteristics, see Fookes *et al.* (1971).

depending on the climate, either process may be dominant. Mechanical weathering, found mainly in arid or cold climates, results in the widening of discontinuities and the formation of new breaks by rock fracture arising from differential expansion and contraction and from the effects of repeated freezing and thawing; mineral grains are loosened and often fractured. The crystallisation of mineral salts brought near to the surface of rock masses by evaporation and capillarity has similar effects, which must be recognised and such zones avoided, particularly in potential aggregates for use in concrete in desert environments.

Chemical weathering may be much accelerated in humid hot climates and penetrate to depths in excess of 300 m, perhaps as much as 1500 m (Fookes *et al.* 1971). It eventually leads to the decomposition of certain of the silicate minerals to clay minerals: the feldspars and ferro-magnesian minerals in igneous and metamorphic rocks are particularly susceptible, though some minerals, notably quartz, are resistant to attack and may survive unaltered. Solution of silica and carbonate cements in sandstones and limestones, in common with chemical weathering generally, commonly results in the progressive discolouration of the rock, and the degree of weathering can often be described and assessed subjectively from the change in rock colour. Of the various published engineering classifications of weathering grade (see reviews by Little 1969 and Sanders and Fookes 1970), those proposed by Weinert (1964), for the field determination of the state of weathering of igneous rocks for use as roadstone in South Africa, and by Fookes *et al.* (1971) for rocks in the Dartmoor area, appear to have general validity. The classification of the Geological Society Engineering Group Working Party Report (1977(b), table 8), is recommended here (Table 2.1).

Weathering tends to become less with depth but may locally extend along discontinuities deep into otherwise fresh rock, and some rock-types may be more susceptible to weathering processes than others. Weathering profiles may also be of great antiquity. They may be repeated in a rock sequence and they may not reflect the response of the rocks to the present climate but to previous environments: the contemporaneous weathering of the upper surfaces of successive lava flows to lateritic residual soils or 'boles' provides an example, only the relatively unweathered lavas being quarried for aggregate, as with the Clyde Plateau basalts near Glasgow and the Antrim Basalts in Northern Ireland. On the other hand, the breakdown of the matrix by weathering in some sources of aggregate, such as breccia, conglomerate or clay-bound sand, may enhance their value by reducing the costs of extraction, processing or beneficiation.

Although discolouration may be an indication of weathering and alteration in rocks, the extent of the deterioration may not be easily distinguished except by means of a full petrographic analysis of a thin section of the rock. Such an examination should be undertaken to help assess the suitability of the rock for particular aggregate uses as indicated in Chapter 6.

2.3 Occurrence of sand and gravel, and of conglomerates

The properties of gravel and, to a lesser extent, of sand largely depend on the rocks from which they are derived, although during their transport by water prior to deposition the weaker particles tend to be selectively removed by attrition so that the resulting material is usually stronger aggregate than the parent rock when crushed (Harris *et al.* 1974, p. 336).

The principal sources of sand and gravel are relatively young and unconsolidated superficial (so-called Drift) deposits which have accumulated in the recent geological past, usually in the northern hemisphere at least, since the onset of the Pleistocene Ice Age. In North America, these deposits are termed surficial deposits or regolith. All older (pre-Quaternary) geological deposits, which form the bedrock, country rock, formation or rock head upon which the softer Drift materials rest, are grouped together as the Solid formations. Drift deposits include marine, beach and lacustrine deposits, alluvial deposits, hill and mountain slope solifluction wastes largely of periglacial origin, as well as those of the glacial deposits resulting from the Pleistocene ice-sheets. Older (Solid) deposits at many levels in the geological column are seldom sufficiently unconsolidated to be regarded as sand and gravel: they exist as pebble beds, conglomerates or pebbly sandstones which may have to be processed by blasting and crushing (Archer 1972).

2.3.1 Alluvial deposits. Of all the aggregate-bearing Drift deposits in Britain the alluvial deposits (including the river terrace gravels) which have been naturally processed by running water, are the most important, especially in south-east England and the Midlands. They generally carry little overburden and, depending on their location in a valley, they may have a high water-table and can be extracted using wet working techniques; their particle size distribution and thickness are more consistent and they usually contain a lower proportion of 'fines'—that is silt and clay grades—than the glacial and fluvio-glacial outwash gravels (Beaver, 1968).

2.3.1.1 River channel deposits. In the normal course of events, the finest grained products of weathering and erosion (clay and silt) are carried farther afield than the sand and gravel, which may come to rest in stream channels to form a significant resource that may be continuously or seasonally replenished, particularly in large river channels and near mountainous tracts. During flooding, sand and gravel as well as the fines, may be spread extensively over the flood-plain, and commonly wide spreads of potentially useful aggregate exist beneath a cover of silty alluvium, particularly in areas where during the recent geological past, the streams were more vigorous and transported greater volumes of coarser material than they do today.

2.3.1.2 River terrace deposits. River terraces represent the dissected remnants of former flood-plain deposits: as a river progressively erodes its valley, it cuts down its channel to lower levels and stepped terraces remain on bedrock benches that are characteristically present on both flanks of a valley (Figs 2.1 and 2.2). The terrace deposits, like those of the flood-plain, are commonly gravelly below, becoming more clayey and silty above, but the higher, older terraces having been subjected to processes of weathering and erosion for longer, characteristically

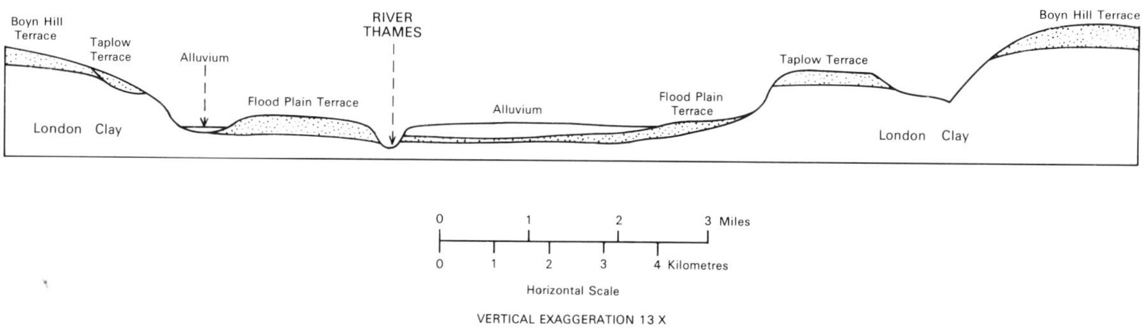

FIG. 2.1. Idealised cross section of the Thames Valley showing the relationship between the major river terrace deposits and the alluvium of the present river. (After Beaver 1968.)

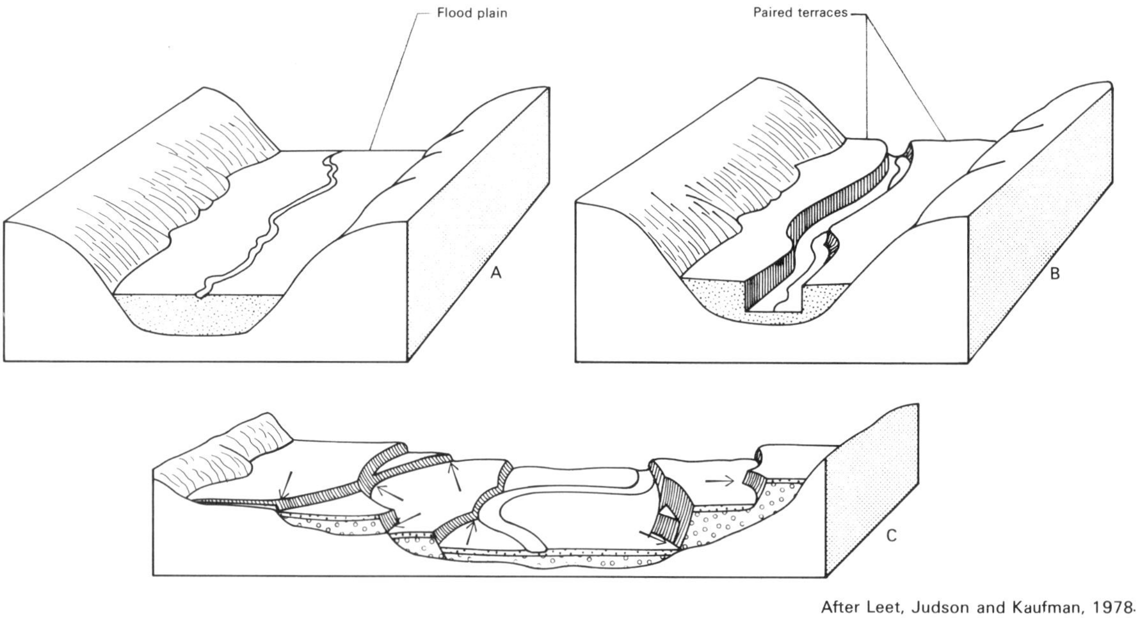

FIG. 2.2. The formation of river terraces. (A) Deposition of valley floor deposits. (B) Erosion of deposits leaving remnants of old flood plain as paired terraces. (C) Unpaired terraces may result from the effects of downward and lateral erosion: the laterial migration of the stream may be arrested when it intercepts resistant bedrock. After Leet, Juston and Kauffman 1978.

carry less alluvial overburden, may contain significantly reduced thicknesses of sand and gravel and the particles may be contaminated by weathering products, such as clay, and iron oxides by leaching; the material in the oldest terraces may be useful as low-quality aggregate.

Although flood-plain and terrace gravels are particularly favoured sources of aggregate, they are nevertheless subject to unpredictable discontinuities and variations in quality: the persistent changes in the location of the river channel after flooding and in response to meandering and braiding of the streams, result in clay, silt, sand and even gravel-filled channels and washouts often of wide extent (Fig. 2.3). Indeed, where a river catchment contains easily degraded rocks, such as mudstone, soft siltstone, sandstone and coarse-grained feldspathic rocks, potentially useful aggregate-bearing deposits may be generally lacking, as for example, in the terraces of the central Wealden rivers of Sussex and Kent. Very commonly, the composition of the sand and gravel in a river basin reflects that of the rocks in the uplands drained by tributaries: the extensively worked gravels of the Thames in the vicinity of Oxford are largely derived from the Jurassic oolitic and shelly limestones of the Cotswolds (Cullimore *et al.* 1976) and so have provided somewhat less attractive concreting aggregates than the predominantly flint gravels of the middle and lower reaches of that river and of many others in south-east England, which drain the chalk uplands or the predominantly flinty drifts that mask them (Squirrell, 1976, p. 4; 1978, p. 6).

2.3.1.3 Alluvial fans. An alluvial fan or alluvial cone is formed when streams carrying large volumes of sand and gravel often from a glacial upland across steep slopes enter an adjacent valley or plain, and although the surface of the fan-shaped body characteristically slopes gently, the proximal deposit of sand and gravel may be thick. Because of frequent shifting of the distributary streams, fan deposits ordinarily contain lenticular beds and tongues of poorly-sorted sand and gravel interbedded with varying proportions of silt and clay: the rock particles are generally subangular to subrounded and in large fans, such as that which covers some 40 km^2 near Wrexham, Clwyd, much of the 'fines' may have been washed far afield (Fig. 2.4).

2.3.2 Glacial deposits. Glacial deposits are commonly much less predictable than alluvial deposits in almost every respect: the shape of the aggregate-bearing bodies associated with glacial till (boulder clay) and fluvio-glacial outwashes, their particle-size distribution, particle shape and composition, thickness and extent, vary widely. Equally, the land-forms associated with the acculation of these deposits and from erosion by glacial ice, are distinctive (Figs 2.5 and 2.6). The part that water has played in the deposition of the sediments (and hence, their usefulness as a source of natural aggregate) is the most important criterion in their grain size classification.

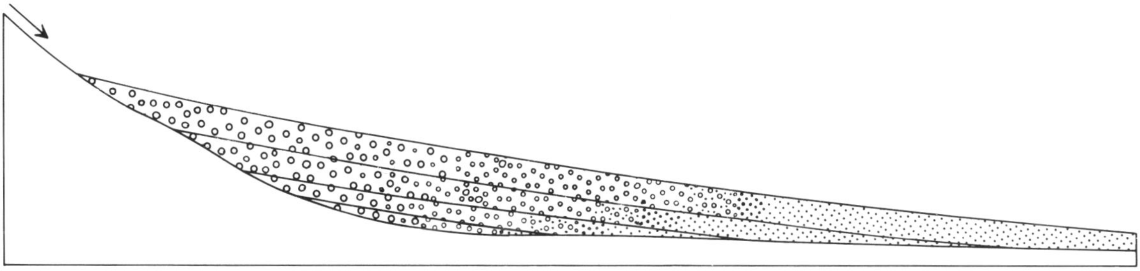

FIG. 2.3. Alluvial cone built up on a plain by a river carrying debris from an upland. Coarse material is deposited proximally. (After Lahee 1952.)

Thus, till, which has been formed directly from the ice without the intervention of water in quantity, is unstratified and unsorted material embodying all size grades from clay to boulders. Because of this mode of origin, till is unlikely to provide a potential source of aggregate, unless the matrix is predominantly sandy rather than clayey (Aitken *et al.* 1979).

On the other hand, outwash sediments (sandar), which have been carried from the ice front by melt-water and then laid down in fairly well-sorted stratified proglacial deposits, provide relatively good prospects (Fig 2.7). Between these two types of deposits, roughly stratified ice-contact deposits (such as kames and eskers) constructed near or within the ice and washed to some extent by melt-water, may also provide potential sources of aggregate.

2.3.2.1 Kames and eskers. Masses of glacial gravels particularly those derived from recent glacial events in or near upland localities close to the ice fronts may take the form of kames and eskers (Fig. 2.8). Kames are commonly a melange of boulders, cobbles and coarse gravel deposited by glacial melt-water in crevasses, kettleholes and between the ice mass and the country rocks, so that the materials typically form arcuate hills and ridges up to perhaps 15 m high, often banked against each other (the Carstairs Kame Belt in Lanarkshire (Laxton and Nickless, 1979)) or a confining valley rock slope. Deposition in a transient body of water, such as an ice-dammed lake, may lead to the formation of kame terraces. Eskers are linear or beaded ridges typically of coarse to very coarse deposits which were emplaced by streams under or within the glacier ice.

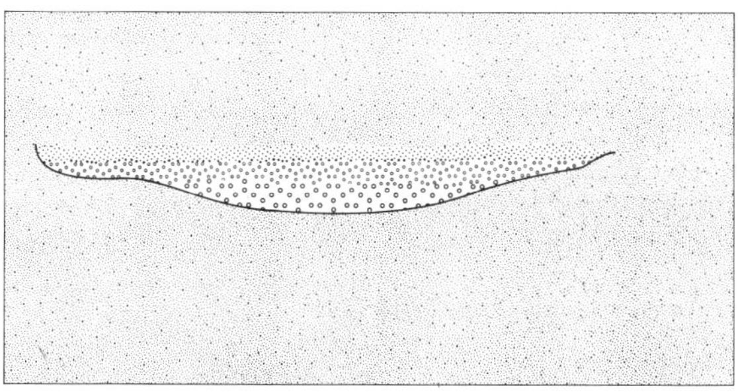

FIG. 2.4. Section of a lens of gravel in a predominantly sandy formation, laid in a stream channel as a 'wash-out'. (After Lahee 1952.)

2.3.2.2 Outwash deposits (sandar). In general, outwash debris varies in particle size with the distance it has been carried from the ice-front and in lithological composition according to the provenance of the ice-sheet and the range of rock types it has incorporated. The extent of the outwash deposits, the degree of sorting and their thickness depend on the scale and duration of the glacial and fluvio-glacial activity, the waxing and waning of the glaciers and ice-fields and the consequential advances and retreats of the glacial and periglacial climatic and depositional environments; tills may therefore pass laterally into sands and gravels of the same glacial episode over several kilometres, may be sandwiched between materials of any of these lithologies from earlier or later episodes, or they may be intermixed and contorted by the effects of glacial erosion or of deep freezing of the ground (cryoturbation). In central southern England, there are extensive spreads of sand and gravel of uncertain age or origin called Plateau Gravel on Geological Survey maps, a term which probably includes glacial, fluvio-glacial and river terrace gravels; locally, other largely periglacial gravels have been termed Pebble Gravel, Downwash Gravel, Hill Gravel and Taele Gravel as well as Head. In East Anglia and Essex, spreads of chalky till in the form of ground moraine are generally though not invariably underlain by outwash sheets or otherwise associated with irregular masses of *glacial sand and gravel* (Nickless, 1971; Clarke, 1975). An impression of the complexity of depositional associations in glacial deposits near an ice front is shown in Fig. 2.9.

2.3.2.3 Periglacial phenomena and deposits. The term 'periglacial' denotes conditions under which frost action is the predominant weathering process. Frost shattering of rocks and deep freezing of the ground result in the downslope movement of the weathering products, some of which may be useful as a source of aggregate. Periglacial processes are associated with very cold climatic conditions such as those near the margins of glacial ice in the polar regions.

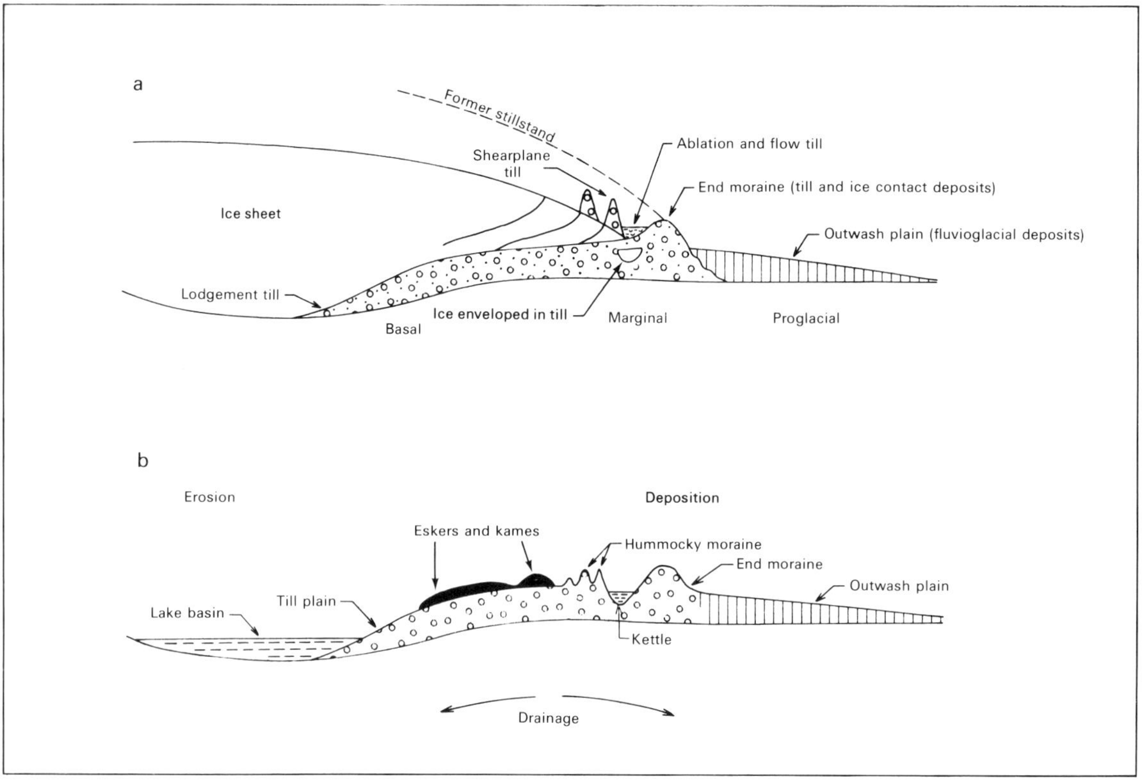

FIG. 2.5. The relationship between the various glacial materials and landforms associated with ice stand-still and retreat (see Figure 2.8 also). After West 1977. (a) The location of basal, marginal and proglacial deposition near the maximum extension of an ice sheet. (b) After the retreat of the ice, till and ice-contact deposits remain.

Approximately one-fifth of the earth's surface is currently underlain by permanently frozen ground (permafrost) and there is evidence in the glacial deposits of the northern hemisphere that during the Pleistocene much more was effected by cold climates.

Most periglacial effects from the Pleistocene glaciations are well preserved in southern and eastern England between the limits of the last (Weichselian) glaciation and the wind-blown silt (loess, brickearth) and sand (cover-sand, dunes) belts of northern and central Europe, but they are not restricted to this area: periglacial phenomena can be found over the whole of Britain because they moved progressively northward as the ice front retreated.

2.3.2.4 Screes, block-fields and head deposits. Repeated freezing and thawing breaks off flakes and fragments from exposed rock surfaces by 'ice-wedging' in fractures and joints and by inducing differential movements in the rock mass. The mechanical disintegration of rocks in upland areas may produce large amounts of predominantly angular rock debris or talus. In the absence of running water and where slopes are steep, large-scale accumulations of debris near to the angle of rest (*screes*) may provide a ready source of coarse aggregate material, though it must be remembered that it may have been adversely affected by long exposure to weathering and any disturbance by working may render the loose material unstable and therefore hazardous. Where the ground is flatter, frost-shattering may give rise to *block-fields* of angular stones which may migrate downslope in response to gravity only slowly, if at all, as the result of heaving by ice crystals, thawing, spalling and the alignment of the fragments into patterned ground.

In soft-rock terrains, when climatic conditions are or have been such that little or no running water is available and the ground is subjected to alternate freezing and thawing, surface and near-surface materials may creep down-slope (solifluct), and form a structureless (unbedded) deposit, termed *head*. Such debris, which may be partly weathered, is the waste from local formations and may contain useful material where it is derived from high level terrace or plateau gravels, for example, the so-called Downwash Gravels in the Aldershot district (Dines and Edmunds, 1929; Clarke *et al.* 1979; 1980). In the vicinity of Chelmsford, Essex, gravelly head of late Pleistocene age is widespread but so heavily contaminated by clay as to be of little use as aggregate, and this is commonly so in present-day periglacial environments. The effects

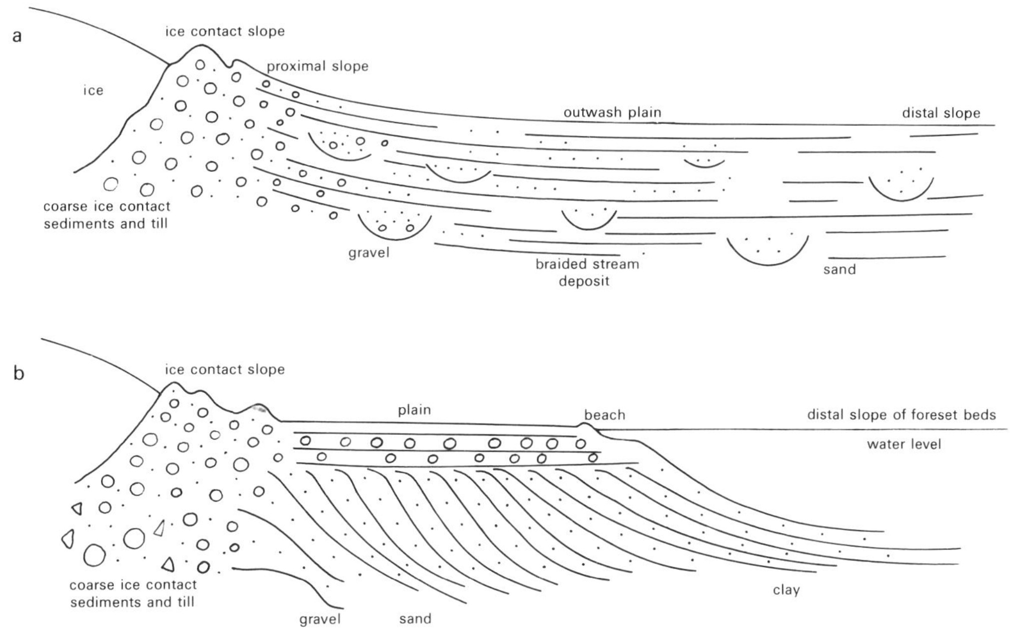

FIG. 2.6. Proglacial landforms and associated deposits. After West 1977. (a) Section through the deposits of an outwash plain. (b) The relations of the deposits in a marginal delta (Compare with Fig. 2.3).

of periglacial freeze-thaw phenomena on the distribution of potentially useful natural aggregate materials may be far reaching; they may generally disrupt and redistribute existing deposits through the development of patterned ground by ice-wedging mix deposits by frost heaving and sub-surface contortions, form deep pockets of gravel and sand in clay formations mobilised by the growth of ice crystals (cryoturbation), and deposits may be spread out on slopes as the result of downhill solifluction and thinning of hill cappings by soil creep. Although evidence of small-scale and post-depositional disturbances of incoherent deposits by frost-heaving is fairly common-place, large-scale diapiric phenomena, affecting the distribution of sand and gravel deposits, are much less so. On Danbury Hill near Chelmsford, Essex, however, a fluvio-glacial outwash sheet overlying soft bedrock (London Clay) has been thrown into large-scale compressional folds of some 30 m amplitude as the result of deep-freezing in periglacial conditions. Erosion has removed much of these structures, which have been outlined by geological mapping for a resource survey (Fig. 2.10).

2.3.3 Coastal deposits. Coastal deposits as distinct from the 'terrestrial' gravels described above, accumulate near the margins of permanent water bodies between the lowest low tide mark and the top of the storm beach as the result of wave action on the rocky or debris-strewn shores of large lakes, estuaries and the sea. Some material derived by denudation from land areas, may however be deposited near a coast and the coarse fraction contribute to beach shingle, where the energy of breaking waves will tend to drive it (Fig. 2.12). The deposits may be augmented by ice-rafted dropstones and by rock ballasting dropped by ships in shallow water prior to taking on cargoes. Because of the erosive and winnowing action of waves and currents, cobbles and pebbles tend to be swept high onto the coast (back-shore) and the sands and finer particles out to sea; beach gravels are therefore commonly strongly localised vertically but laterally extensive on the shore and they are characteristically well-rounded and sorted; they also reduce in grain-size from the top of the beach into deep water so that on exposed shores, a stable beach will rarely contain material finer than medium sand (about 0.2 mm). Movement of materials also takes place along the shore by the process of long-shore drifting and deposits accumulate in ridges and bars, which in particular circumstances may coalesce to form extensive shingle ridge deposits (such as that at

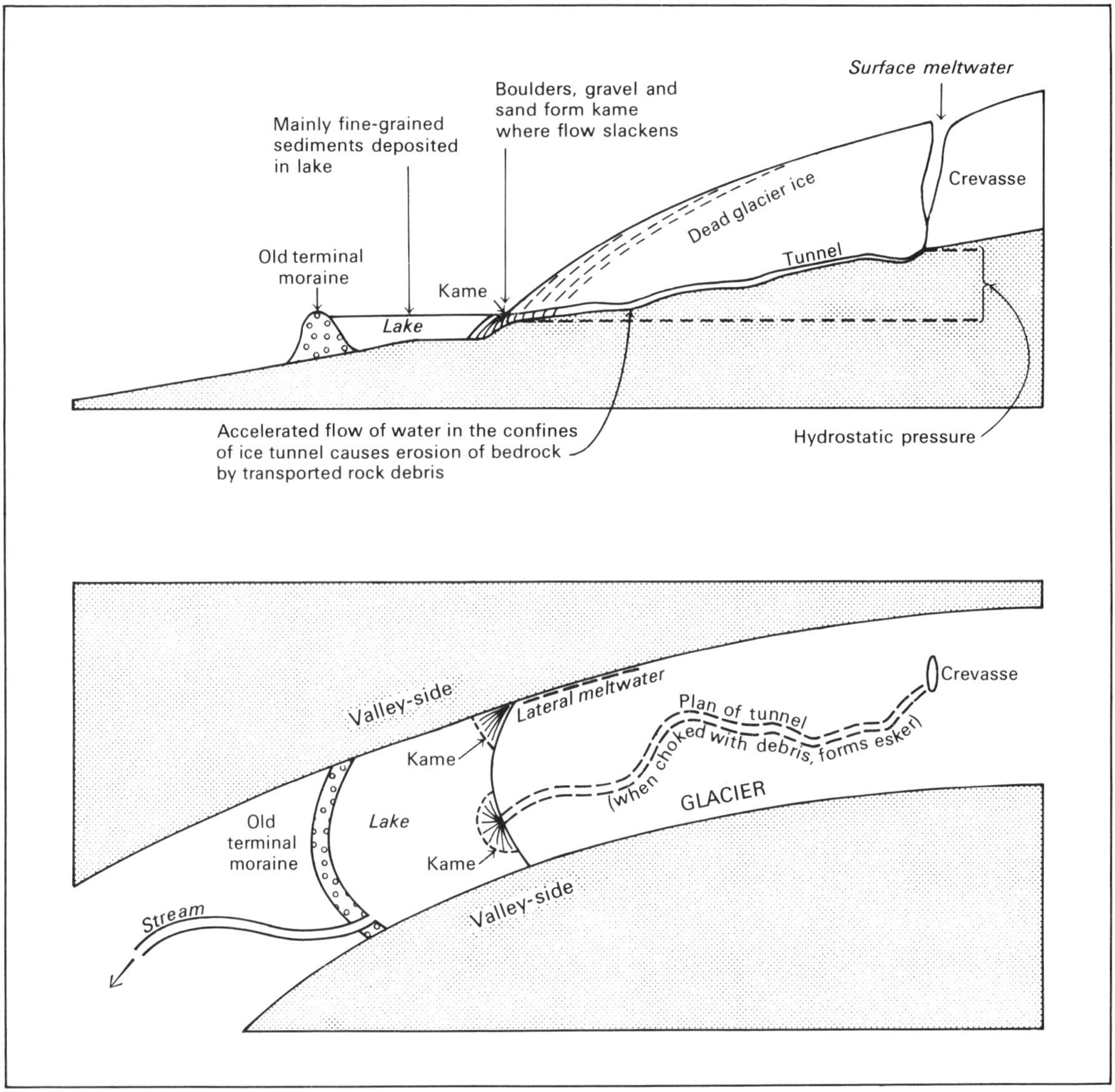

FIG. 2.7. Formation of kames and eskers. (After Gresswell and Lawrence 1978.)

Aldeburgh, Suffolk, and the complex at Dungeness, Kent) or, influenced by long-shore drifting and winnowing, they may demonstrate a progressive gradation from coarse to fine-grained deposits along a beach. For example Chesil Beach extends 24 km from the Isle of Portland to Burton Bradstock, Dorset, grading from boulders to fine gravel and sand westward.

The extraction of aggregate from beach and near-shore deposits must be approached with care; there may exist a very delicate balance between erosion and accretion which might be upset if the supply of material is changed whether by the construction of sea defences (such as a sea wall, or the emplacement of groynes) or by its removal by digging from beaches, foreshore or off-shore zones.

2.3.4 Marine deposits. It is likely that gravels found well away from the present coastlines beneath the shallow seas were originally laid as fluvial or fluvio-glacial deposits which have been drowned by the general post-glacial rise in sea level. Whilst sand may migrate as sand-waves on the shallow sea floor (to depths of 180 m or so) under the influence of currents with velocities in excess of about 2.7 kilometres per hour, it seems unlikely that pebbles will do so on a large scale in currents of less than 7.5 km/h though it is

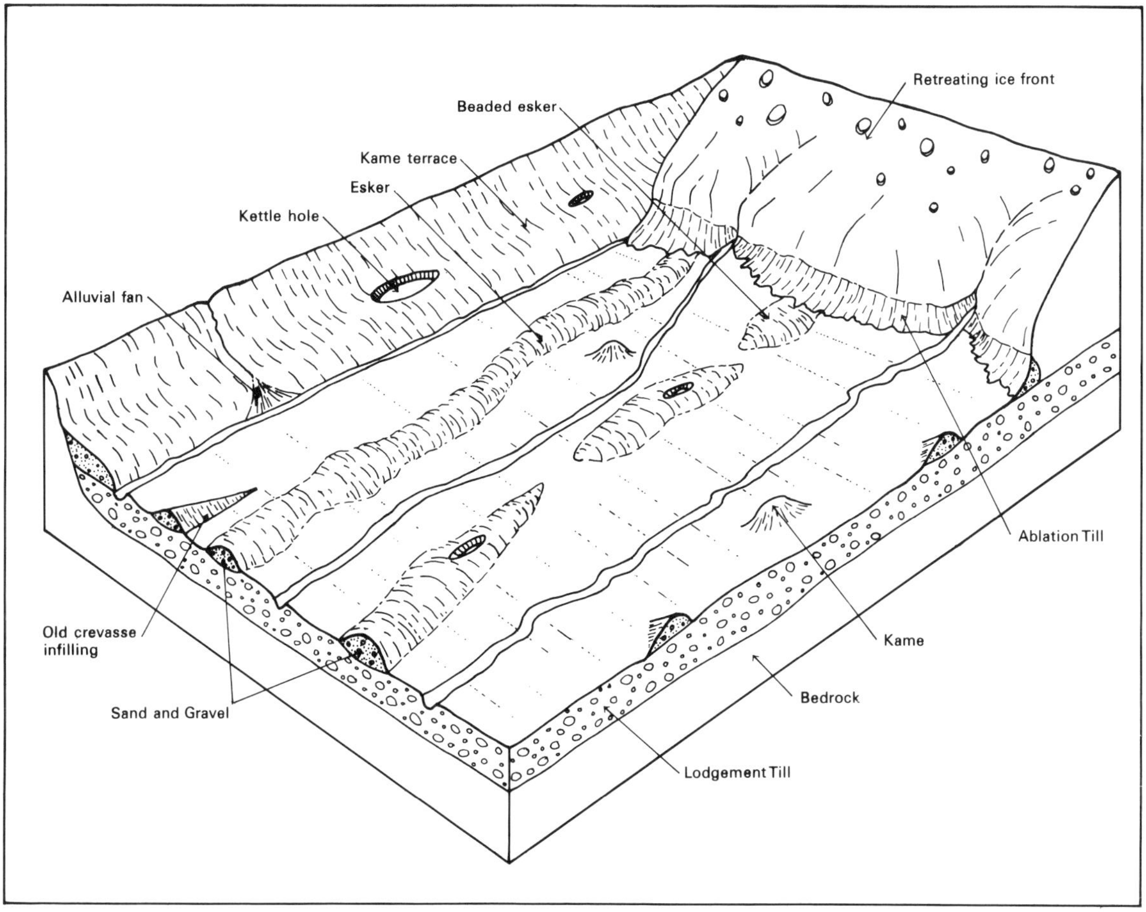

FIG. 2.8. Idealised Fluvioglacial and Ice Contact Landforms. (After Fookes, Gordon and Higginbottom 1978.)

possible that concentrations of gravel may result from submarine erosion of till or other gravelly deposits. Offshore deposits may be replenished, often seasonally, by debris from the marine erosion of coastal deposits and from subaerial and glacial erosion of near-shore land areas of high relief, which may be transported and distributed over considerable distances by fluvial and tidal currents.

2.3.5 Gravels in the solid formations. Although modern shoreline gravels account for only a small proportion of available resources, because they are confined to relatively narrow linear tracts, many of the sand and gravel deposits in the 'solid' rock sequence take the form of basal pebble beds whose origin has been traditionally attributed to the transgression of shoreline gravels with a rise in sea level, resulting in the deposition of a relatively thin sheet of basal gravel, which may be indurated or cemented to form a conglomerate. The Tertiary Blackheath Beds in North Surrey and the Bovey Beds near Newton Abbot, Devon, which have been worked on a small scale, are examples. North of Glasgow, several small discrete lenticles of quartz conglomerate are worked from Carboniferous sandstone.

Conversely, when deposits are laid down on a slowly rising sea-floor, successive layers are related such that coarser grades come to rest on the finer parts of older layers, as the sea regresses from the land. The record of marine transgression and regression as demonstrated in a borehole core is shown in Fig. 2.11.

Perhaps the largest volumes of gravelly sediments have formed as alluvial fans or deltas in arid or semi-arid regions, particularly where gravel-carrying torrents debouch into an intermontain tract. The Triassic 'Bunter' Pebble Beds of the Sherwood Sandstone Group of the north Midlands and the Budleigh Salterton Beds of approximately the same age in east Devon are the main source of gravel from pre-Quaternary formations in Britain. The proportions of conglomeratic material to largely pebble-free sandstone varies widely and the workings tend to exploit only the most pebbly developments: sand is

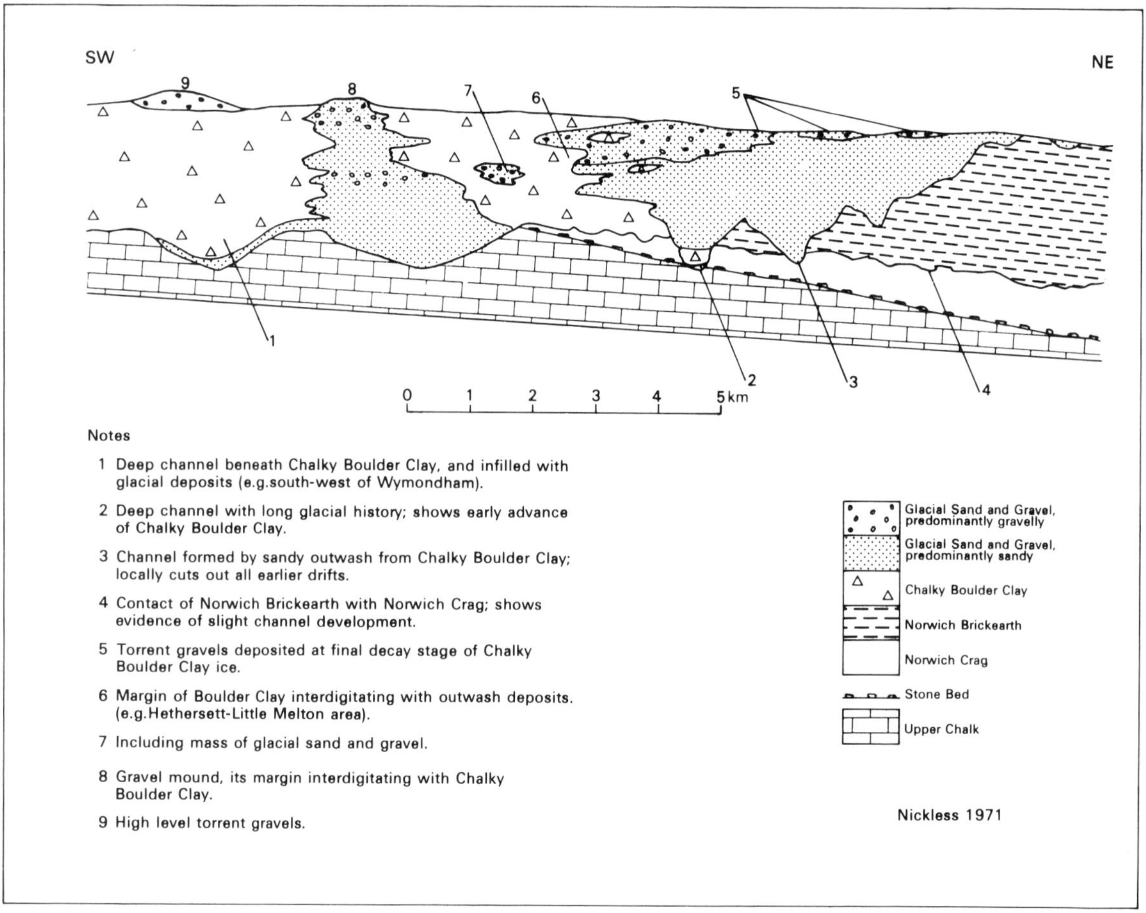

FIG. 2.9. Schematic section across the Chalky Boulder Clay ice front, showing the stratigraphy of the Norwich area. (Nickless 1971.)

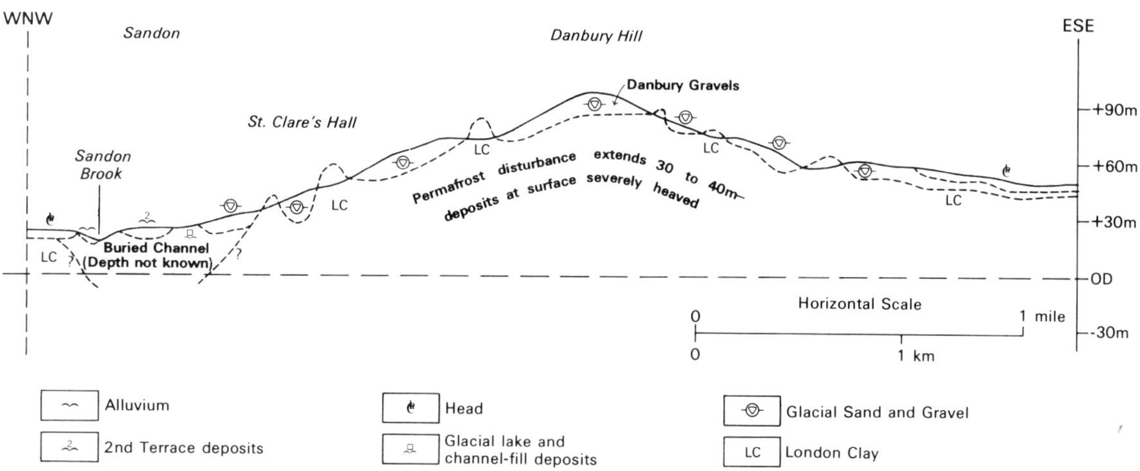

FIG. 2.10. Geological sketch section across the area of Danbury Hill, E of Chelmsford, Essex showing the effect of large-scale frost heaving on the distribution of the deposits. (Clarke 1975.)

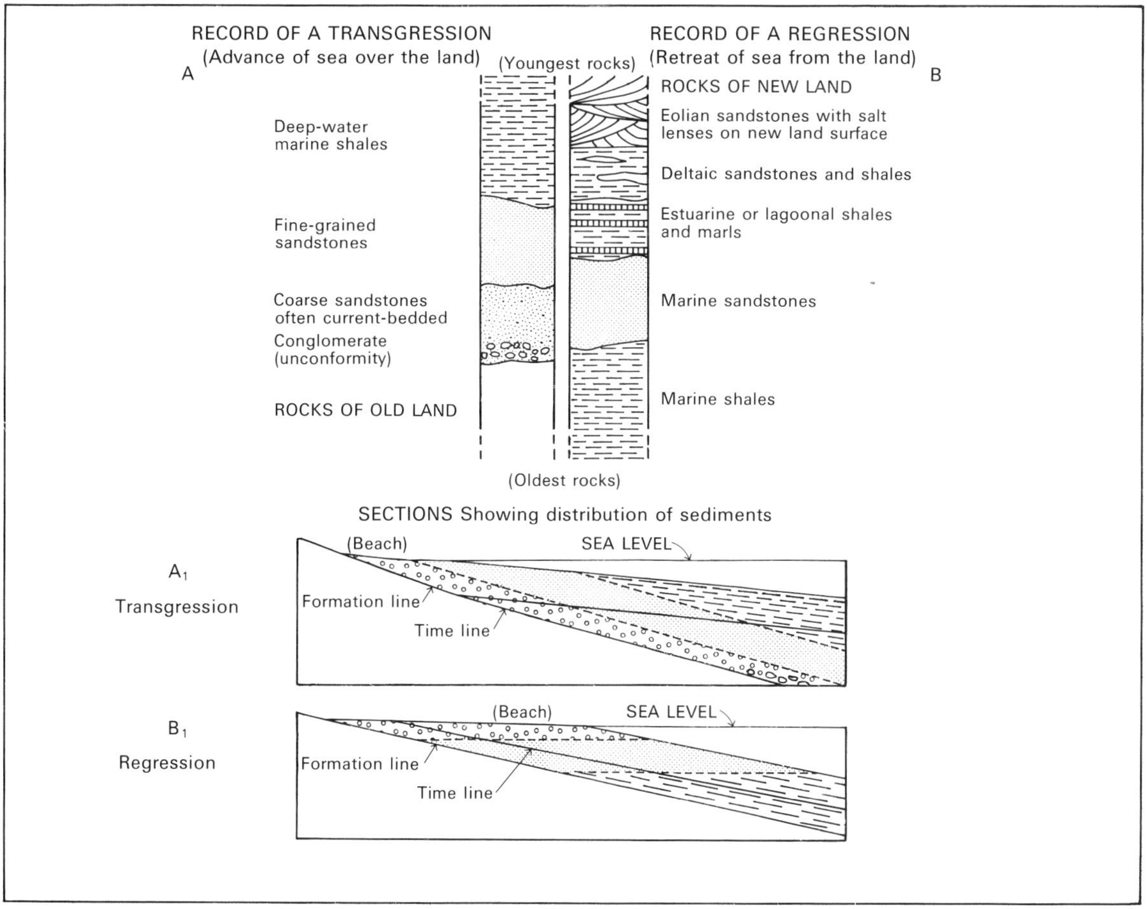

FIG. 2.11. Sea-board deposits resulting from the transgression and regression of the sea, as seen in borehole cores (A and B) and in idealised horizontal sections (A_1 and B_1). Note that in the sections, lines parallel to the sea-bed are time lines as they join sediment deposited contemporaneously; lines essentially parallel to gravel, sand and clay deposits are formation lines, and are those commonly shown on geological survey maps.

generally regarded as a secondary product. The predominant pebbles are very hard, well-rounded, buff to liver-coloured quartzite held in a loosely cemented sandy matrix (Cement, Lime and Gravel, 1963; Thompson, 1970). They are processed by light crushing and washing to remove the fines, which occur mainly as green marl seams, and sand.

2.4 Aggregates from the sedimentary rocks

As the result of the denudation of the rocks of the land surface through processes of weathering (mechanical and chemical, see 2.2.2) and of erosion during transport by moving water and ice, and by wind, large amounts of detrital and chemical including biochemical materials become available to form new sedimentary deposits. In addition, some sediments originate as ash, dust and rock debris from volcanism and others, of much smaller volume, are principally organic residues, like peat and coal and are not derived from pre-existing rocks.

Sedimentary rocks are generally divided into clastic or detrital rocks on the basis of a convenient grain-size classification (giving pebbly, sandy and muddy categories) and chemical and bio-chemical (organic) rocks, on the basis of the predominant raw material (Table 2.2). Sources of aggregates in the unconsolidated sediments (the drift deposits) have been described above (in 2.3); materials suitable for aggregate in the compacted, indurated and cemented sediments, which are part of the so-called solid deposits, are the limestones and sandstones, described below.

2.4.1 Limestone. This term is applied to the large and petrographically diverse group of sedimentary

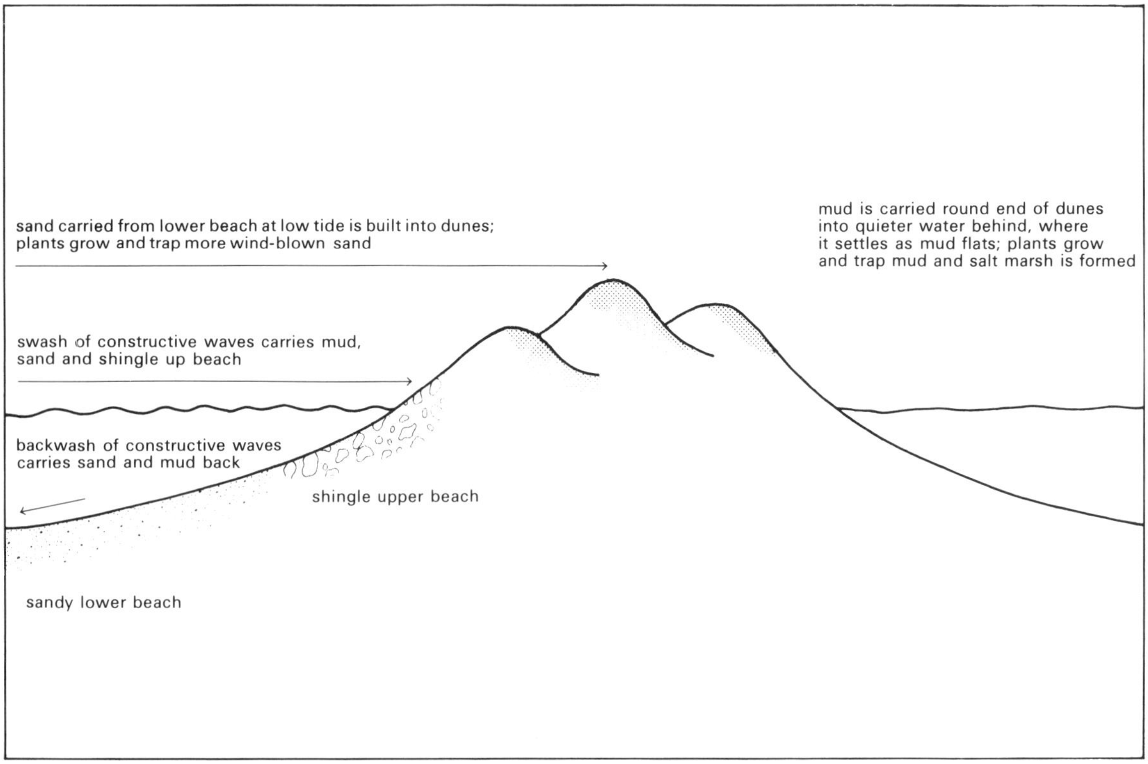

FIG. 2.12. Shoreline depositional features. (After de Boer 1973.)

TABLE 2.2. *Outline classification of sedimentary rocks*

Main character	Raw material (unconsolidated)	Equivalent rock (consolidated)
Detrital (mechanically sorted)		
PEBBLY	Pebbles, gravel (over 5 mm)	Conglomerate*
	Scree (talus)	Breccia*
	Boulder clay	Tillite
SANDY	Sand (0.06–5.0 mm)	Sandstone*, with varieties according to cement (siliceous, calcareous, clayey, ferruginous) or admixture of minerals other than quartz (feldspar, lithic fragments = Arkose*, Greywacke*)
	Silt (0.002–0.06 mm)	Ganister, brickearth, loess
MUDDY	Clay (less than 0.002 mm)	Shale, mudstone
Chemical and biochemical (organic)		
CALCAREOUS	Shells, corals, crinoids	Limestones* (shelly, coral crinoidal)
	$CaCO_3$ precipitated	Oolitic limestone, stalacetite
	$CaMg(CO_3)_2$ precipitated or replacement	Dolostone* or dolomitic limestone*
SILICEOUS	Silica gel, siliceous fossils	Flint*, chert*, diatomite
SALINE	Salt (lake) precipitates	Evaporites, duricrusts*
CARBONACEOUS	Peat, liquid hydrocarbons	Coals, oil-shale
FERRUGINOUS	Iron carbonate, iron silicate	Siderite charmosite ores

* Rocks most likely to be useful for aggregate (after Blyth & de Freitas 1974).

rocks in which the carbonate fraction (very commonly as calcium carbonate) exceeds the non-carbonate. Thus, if more than 50% of sand-sized quartz or of mudstone is present, the rock is calcareous sandstone or calcareous mudstone, rather than limestone. Similarly, the progressive admixture of magnesium carbonate as dolomite ($CaMg(CO_3)_2$) gives rise to magnesian limestone (up to 10% dolomite), dolomitic limestone and finally dolostone, in which more than 50% of dolomite is present. As limestones and dolostones are commonly the most indurated members within the younger sedimentary sequences, they may yield good aggregate provided they have a dense crystalline or cemented fabric and are not weakened by the existence of cavities or high porosity.

The limestones and dolostones, which are the main carbonate rocks of importance for aggregates, form a diverse, polygenetic group—some are detrital (particulate) sediments, others chemical or biochemical. They tend to be soluble and easily recrystallisable, so that their original characteristics may be considerably changed after deposition and burial.

The detailed petrographical description and classification are complex, and simpler schemes more appropriate for use in engineering contexts have been proposed, for example, by Fookes and Higginbottom (1975) and Burnett and Epps (1979), to which the reader is referred. For a comprehensive review of the availability, exploitation and use of limestone and dolomite, including their occurrence and applications in the aggregates field, refer to Mineral Dossier No. 23 (Harris, 1982).

Limestones, which commonly range in colour from dark grey to white and may be tinted by small amounts of ferruginous or other components, are characteristically bedded rocks often containing many fossils; they are readily scratched with a steel blade and, with the exception of dolostones, they effervesce in cold dilute hydrochloric acid. Many limestones may be described as 'shelly', that is they contain the debris and more rarely well-preserved examples of fossil shells, such as bivalves, corals and crinoids ('sea-lilies'), which may constitute a large part of the rock: algal, coral, brachiopod and crinoidal limestones take their names from prominent fossil constituents. The reef limestones from mounds or pavements (reef-knolls and bioherms) of largely unbedded rock of more limited occurrence, and the oolitic limestones, which may also contain much shelly and clayey material, are characterised by the predominance of concentrically layered grains of sand-grade calcium carbonate resembling in appearance the hard roe of a fish. All of these rocks are more durable and useful for aggregate when they are indurated, crystalline, fine-grained in texture and particularly when the matrix which holds the grains together is itself crystalline.

Although the softer more porous post-Triassic limestones in southern and eastern England will not generally satisfy the specifications for good-quality aggregate, samples of Middle Jurassic crushed limestone aggregate with water absorption values up to 9% have yielded ten per cent fines values above 50 kN in laboratory tests (R. J. Collins pers. comm.) and might be utilised in appropriate applications. Some Permian limestones are acceptable and most older limestones are entirely satisfactory for good-quality aggregate and are quarried widely. In the UK, Carboniferous Limestone currently provides about 60% of national crushed-rock aggregate production (Institute of Geological Sciences 1982) and it probably represents the largest resource of good-quality aggregate in British sediments. The limestones, which are commonly thickly bedded and structurally simple, give rise to fairly extensive well-drained uplands in which large quarrying operations can be established economically. The main occurrences and centres of extraction (Ashworth, 1975) are in the Mendip Hills, (Somerset County Council, 1971; Quarry Management and Products, 1975), around the northern rim of the Bristol coalfield, in the uplands at the periphery of the South Wales coalfield and in North Wales, in the Derbyshire Dome, on the borders of the Lake District, in North Yorkshire and Lancashire and to a minor extent in Scotland. In the north Pennines, several thin limestone beds are interbedded with soft sandstone and shale (Yoredale facies). One of the more prominent of these, the Great or Main Limestone, extends into Durham and Northumberland.

Other major limestones being worked for aggregates include parts of the Permian Magnesian Limestone in Nottinghamshire, Derbyshire, Yorkshire and Durham, the Devonian limestones near Torquay and to a lesser extent the Silurian limestones in the Welsh Borderlands. Certain of the more indurated Jurassic and Cretaceous limestones are quarried locally for building and ornamental dimension stone but mainly for 'fill' and low-strength mass concrete applications. In southern and eastern England however, some potentially useful good quality older limestones and other aggregate materials may lie at no great depth beneath the Mesozoic cover (for example, Carboniferous Limestone beneath Cambridge), and in due course, they might become economically exploitable by mining (Verney, 1976, chapter 9), as they are in Kansas, U.S.A. (Harris, 1976).

2.4.2 Sandstone. Sandstone (or arenite) is a lithified accumulation of sand grains which by most authorities are acknowledged to range in size (or grade) from about 0.06 to 2 mm in diameter, though exceptionally particles up to 5 mm in size and may be included (Chapter 6). The grains, which are predominantly of quartz, were produced by weathering and attrition of pre-existing rocks and were sorted, abraded and transported by moving ice, air and water from sources outside the basin of deposition, where they became consolidated (commonly by cementation) into a coherent rock. Coarse-grained sandstones with angular or subangular particles are termed gritstones and the admixture of other materials, in addition to quartz, gives rise to a range of arenaceous rocks, for example, a feldspathic sandstone is termed arkose and the presence of lithic (rock) fragments and clay minerals produces the so-called 'dirty' sandstones or greywackes.

The use of sandstones, arkoses, greywackes and their low-grade metamorphic derivatives as aggregate requires that they should have high strength, low porosity and, for pavement wearing course applications, that they should be resistant to polishing (Harris, 1977). A review of the suitability of some British arenaceous rocks for roadstone (Hawkes and Hosking, 1972, 1973) relates the petrography to the results of physical and mechanical tests on 86 samples specially collected from a range of pre-Permian formations, exposed mainly in disused and working quarries, and demonstrates that high-quality road surfacing materials exist widely in the Carboniferous sandstone of the West Country and South Wales. They are also extensively represented in the Lower Palaeozoic sediments of Wales, Cumbria, North Yorkshire and the southern uplands of Scotland. To be useful, the quarried material should not contain excessive amounts of weathered constituents and mica or exhibit small-scale bedding laminations or jointing discontinuities, as they may result in the crushed product becoming unacceptably flaky and weakened by the alignment of the constituent grains.

Although quartzites occur fairly widely, their outcrops are comparatively much less extensive both laterally and vertically. Examples which are exploited for aggregate include the Cambrian quartzites of Hartshill, near Nuneaton, and the pre-Cambrian Holyhead Quartzite in Anglesey. The Carboniferous sandstones of the Millstone Grit and Lower Coal Measures of Northern England, which have been quarried traditionally for dimension stone, now commonly provide aggregate for relatively undemanding applications, because they tend to be porous and weak. On the other hand, the Pennant Sandstone in the Upper Coal Measures of South Wales (Quarry Managers' Journal, 1972), and the Carboniferous and adjacent Devonian rocks of north Devon are strong and have very good polish-resistance properties, but, like the greywackes, they often have to be worked selectively from amongst substandard interbedded materials.

In south-eastern England, an area virtually without indigenous hard rock aggregate resources, a minor source of hard calcareous sandstone and sandy limestone, the Kentish Ragstone, forms part of the Hythe Beds of the Lower Greensand; a similarly hard fine-grained sandstone in the top of the Jurassic Portland Beds is worked as a by-product from gypsum mining in southern Sussex (Quarry Management and Products, 1979). Both are used for roadstone.

In addition to these granular quartz rocks, chert and flint are important aggregate materials. Chert is essentially, though not exclusively, siliceous material of sedimentary, often organic origin (such as sponge spicules, diatoms) lithified by a cement of redistributed, minutely crystalline silica, to form nodules and beds in rock formations of almost every geological age. The term flint is applied exclusively to often high purity silica concretions which form bands and nodules on bedding planes and joints in the White Chalk of the Upper Cretaceous of western Europe and elsewhere. The flints are typically hard compact and durable and they are the predominant constituent of the high-quality glacial and fluvial gravels of eastern and southern England.

2.4.3 Other sedimentary rocks and natural waste materials. Although the fine-grained (muddy) sediments (which comprise possibly widely varying mixtures of clay, quartz (silt-grade) and carbonate) are not useful for aggregate, their indurated derivatives such as claystone, siltstone, argillite and slate (Fookes and Higginbottom, 1975), may be acceptable in applications for which specifications are relatively undemanding. Except where good-quality aggregates are unavailable, it is unlikely that workings for the supply of such materials would be specially commissioned; waste-materials from collieries, slate quarrying and metal mining are more usual sources of these so-called secondary aggregates. Colliery shale is amongst the most widely available in Great Britain and has been used as sub-base and base material in the construction of lightly trafficked roads (Gutt *et al.* 1974) particularly as a cement-bound granular material (Kettle and Williams, 1969). Boulders of slate waste from slate quarrying has been used for armour stones in sea defence works on the North Wales coast.

2.5 Aggregates from the igneous and metamorphic rocks

2.5.1 Igneous rock. Although igneous rocks (that is, rocks which have solidified from the molten state) exhibit a very wide range of chemical composition, as reflected in the intricacies of their nomenclature, their suitability for aggregate depends also on their mineral constituents their crystalline fabric and texture, and these in turn relate to their modes of occurrence (Figs 2.13, 2.14). The plutonic rocks formed at depth in large masses (granite, diorite) are typically coarsely crystalline, whereas smaller bodies which may have been intruded as dykes, sills and bosses into country rocks have a medium to finely crystalline texture (microgranite, dolerite) perhaps, in some examples, with a scattering of larger crystals (porphyries). Extrusive igneous rocks have solidified rapidly as lava flows on the surface or under the sea and they may be of glassy or micro-crystalline texture (basalt, andesite). Lava flows may be of little use for aggregate if they include flow-banded strongly jointed, vesiculate or brecciated members, and pyroclastic materials such as dust, ash and rock debris ejected from volcanoes also may not be useful unless they have become indurated into hard rock by heating (welded tuff) or compacted by burial or cementation.

On the other hand, pumice, a highly vesicular sometimes froth-like lava resulting from the expansion of gases during solidification, is a naturally-occurring light-weight aggregate much favoured by the construction industry for making light-weight concrete blocks for inner leaf insulation.

From a chemical point of view the vast majority of igneous rocks are made up by combinations of only eight elements. Of these, oxygen is dominant, next is silicon and then aluminium, iron, calcium, sodium, potassium and magnesium. In terms of oxides, silica (SiO_2) is by far the most abundant, ranging from 40% to 75% of the total, and the silica percentage forms the basis of a three-fold classification into acid, intermediate and basic categories.

In general, silica-rich (acid) igneous rocks which typically contain free silica in the form of clear, colourless quartz crystals and opaque but pale-coloured feldspars, are generally paler and lighter in weight than the intermediate and basic rocks, which have increasing amounts of dark-coloured ferromagnesian minerals of higher density and only very rarely any free quartz. An outline classification of igneous rocks according to these features is shown in Table 2.3.

Most of the crushed igneous rock aggregates in Britain are produced from fine-grained basic rocks (dolerite and basalt), and to a lesser extent from fine-grained acidic rocks (feldsites, porphyries and micro-granites) and some from certain of the coarse-grained rocks (granites and diorites). Many quarries are located in small intrusions of dolerite and basalt, for example in Wales and the Welsh Borders, in Devon, in the Coal Measures at Rowley Regis in Staffordshire and in the Carboniferous Limestone of the Derbyshire Dome. The Whin Sill and related intrusions are sources in Northern England and in the Midland Valley of Scotland. There is a large production from the Clyde Plateau Lavas, the Antrim Basalts and the Old Red Sandstone basalts and andesites in Fife and Tayside, in Scotland (Quarry Management and Products, 1977(a)). Some coarse-grained rocks may be less attractive for aggregate because of their relatively low crushing strength which

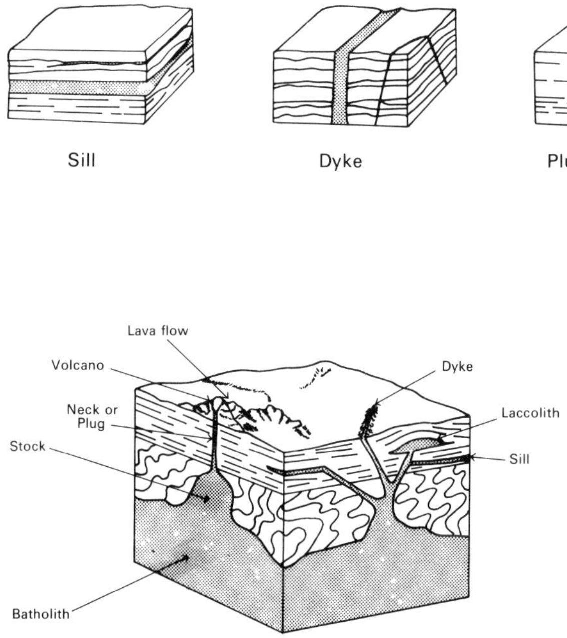

Diagrammatic representation of the relationships between bodies of igneous rocks.

FIG. 2.13. Igneous rock bodies. (After Ollier 1975.)

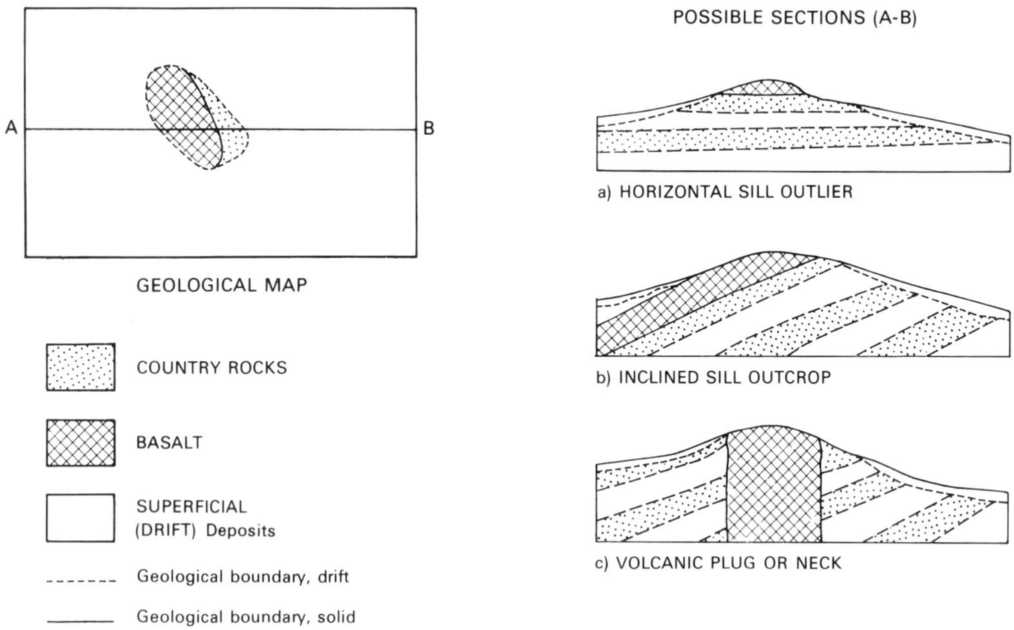

FIG. 2.14. Geological map of an occurrence of igneous rock with possible subsurface extensions of the rock body shown in geological sections. (After Tomkeieff 1964.)

arises from fracturing along the cleavages of the coarse crystalline constituents.

Igneous rocks are unevenly distributed. In south-east England, where the demand for aggregate is greatest, such rocks do not crop out. There are only small outcrops of igneous rocks in the Midlands but they are extensive in south-west England, west and north-west Wales, parts of northern England and in Scotland (Harris, 1978). In Northern Ireland, igneous rocks are the source of more than half the aggregate produced (Institute of Geological Sciences, 1980).

Though many granites yield good aggregate, the quality is not uniform: for example, the fine-grained core of the Land's End granite in Cornwall is exploited, rather than the coarse-grained material which comprises the greater part of the intrusion. The feldspars in many, such as the nearby St. Austell granite and the Arran granites in Scotland are commonly deeply weathered or affected by kaolinisation; they are prone to serious attrition during normal handling and transport. The igneous rocks of Leicestershire include granodiorite at Mount Sorrel and several smaller intrusions of diorites which have assumed considerable economic importance because of the rarity of hard rock in southern England (Quarry Management and Products, 1974). Several Ordovician porphyritic felsites and microgranites in North Wales, the Threlkeld microgranite (Cumbria) and the Newer Granites of the Scottish Highlands are examples of sources of good quality aggregates a few of which are distinctively coloured and therefore useful for decorative finishes and road surface dressings.

2.5.2 Metamorphic rock. Since metamorphic rocks result from the alteration of existing igneous metamorphic and sedimentary rocks by heat, pressure and chemical activity, they comprise a very complex range of rock types (Blyth and de Freitas, 1974; McLean and Gribble, 1979). Thermal or contact metamorphism takes place when rocks are heated by the intrusion of a mass of molten igneous rock. Clayey rocks, such as shales and mudstones may recrystallise into hard, dense, flint-like rock, termed hornfels,

TABLE 2.3. *A classification of igneous rocks on silica percentage and grain size*

ACID	INTERMEDIATE	BASIC
COARSE-GRAINED (PLUTONIC) ROCKS (grain size larger than about 5 mm; liable to be brittle from presence of large crystals)		
Granite	Syenite	Gabbro
Granodiorite	Diorite	Norite
MEDIUM-GRAINED (*HYPABYSSAL*) ROCKS (grain size between 5 and 1 mm; crystals commonly intergrown—good roadstones)		
Microgranite	Porphyry	Dolerite
Granophyre	Porphyrite	Diabase
FINE-GRAINED (VOLCANIC) ROCKS (grain size less than 1 mm, i.e. below the range of the unaided eye; some rocks liable to be brittle and splintery—otherwise good aggregate material)		
Rhyolite, pumice	Trachyte	Basalt
Felsite	Andesite	Spilite
← Continuous variation in properties and composition →		
Light colour Low relative density (2.6) High silica percentage, orthoclase feldspar		Dark colour High relative density (2.9) High ferro-magnesian (mafic) minerals

(after Fookes 1975)

whilst quartz-rich rocks, the greywackes and sandstones, fuse or recrystallise into quartzites which can provide high-quality aggregates (Table 2.4).

The effect of progressive regional or dynamic metamorphism, is first to impart foliations; where local pressure-induced movement occurs, the rocks become sheared. Platy or foliated structures in metamorphic rocks indicate that high shearing stresses have been the principal agency in their formation, and the aggregate potential of such rocks is reduced: the foliation may not always be visible to the naked eye, though individual mineral grains may exhibit strain lines when seen under the microscope. More intense metamorphism gives rise to widespread recrystallisation of minerals and in extreme conditions many rock constituents melt and may be intruded as an igneous

TABLE 2.4. *Generalised chart showing the origin and classification of common metamorphic rocks*

Original rock	Metamorphic environment—temperature and pressure conditions: Low grade—shallow burial	Medium and high grade—deep burial
	FOLIATED PLATY ROCKS (effects of pressure paramount)	
clay, shale and volcanic tuff	slate	phyllite, schist, gneiss*
clayey sandstone	slate and quartzite*	quartz-mica schist, fine-grained gneiss
clayey limestone	slaty marble*	calcareous schist
granite, quartzitic tuff	sheared granite, slate	granite-gneiss*, quartz-mica schist.
basalt, basic volcanic tuff	green (chlorite) schist	amphibolite (amphibole schist) hornblende gneiss*
	NON-FOLIATED, MASSIVE ROCKS (effects of increased temperature paramount)	
any parent rock		hornfels* (some recrystallisation but often original features still present)
quartzose sandstone		quartzite* (includes mica if impure) granulite*
limestone and dolostone	marble* (if impure, contains a wide range of calcium and magnesium silicates with increasing metamorphic grade)	

* Rocks most likely to be useful for aggregate.

body. In these conditions clay-rich sediments first form slates, which are characterised by closely-spaced cleavage planes perpendicular to the direction of greatest pressure, then phyllites (foliated rocks containing much white mica) and finally schists, in which the constituent minerals tend to be irregularly aligned imparting an uneven, wavy, 'schistose' texture. Gneisses are formed from igneous, metamorphic and sedimentary rocks in more extreme conditions. They are typically coarse-grained and banded in structure, reflecting differences in original composition accentuated by segregation of the constituents during recrystallisation and magmatic flow under great pressure deep in the crust. Striped gneisses resulting from the injection of thin sheets of molten material (commonly rich in quartz and feldspar) along parallel planes in a variety of country rocks are termed injection gneisses or migmatites. Such rocks are commonly massive and granular and can yield good quality aggregate. The main metamorphic rocks derived from common rock materials are classified in Table 2.4.

The wide diversity of metamorphic rocks arises from the variety of metamorphic processes and pre-existing rocks that are available for metamorphism, and their usefulness for aggregate is equally variable. It is probably generally true to say that the originally clayey, foliated or platy rocks must have been subjected to extreme or high grade metamorphism and to have undergone recrystallisation to render them potentially useful as aggregate and that the massive quartzitic metamorphic rocks probably provide the best prospective aggregate sources.

Few metamorphic rocks are of more than local importance as aggregates in the UK as they are commonly remote from the main centres of population. However, some thermally metamorphosed rocks such as the hornfels adjacent to the Shap granite in Cumbria, are of value because of their high crushing strength, and in Cornwall, Anglesey and northern Scotland metamorphic rocks contribute to local aggregate production.

2.6 Aggregate materials in desert regions

The effects of climate and ground-water conditions on the geological materials in an arid region, such as the Middle East where much experience on the use of indigenous aggregates has accumulated (Fookes and Higginbottom, 1980), are mechanical (thermal fracturing, wind erosion) and chemical. The widespread excess of evaporation over rainfall induces persistent upward leaching of salts, particularly carbonates, sulphates and chlorides, causing the mechanical disintegration of porous rocks through the growth of crystals in the pores. Where a high saline water table is present as in the coastal sabkhas and salinas of the desert interior, 'duricrusts', composed of substances transported by solution in the pore fluids may be developed. The most widespread crusts consist of calcium and calcium-magnesium carbonates ('calcrete' and 'dolocrete'), calcium sulphate ('gypcrete', or 'gypcrust' where weakly cemented) and sodium chloride ('salcrust'). Although the development of a duricrust may effectively strengthen the uppermost few metres of an originally weak material, such as the porous Eocene limestones on the western and southern shores of The Gulf, the immediately

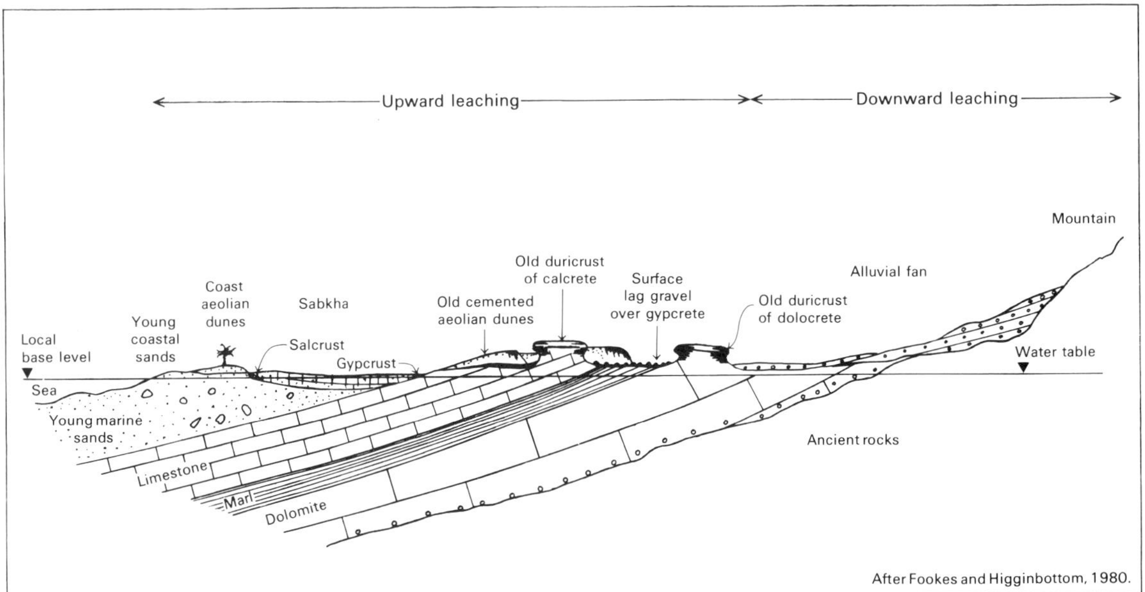

FIG. 2.15. Idealised section from desert coast to mountain interior showing typical associations of landforms and superficial deposits. (After Fookes and Higginbottom 1980.)

underlying leached layer may be rendered weaker and more porous than the unaltered parent material. More commonly in cement-bound materials the build-up of salts is deleterious in aggregates for use in concrete, although they may be usable in unbound applications and in bituminous mixes. The typical distribution of some common duricrusts is illustrated in the idealised section shown in Fig. 2.15 which also shows the distinction between areas of upward and downward leaching. Downward leaching commonly takes place where periodic flooding in the mountains and foothills and associated wadis effectively prevents the accumulation of sulphates and chlorides, though cementation by the less soluble carbonates may still occur in these zones of modern deposition, particularly in the coarse deposits which are not moved by the periodic floods. Thus, the active wadis in the mountains and parts of their alluvial fans where they open out on to the piedmount are the main sources of well-graded relatively clean sand and gravel: the material is in general poorly sorted, torrent-bedded localised spreads, which, however, can yield the most acceptable 'all-in' natural aggregate in a desert region.

Apart from the wind-blown dune sands of the deep interior of deserts, which consist predominantly of silica and silicate lithic grains but are commonly too fine-grained, rounded and dusty to be attractive for aggregate uses, the carbonate sands of coastal origin, which comprise the water-worn debris of modern marine organisms, are more acceptable, despite the frequency of porous, sometimes hollow shells contaminated by evaporitic substances. In some areas, the carbonate sands of the coastal dunes have contributed, through the action of on-shore winds, to the aeolian dunes farther inland, where, with increasing distance from the sea-board sabkhas, contamination by sulphates and particularly by chlorides tends to decrease.

Selective working of the lower, more coarse-grained parts of these dunes and blending dune sand with other material, such as crushed-rock sand, can yield commercially acceptable fine aggregate, provided that the crushed component possesses reasonably good particle shape, does not increase the concentration of chemical contaminants and does not contain excessive amounts of fines.

Of the hard rock sources in deserts, many of the calcretes suffer from the variable and random nature of the hardening process as well as potentially serious chemical problems; dense crystalline nodules may be surrounded by weak absorptive stone or contain veins of clay, which is commonly carbonate-rich and must be removed from the aggregate. In semi-arid areas, red clay soils ('terra rossa') may cover the hardened ground and invade solution-widened joints in limestone terrains. Generally speaking, the most promising hard rock sources in arid regions are the igneous and metamorphic rocks, which have good soundness, aggregate crushing and absorption characteristics, when unweathered.

2.7 Aggregate materials in tropical regions

In hot, wet areas, the decomposition of rocks through chemical weathering is intense and generally exceeds the effects of physical disintegration. In the inter-tropical zone weathering of rock produces clay-rich weathering mantles generally up to 100 metres deep and exceptionally very much more (Fookes *et al.* 1971), and fresh, unweathered and unaltered bedrock which might be expected to yield aggregate is commonly not readily accessible (but see below).

The distinction between soil and the weathering mantle is hard to draw—the weathered rock that is still in place tends to retain features of the unweathered rock and is separated from the overlying loose mantle of more or less transported material by stone lines of varied composition and controversial origin. In strictly residual soils, weathering attacks the rock by

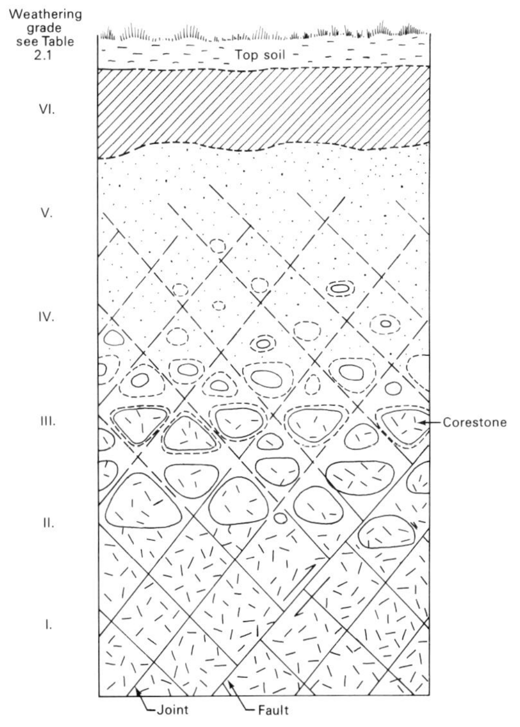

FIG. 2.16. Diagrammatic weathering profile of a jointed and faulted (igneous) rock body. (After Deere and Patton 1971.) The weathering grade is indicated at left (see Table 2.5).

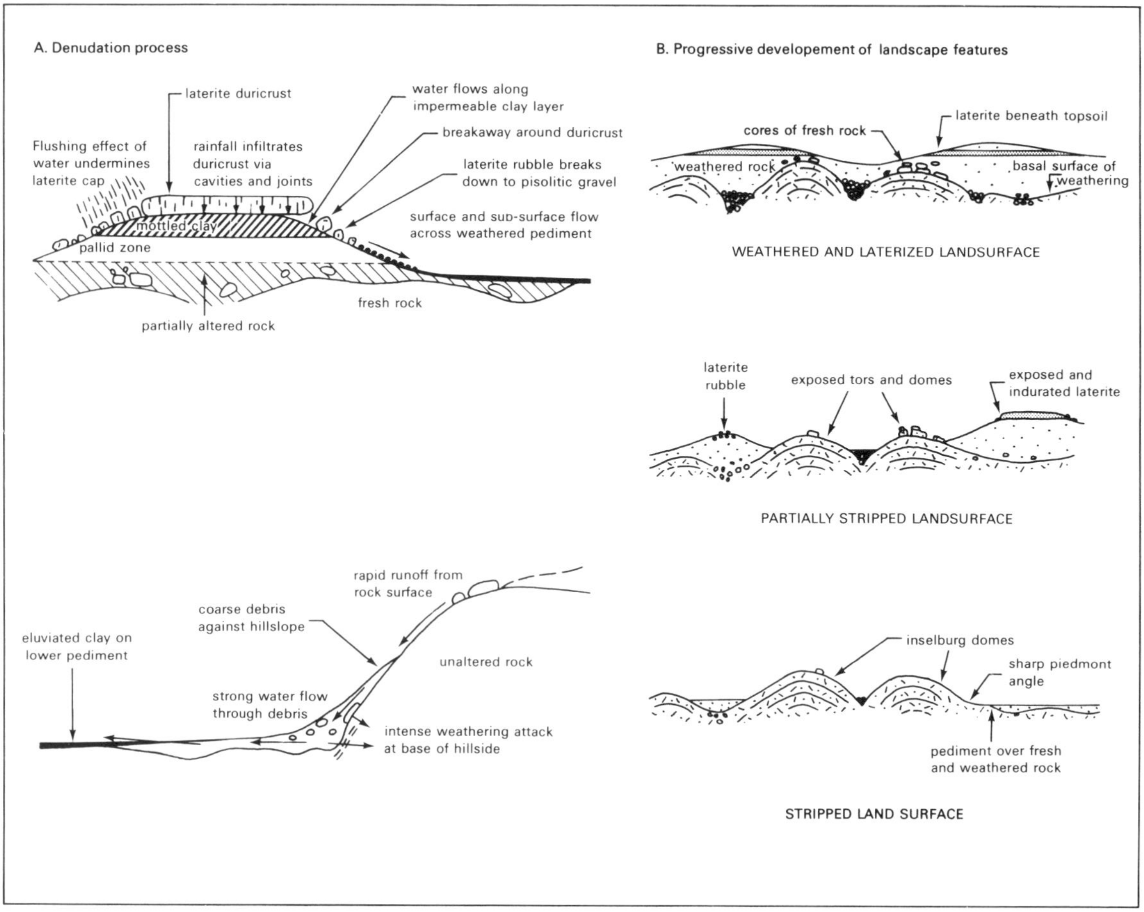

FIG. 2.17. Possible aggregate sources in a landscape subject to a seasonal tropical climate (savanna type, between the afforested humid tropics and the deserts). (After Thomas 1971.)

progressively permeating joint-bounded blocks, leaving kernels of fresh material within envelopes of decomposed rock. Further attack reduces the kernels by spheroidal wasting until total decomposition results in the formation of red earth residual soils (Fig. 2.16). In areas with heavy rainfall, dissolved chemicals are leached out and bases such as calcium and magnesium (and even silica) may be removed, leading to higher concentrations of alumina and iron oxides and hydroxides and the formation of clay minerals, such as kaolinite. Layered soils with an iron-rich layer, which may harden to ironstone gravel (ferricrete), are the lateritic soils (West and Dumbleton, 1970). In regions where erosion has followed laterite formation, flat-topped relict hills capped by lateritic ironstone may be common in the landscape, particularly on the borders of tropical forests, where periodic drying out can occur (Fig. 2.17).

Rain forest is developed on a variety of rock types and in a wide range of topographic situations. High plateaux and gently sloping uplands of Africa, the East Indies, Malaysia, the Guianas, Brazil and other parts of South America (Sanders and Fookes, 1970) may be deeply though irregularly weathered, both in depth and laterally. Occasional hills of fresh rock (inselbergs) may rise sharply from a surrounding plain to a height of tens or even hundreds of metres; limestones in particular, in tropical forests, may also give rise to a distinctive landform—the so-called 'haystack hills', and may contain much potentially useful aggregate.

References

AITKEN, A. M., MERRITT, J. W. & SHAW, A. J. 1979. The sand and gravel resources of the country around Garmouth, Grampian Region. *Miner. Assess. Rep. Inst. geol. Sci.* No. 41.

ARCHER, A. A., 1972. Sand and gravel as aggregate. *Miner. Dossier Miner. Resour. consult. Comm. No. 4.* H.M.S.O.

ASHWORTH, K. A., 1975. Limestones in the UK. *Ind. Miner.* **90,** 15–28.

BEAVER, S. H., 1968. The Geology of Sand and Gravel. Sand and Gravel Assoc. of Great Britain, London.

BLYTH, F. G. H. & DE FREITAS, M. A. 1974. A Geology for Engineers. 6th Edn. London: E. Arnold.

BRITISH STANDARDS INSTITUTION, BS 5930:1981 Code of Practice for Site Investigations (formerly CP 2001).

BURNETT, A. D. & EPPS, R. J. 1979. The engineering geological description of carbonate suite rocks and soils. *Ground Eng.* 41–48.

CEMENT LIME AND GRAVEL, 1963. Working the Bunter Pebble deposits at Weeford, West Midlands. *Cem. Lime Gravel,* **38,** 317–322.

CLARKE, M. R., 1975. The sand and gravel resources of the country east of Chelmsford, Essex. *Miner. Assess. Rep. Inst. geol. Sci. No.* 13.

CLARKE, M. R., DIXON, A. J. & KUBALA, M. 1979. The sand and gravel resources of the Blackwater Valley (Aldershot) area. *Miner. Assess. Rep. Inst. geol. Sci. No. 39.*

CLARKE, M. R., RAYNOR, E. J. & SOBEY, R. A. 1980. The sand and gravel resources of the Loddon Valley area. *Miner, Assess. Rep. Inst. geol. Sci. No.* 48.

CULLIMORE, R. N., PIKE, D. C. & JORDAN, J. P. R. 1976. Jurassic rock gravels as concrete aggregates. Tech. Paper 9, Sand and Gravel Assoc., London.

DEARMAN, W. R. 1974. Weathering classification in the characterisation of rocks for engineering purposes in British practice. *Bull. Int. Assoc. Engng. Geol.* **9,** 33–42.

DE BOER, G. 1973. Land created by sea. *Geogr. Mag. XLV,* **5,** 377–382.

DEERE, D. U. & PATTON, F. D. 1971. Stability of slopes in weathered rock. 4th Pan. Am. Conf., 87–163.

DINES, H. G., & EDMUNDS, F. H. 1929. The Geology of the country around Aldershot and Guildford. *Mem. geol. Surv. G.B.*

FOOKES, P. G. 1975. Geology for Engineers. Chap. 7, Civ. Engng. Practice (3rd Edn. Ed. Blake) 1975.

FOOKES, P. G., DEARMAN, W. R. & FRANKLIN, J. A. 1971. Some engineering aspects of rock weathering with field examples from Dartmoor and elsewhere. *Q. J. eng. Geol. London*, **4,** 139–185.

FOOKES, P. G. & HIGGINBOTTOM, I. E. 1980. Some problems of construction aggregates in desert areas, with particular reference to the Arabian peninsula. I—Occurrence and special characteristics II—Investigation, production and quality control. *Proc. Instn. Civ. Engrs. pt. 1,* 68 Feb. 1980, 39–90.

FOOKES, P. G. & HIGGINBOTTOM, I. E. 1975. The classification and description of near-shore carbonate sediments for engineering purposes. *Geotechnique. London* **25** (2) 406–411.

FOOKES, P. G., GORDON, D. L. & HIGGINBOTTOM, I. E. 1978. Glacial landforms their deposits and engineering characteristics in The engineering behaviour of glacial materials; *Proc. Symp. Midland Soil Mech. Foundn. Engng. Soc., Geo-Abstracts, Norwich.*

GEOLOGICAL SOCIETY ENGINEERING GROUP REPORT, 1977 (b) The description of rock masses for engineering purposes (Rep. Geol. Soc. Engng. Group Working Party). *Q. J. eng. Geol. London*, **10,** 255–388.

GRESSWELL, R. K. & LAWRENCE, G. R. P. 1978. The work of the Sea, Rivers and Ice. Hulton Educ. Pub.

HARRIS, P. M. 1976. Underground limestone mining in the U.S.A. *Trans. Instn Ming Metall. A.,* Vol. **85,** 75–84.

HARRIS, P. M. 1977. Sandstone. Mineral Dossier No. 17, *Miner. Resour. consult. Comm.*, H.M.S.O., London.

HARRIS, P. M., 1978. Igneous and metamorphic rock. No. 19, *Miner. Dossier Miner. Resour. consult. Comm.,* H.M.S.O., London.

HARRIS, P. M. 1982. Limestone and Dolomite. No. 23, *Miner. Dossier Miner. Resour. consult. Comm.* H.M.S.O., London.

HARRIS, P. M., THURRELL, R. G., HEALING, R. A. & ARCHER, A. A., 1974. Aggregates in Britain. *Proc. R. Soc. London,* **A339,** 329–353.

HAWKES, J. R. & HOSKIN, J. R. 1972. British arenaceous rocks for skid-resistant road surfacings. *Transp. Rd Res. Lab. Rep. No. LR 488.*

HAWKES, J. R. & HOSKING, J. R. 1973. British arenaceous rocks for skid-resistant road surfacings. *Quarry Managers J., London* **57,** 39–47.

INSTITUTE OF GEOLOGICAL SCIENCES, 1982. United Kingdom Mineral Statistics 1981. H.M.S.O., London.

LAHEE, F. H. 1961. Field geology. 6th Edn. McGraw-Hill.

LAXTON, J. L. & NICKLESS, E. F. P. 1979. The sand and gravel resources of the country around Lanark, Strathclyde Region. *Miner. Assess. Rep. Inst. Geol. Sci. No. 49.*

LEET, L. D., JUDSON, S. & KAUFFMAN, M. E. 1978. Physical Geology 5th Edn. Prentice-Hill, New Jersey, USA.

LITTLE, A. L. 1969. The engineering classification of residual tropical soils. *Proc. Intern. Conf. Soil Mech. and Found. Eng. 7th, Mexico* **1,** 1–10.

MCLEAN, A. C. & GRIBBLE, C. D. 1979. Geology for Civil Engineers: Allen and Unwin, London.

NICKLESS, E. F. P. 1971. The sand and gravel resources of the country south-east of Norwich, Norfolk. Description of 1:25 000 resource sheet TG 20. *Rep. Inst. geol. Sci.* London 71/20.

OLLIER, D. 1975. Weathering. 304 pp. Longman, Edinburgh.

QUARRY MANGERS' JOURNAL, 1972. Welsh gritstone for premium road materials. Developments at Craig-yr-Hesg Quarry of ARC. *Quarry Managers J. London,* **56,** 409–416.

QUARRY MANAGEMENT AND PRODUCTS, 1974. Buddon Wood Quarry. A major new development in Leicestershire by Radland Roadstone. *Quarry Manage. Prod. London* **1,** 161–176.

QUARRY MANAGEMENT AND PRODUCTS, 1975. Developing and operating a mammoth quarry. *Quarry Manage. Prod. London* **2,** 12.

QUARRY MANAGEMENT AND PRODUCTS, 1977 (a) New processing plant at Scottish basalt quarry. *Quarry Manage. Prod., London* **4,** 117–121.

QUARRY MANAGEMENT AND PRODUCTS, 1979. Road materials from Underground. *Quarry Manage. Prod., London* **6,** 246–248.

READ, H. H. & WATSON, J. 1966. Beginning geology. MacMillan, Allen and Unwin, London.

SANDERS, M. K. & FOOKES, P. G. 1970. A review of the relationship of rock weathering and climate and its significance to foundation engineering. *Eng. Geol.* **4,** 289–325.

SIMPSON, B. 1966. Rocks and minerals. Pergamon, London.

SOMERSET COUNTY COUNCIL, 1971. Quarrying in Somerset. Taunton, UK.

SQUIRRELL, H. C., 1976. The sand and gravel resources of the Thames and Kennet Valleys, the country around Pangbourne, Berkshire. *Miner. Asses. Rep. Inst. geol. Sci., No. 21.*

SQUIRRELL, H. C. 1978. The sand and gravel resources of the country around Sonning and Henley, Berkshire, Oxfordshire and Buckinghamshire. *Miner. Assess. Rep. Inst. geol. sci., No. 32.*

THOMAS, M. 1971. Savanna lands between desert and forest. *Geogr. Mag. XLIV,* **3,** 185–189.

THOMPSON, D. B. 1970. Sedimentation of the Triassic (Scythian) Red Pebbly Sandstones in the Cheshire Plain and its margins. *Geol. J.*, **7,** 183–216.

TOMKEIEFF, S. I. 1964. The economic geology of quarried materials—2 Igneous and metamorphic rocks. *Quarry Managers J. London* **48,** 85–94.

VERNEY, SIR, R. B. (Chairman) 1976. Aggregates: the way ahead. Report of the Advisory Committee on Aggregates. (H.M.S.O., for Dept. of the Environment).

WEINERT, H. H. 1964. Basic igneous rocks in road foundations. *S. African Council Sci. Indust. Res., Report No. 218, Nat. Inst. Road Res. Bull,* **2,** 1–47.

WEST, G., & DUMBLETON, M. J. 1970. The mineralogy of tropical weathering illustrated by some west Malaysian soils. *Q. J. eng. Geol., London*, **3,** 25–40.

WEST, R. G. 1977. Pleistocene geology and biology. 2nd Edition, Longmans, London.

Chapter 3. Field investigation of deposits

3.1 Introduction

The location of aggregate materials through the appraisal of geological and geotechnical information and the eventual selection of sites from which to extract them, requires the collation of data from many sources through desk studies, field reconnaissances and the evaluation of prospects to the necessary level of geological assurance. Following the discovery of a possibly useful deposit, further, more specific investigations of its physical and mechanical characteristics may confirm that it is potentially useful as aggregate—it is a resource (McKelvey, 1972)—and that certain parts of it may be capable of being worked at a profit and may be regarded as a reserve (3.7.2).

Whilst the main objective of this section is to outline methods for investigating and reporting on possible sources of aggregate, some attention is also given to operational, commercial, environmental and planning matters which, along with the geological factors, have to be considered by the engineer or geologist during the assessment of reserves for a quarrying prospect.

3.2 Desk study

Although the assembly of published and unpublished data will draw primarily on the collation of available previous geological and geotechnical work, particularly geological maps and records, and site investigation reports, many other sources may give a lead to the presence of aggregate (Dumbleton and West, 1976; Rose, 1976; Eastaff *et al.* 1978). Thus, previously unreleased records of mining and quarrying operations, former aggregate workings or of investigations for them or for civil engineering works may be relevant, along with land use or soil surveys for agricultural purposes, old topographic maps and even local oral tradition from well sinkers and the local population (Chaplow, 1975).

3.2.1 Topographic maps. The prime value of contoured topographic maps lies in their portrayal of physiographic features (land forms) which may give an indication of the presence of aggregates and their accessibility (2.3.2). At the reconnaissance stage, small-scale maps at 1:250 000 to 1:50 000 scales with contouring at 10 m intervals or more may be adequate, but at later stages in an investigation large-scale maps for plotting the locations of all relevant information will be needed for prospect evaluation. Working maps down to 1:500 scale demonstrating such details as property boundaries, drainage lines and directions, the location of specimen trees and intervisibility information in support of submissions for planning approvals for site development, quarry management and the eventual restoration of the workings, will be indispensable.

3.2.2 Geological maps. The essential prerequisite when prospecting for aggregates is a geological map. If none exists, geological mapping specifically for mineral aggregate deposits should be undertaken using appropriate supporting investigative techniques including ground geophysics (Zalasiewicz *et al.* 1985), geomorphological mapping (Doornkamp *et al.* 1979), and drilling and sampling; in addition, particularly in little-known terrains and in developing countries, remote geophysical and optical sensing from aircraft (or even from spacecraft) may be useful. There is no doubt that, at the site investigation stage, the evidence of good geological mapping can be used to reduce the frequency and cost of drilling and sampling by tying point data into an interpretative pattern, which is the crux of the mapping exercise (Himus and Sweeting, 1968, Geological Society, 1972). Conversely the geological mapping is constantly refined to accommodate new evidence on the variability of deposit thickness and quality until the interpretation is judged to be good enough for aggregate assessment at whatever degree of sophistication, whether for measurement of regional resources or proof of reserves for extraction. The exercise of this judgement is the particular expertise of the geologist concerned with aggregates (Fox, 1979).

In the United Kingdom, the New Series 1:50 000 geological maps together with the detailed descriptive memoirs of the British Geological Survey (BGS)—[established as the Geological Survey of Great Britain in 1835 and known as the Institute of Geological Sciences (IGS) from 1965 until 1983]—are interpretations and summaries derived from the national archive of subsurface records and 1:10 000 scale, formerly 6-inch to 1 mile, field survey maps which have formed the basis of geological surveys since before the turn of the century. These detailed maps are published for most coalfields and some other areas, and are available for consultation in manuscript or can be purchased as photocopies or prints from dye-line masters. Approximately four-fifths of the UK are mapped on this scale, and although the quality of the interpretation embodied in some older maps varies considerably (Sabine, 1977; Gordon *et al.* 1982), as illustrated in Appendix A3.1, but they provide invaluable guidance on the occurrence and distribution of formations that are potentially useful for aggregate. The main scale of publication however, is now 1:50 000, and the sheet boundaries and availability of these maps (and of the one-inch to the mile maps, which they are replacing) are given in Appendix A3, Fig. A3.2. Over 400 sheets are in print: updated index maps and lists are published in the annual map catalogue of the Ordnance Survey and in the Annual Reports of the BGS. Other specialist publications of the Survey may also be useful in the search for sources of aggregates, for example, engineering geology maps such as those for Belfast and Milton Keynes, geophysical maps, both gravity and aeromagnetic, hydrogeological maps and aggregate resource maps.

3.2.3 Aggregate resource maps in the UK. In recent years, drilling and sampling surveys designed specifically to appraise the amount and main physical characteristics of the major occurrences of sand and gravel in England, Wales and Scotland have been commissioned by the Department of the Environment, and the results, in the form of 1:25 000-scale resource maps with reports, are published in the Mineral Assessment Reports of the Geological Survey. To date more than one hundred and forty reports have been published (Appendix A3, Fig. A3.3). The objective of these surveys (which have been extended to include the identification of chemical-grade limestone resources, as well as those remaining for use primarily as aggregate, conglomerate and sandstone) is to provide information at the 'indicated' level (Bureau of Mines and Geological Survey, 1948, 1976) primarily for land use and mineral planning purposes by local and central authorities and by industry. The investigations add significantly to the national geological archive particularly in regard to glacial and fluvial deposition, Quaternary stratigraphy, and the refinement of the mapping of the drift deposits and hydrogeology. For the purposes of these regional sand and gravel assessments, the BGS has adopted a metric grain-size classification (Archer 1970) based on that of Willman (1942) and after consulting widely, has fixed arbitrary physical criteria in order to define on the resource maps and in the borehole logs, the occurrence of 'potentially workable' sand and gravel, that is, 'mineral' (Dunkley, 1979 and Fig. 3.1).

3.2.4 Small-scale maps. Examples of these maps showing the general distribution of natural aggregate and hard rocks suitable for crushing in Great Britain accompanied the Survey's contribution on national resources to the Report of the Advisory Committee on Aggregates (Verney, 1976). The Ministry of Housing and Local Government published 1:625 000-scale maps showing occurrences of gravel and limestones in Britain in 1965, and more recently the official Quaternary map of the United Kingdom was published at this scale (Institute of Geological Sciences, 1977). A wide range of geological and specialist maps and publications relating to many parts of the world including many journals and papers are available for reference in the main BGS libraries at Keyworth, Nottinghamshire, in South Kensington, London, in Edinburgh and Belfast and in smaller libraries in the Exeter, Aberystwyth and Newcastle offices. Most of the maps are at scales smaller than 1:50 000.

3.3 Field investigation

Geological mapping will usually provide basic information on the nature of the potential aggregates, but in unknown terrains and before any fieldwork is contemplated, it may be helpful in identifying possible targets for more detailed investigation to utilise some form of remote sensing, such as satellite imagery, air photographs and possibly airborne geophysical surveys (Norman, 1968, 1969; Beavan, 1976). Ground geophysical techniques can usefully augment direct ground exploration since most have to be interpreted against proven ground conditions. Ground geophysics in the aggregates field is a valuable aid in confirming rock associations between sampling points, usually boreholes (Cratchley, 1976; Clarke, 1983; Cannell and Crofts, 1984).

3.3.1 Satellite imagery. Data from Landsat satellites either in digital form or as imagery can be enhanced for regional geological studies but the spatial distribution is generally too coarse for mapping aggregates. However, Landsat false colour composite imagery has been used successfully to locate and map areas of indigenous construction material, for example in Botswana. Although detailed studies were limited by Landsat ground resolution, the imagery was effective in the location of calcrete occurrences over large areas at a reconnaissance scale (Beaumont, 1979). Multispectral satellite photographs, taken from Skylab, have higher resolution than Landsat and have been used to identify the courses of buried channels on the Plain of Venice in northern Italy. These sand and gravel deposits could be delineated by changes in reflectivity caused by variations of soil moisture and vegetation (Cassinis, 1977). The use of Skylab photographs was estimated to reduce the need for ground control by ninety per cent.

3.3.2 Aircraft remote sensing. Though much can be done by the non-specialist, photogeological interpretation of stereoscopic photography is a field of expert study (Latham, 1965; Allum, 1966; Avery, 1968; Dumbleton and West, 1970) which can lead rapidly to the identification of lineaments, soil and rock types and landforms for the production of outline geological mapping and to provide a guide to the occurrence of aggregate materials. Thermal imagery, obtained from an airborne infra-red linescanner, records emitted radiation from the earth's surface. It is becoming increasingly used to discriminate between

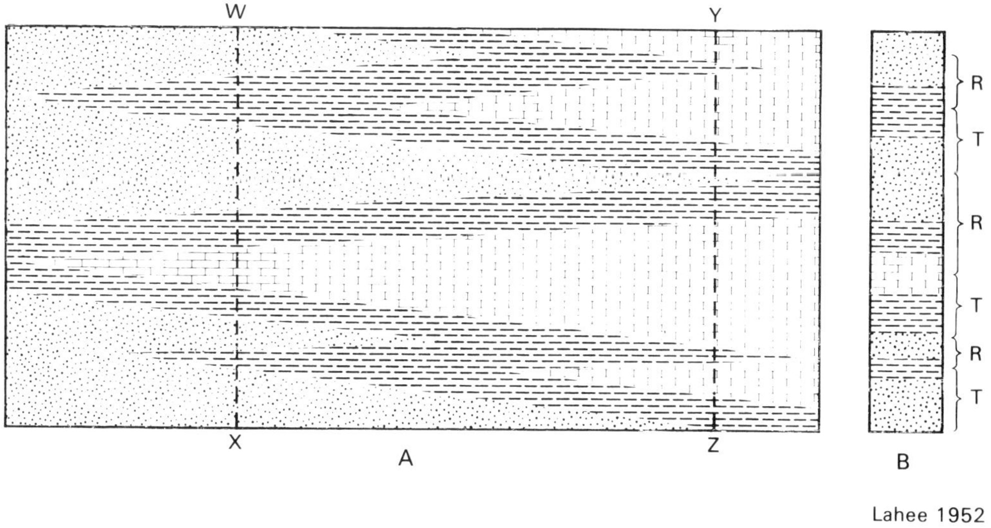

Diagrammatic section, A, of deposits laid down during successive transgressions and regressions of the sea, demonstrating that a formation may be characterised by different rock-types in space (laterally) and time (vertically). Land-derived materials have come from the left (sandy deposits grading into clayey deposits in the deeper water) whilst carbonate sedimentation (limestone) predominated out to sea (to the right)

The columnar section, B, taken at WX, shows transgressive (T) and regressive (R) phases of sedimentation; the same phases at YZ produced quite a different lithological succession.

FIG. 3.4. Sediments deposited contemporaneously in different parts of a sedimentary basin may vary widely in lithology (Lahee 1952).

emplacement. The presence of discontinuities (bedding, shear and fault-planes, cleavages, fissures and joints that are open or openable) and the effects of deep-seated alteration and superficial weathering have a profound influence on the potential suitability of a rock mass for aggregate. In wet humid climates, the effects of tropical weathering may extend to great depths and render most of the readily accessible rock bodies useless for aggregate, whereas feint physical weathering in some cool temperate conditions, for example, may enhance the potential of a deposit by leaving the aggregate material loosened and more easily extractable. The opening of discontinuities by gentle physical weathering may improve the rippability of some aggregate materials.

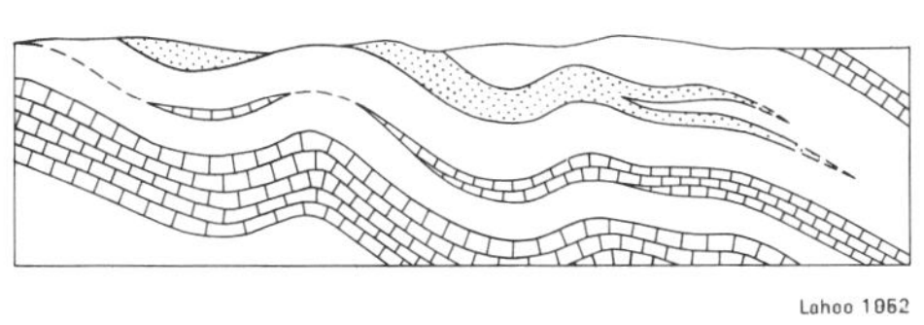

FIG. 3.5. The lensing (or 'pinching') of beds in a gently folded rock succession (Lahee 1952).

3.5 Sample collection

Many of the geological factors in paragraph 3.4 can be partly explored, though perhaps somewhat subjectively and interpretatively during the geological mapping and literature study of the prospect. But no matter how skilled the observer, the necessary detailed objective appraisal of the aggregate potential will require specific data on the subsurface extent of the occurrence and of the quality of the rock material obtained from the study of representative samples, which can only be obtained from a programme of drilling, pitting, trenching or excavation, or of sampling from exposures. Advisory literature on the practicalities and pitfalls involved in appraising aggregate sources is scattered, but the guidelines of the Sand and Gravel Association (1967) and of Dunn and Cutcliffe (1971) include advice and recommended field procedures for exploration, logging and sampling, the latter based on experience in the USA. The following sections are drawn from British practice (for example, Glossop, 1968). Further advice on drilling and sampling methods will be found in Section 19 of the Code of Practice for Site Investigations (BS 5930: 1981).

3.5.1 Drilling for sampling, hard rock. Drilling for sampling of hard rocks may be undertaken by rotary core drilling machines using diamond-studded crowns and double core-barrels with water, mud or air as flushing media. The objective in all investigations of hard rocks is to obtain total recovery of the rock cores for description and testing; the core diameter should be large enough to provide sufficient material for the common index testing procedures to be undertaken to British Standard specification (BS 812: 1975) and for a vertical slice or slab of the core to be retained for future reference, in such condition that original depth interval markings are preserved (Chapter 7). Comprehensive logging should be undertaken according to recommended procedures (Geological Society, 1970). A core diameter of 100 mm, 70 mm (H size or equivalent) or exceptionally 54 mm (NX size or equivalent) may be adequate (BS 4019: 1974—Rotary Core Drilling Equipment).

The drilling and sampling of compacted, indurated and cemented pebble beds and conglomerates, are notoriously difficult to achieve by rotary methods because detached pebbles tend to foul, jam and seriously abrade the drilling tools. Often strongly-cemented conglomerates can be successfully cored.

By means of down-the-hole hammer drilling using a button-bit and air, or air and water as the flushing agents, comminuted samples can be collected in a cyclone with sufficient precision for direct lithological correlations to be made between particle-size analyses of the samples and photographic records of the borehole walls. This method, which follows some of those used by the extractive industry, has been successfully tried in assessing the conglomerate resources of the Sherwood Sandstone Group ('Bunter' Pebble Beds) in Staffordshire (Piper and Rogers, 1980, Rogers *et al.* 1982).

3.5.2 Shallow pits and trenches. Trial pits and trenches may be conveniently dug by means of track-laying hydraulic bucket excavator in dry deposits (A4.1.5) and can rapidly provide bulk samples of sand and gravel from depths of 6 m or so. The diggings, unlike boreholes, permit *in situ* examination of deposits in cross section; if necessary, large samples of truly representative material including the fines and large stones, which may be lost from drill-hole samples, can be obtained for trial processing in an existing commercial plant. Hand-dug pits can be very useful where labour is available.

3.5.3 Drilling for sampling, sand and gravel.

3.5.3.1 Hand auger boring. Using light-weight rods and spiral, spatulate or bucket samplers up to 200 mm diameter, samples may be taken to 5 to 6 m in suitable ground conditions without the aid of a tripod or casing. The strata must be self-supporting, soft and with only very few pebbles. Sampling of gravelly material and from below the water table is rarely satisfactory. Hand auger sampling is nevertheless economical and rapid and it provides an ideal first look at a prospect. Hand augering is, of course, a basic tool in geological mapping in soft-rock terrains.

3.5.3.2 Power auger machines. These range from small mobile drills powered by small-capacity petrol engines capable of reaching depths of 10 m or so at 75 mm diameter and operated by one or two men, to large lorry or tractor-mounted rigs which can reach depths exceeding 25 m at 200 mm diameter and more. Sampling may be achieved by means of continuous auger-flights which are attached throughout the whole length of the drilling string or by single flight auger which is mounted only at the leading end of the drill string and must be withdrawn from the borehole for sample retrieval. As only very few auger machines are capable of advancing borehole casing whilst drilling within it, open-hole power auger drills are commonly used but they have numerous disadvantages when precise sampling is required. Unconsolidated material, particularly beneath the level of the water table, tends to collapse into the borehole and renders progress difficult and ultimately impossible; the rotary action of the auger tends to churn the strata, leading to contamination of one layer with another so that accurate sampling is generally very difficult to achieve. The method, nevertheless, can provide good guidance on the thickness of overburden to depths of about 10 m and in dry ground conditions samples of sand and gravel may be brought to the surface though they will have been severely mixed and contaminated with other deposits. As a reconnaissance tool a power auger has a distinct advantage over other types of drill in that it can quickly penetrate unstable strata without the need for lining tubes. Extra flighted auger lengths can be added at the surface as drilling proceeds but the rate of progress of the drill through the various strata can only be inferred by a skilful operator: the depth from which samples are derived cannot be accurately fixed. Auger machines which can sample from within a casing are much to be preferred both in terms of sampling accuracy and speed (Hill McDonald, 1972) though they are not commonly available.

3.5.3.3 Hollow-stem augers. These can penetrate unstable strata without the need for lining tubes and sampling can be done within the hollow stem by rotary coring, augering or by percussive methods to depths of perhaps 50 m in glacial deposits. Although sampling can be very rapid and reliable, these machines are expensive to buy, costly to run and their great size and weight may pose problems of access to land.

3.5.3.4 Light cable percussion boring. This is an

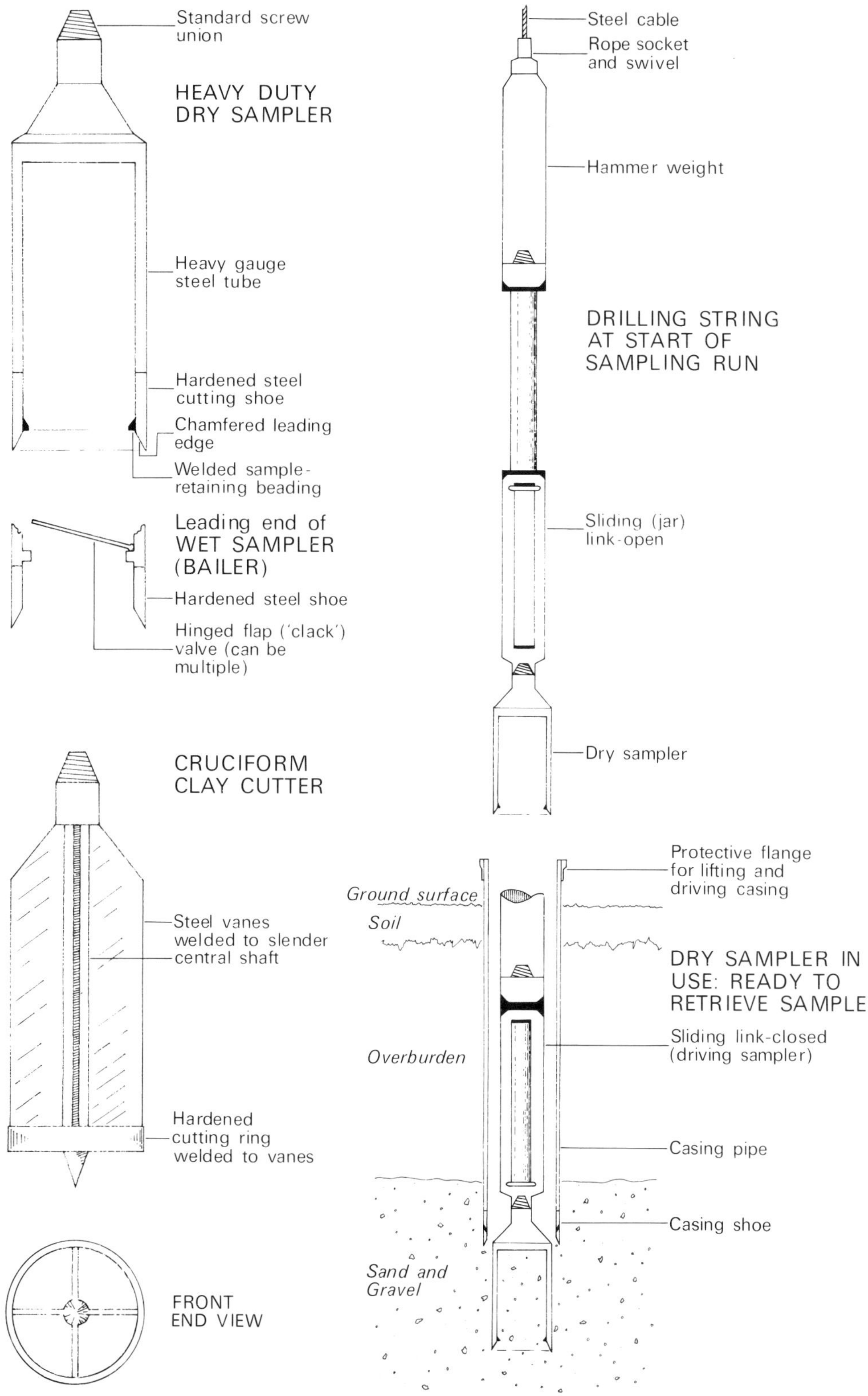

FIG. 3.6. Drilling tools for use with a cable percussion drilling rig for sampling sand and gravel from drift sequences (Thurrell 1981).

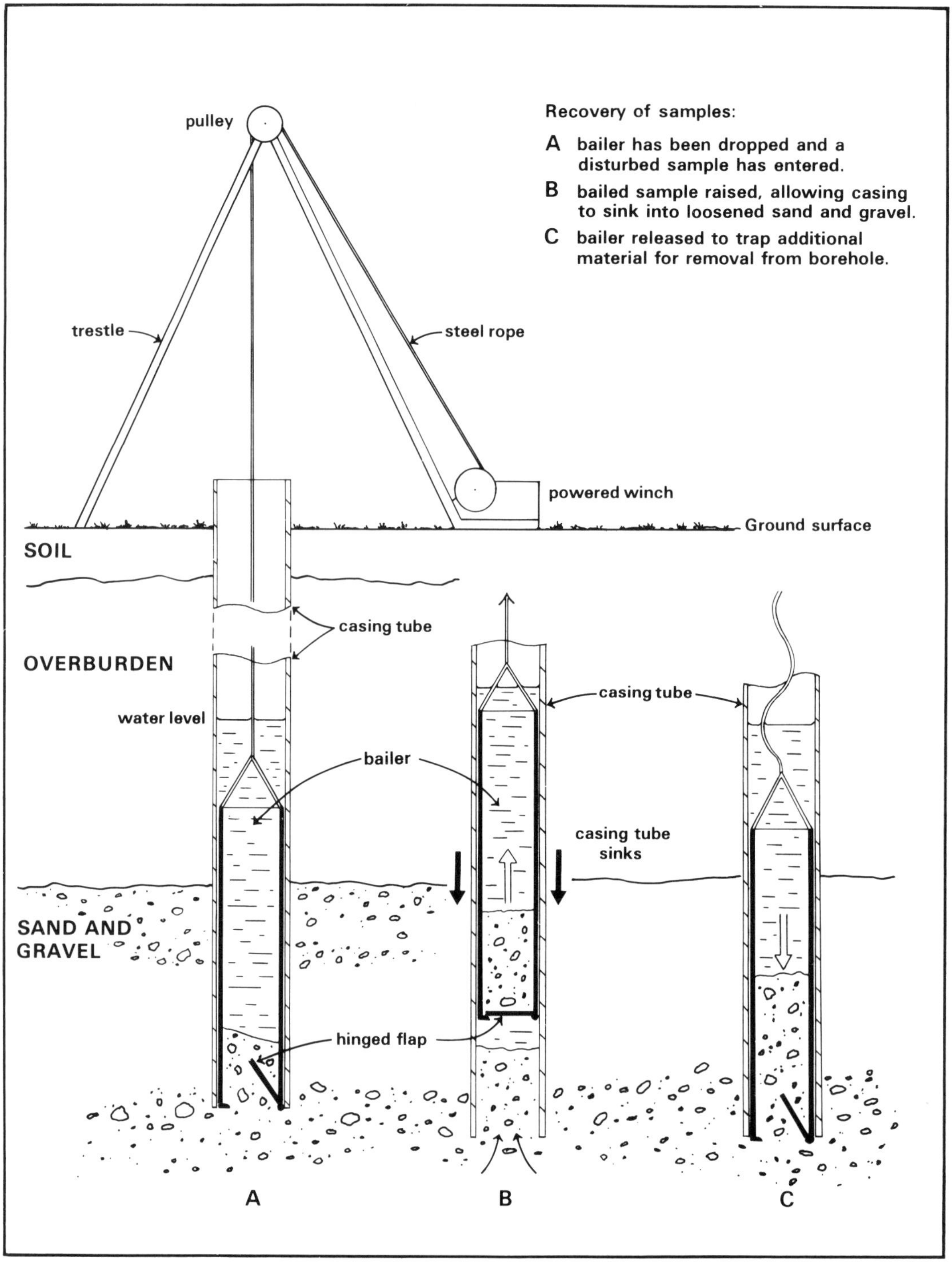

FIG. 3.7. Mode of operation of cable percussion rig for bailing unconsolidated material.

adaption of standard well-boring methods and commonly utilises clay cutters, chisels (for breaking up hard rock obstructions) and bailers (for removing water-laden samples from the borehole) working within lining tubes (casing) which enter the ground under gravity as the boring proceeds. To ensure that samples of sand and gravel are truly representative, it is necessary to take samples in dry and if possible undisturbed conditions within the borehole and this can be achieved by driving a sampling tube ahead of the leading edge of the casing by the use of a sinker bar with sliding hammer (Fig. 3.6). In this way, all of the material from within the ideal daimeter of the borehole can be sampled dry (Hill McDonald, 1972). Slightly clayey and sandy drift deposits containing gravel and even cobble grades (2 mm to 200 mm) can be sampled from depths to 30 m using a 2-ton percussion rig without seriously breaking the particles, provided tools up to 300 mm diameter are employed and the deposits are not unduly consolidated. A limitation on the depth of drilling is commonly imposed by the capacity of the rig to extract the lining tubes from the ground, though this may be overcome to some extent by the insertion of a reduced diameter tube within the main casing tube (and the substitution of appropriate sampling tools). In waterlogged conditions the tendency for groundwater to rise rapidly up the borehole carrying with it material from beneath the sampling level effectively prevents further sampling unless the casing can be advanced into a watertight stratum. Increasing consolidation of the drift deposits at depths exceeding about 30 m in most terrains, calls for the alternate use of the chisel and sampler. The drilling then proceeds very slowly and the samples are commonly severely broken by the percussive force of the chisel; other methods of boring using high-powered machines are to be preferred in these circumstances.

The cable percussion method is nevertheless to be strongly preferred where sampling is the major objective because progress cannot be maintained unless casing is inserted and this ensures that the samples remain uncontaminated (Fig. 3.7). Also, the machinery is relatively unsophisticated, cheap to buy and simply maintained; its modest size is advantageous in gaining easy access to drilling sites (Thurrell, 1981).

3.5.3.5 Reversed circulation drills. These are often of large size, operate with a double-wall usually non-rotatory drill-pipe system: the flushing medium, which may be air or foam, is forced down the annulus between the outer and inner tubes to a rotating drill bit where it sweeps the sample or rock cuttings back up through the central tube at high velocity, to be collected in a pneumatic separator at the surface. Sampling is uncontaminated and entirely representative; in most conditions the depth of sampling can be accurately located, though the speed of the drilling operation is such that refined sampling and logging may not be achieved unless many hands are available. For economical deployment of such machines, full advantage must be taken of their high output and this can normally be ensured only where extensive prospecting is to be undertaken in relatively open terrains.

3.6 Marine exploration

Submarine occurrences may represent inundated and possibly reworked fluvial, glacial, and shoreline deposits with a variable cover of modern sediments deposited from rivers or derived from erosion of coastal deposits. In certain conditions the fine-grained unconsolidated sediments may migrate as sand waves of considerable amplitude (Veenstra, 1969).

Whilst inspection and sampling of sea-bed deposits by skin divers may provide useful preliminary guidance, more sophisticated follow-up methods will normally be deployed. Prospecting must include investigation of the bathymetry of the sea floor as well as the distribution, thickness and quality of the potential aggregate materials.

3.6.1 Echo sounder. The returning signals can be displayed to given an analogue record of the sea-bed or stored on tape for plotting and contouring. The accuracy of the record can be improved by mounting the transmitter in a towed fish rather than in the hull of the ship, but unless the lines of survey are very closely spaced there is a liklihood that significant variations in sea-bed topography may be missed.

3.6.2 Side-scan sonar. This system involves the tranmission of a beam of sound from a towed fish and the collection and display of return signals reflected by irregularities of the sea floor or the texture of deposits, extending over a survey tract up to 1500 metres wide (Simpson, 1979). Areas of soft sand, silt and clay give very weak return signals and show as lighter areas on the display, whereas more strongly reflective granular material, such as gravel and boulder clay, gives a stronger return and exposures of bedrock (and the presence of obstructions, such as wrecks) produce the strongest (darkest) reflections.

3.6.3 Sampling. Although these reconnaissance techniques give a general indication of the shape and likely physical constitution of the surface of the sea-bed sediments, positive identification must depend on direct sampling. Samples weighing a few kilogrammes may be collected from the surface of the sea-bed using small spring-loaded or mechanically actuated grabs lowered from a ship; larger samples may be dredged. Sampling below the sea-bed can be achieved by dropping a weighted coring tube under gravity. Cores up to 100 mm in diameter and 6 metres long can be recovered from soft sediments by this method, but in more consolidated materials, like boulder clay, sand and gravel, sampling tubes can be much more effectively advanced by pneumatic or electrically driven hammer drills (vibro-corers). Deeper sampling to depths approaching 20 metres in

100 metres of water can be achieved by means of an air-lift, hydraulic-jetting system combined with a driving hammer. The speed of the sampling operation and the volume of the material collected by this prospecting drill require special arrangements for splitting and storing the samples on board ship; the average daily performance is about 10 boreholes of 6 m and 10 m in heavy seas with swells up to 2 m or 3 m (Hess, 1971).

3.6.4 Seismic and acoustic prospecting techniques. The high cost of prospecting at sea dictates that rather less sampling is likely to be undertaken than on land, and it is normal practice to interpolate between boreholes or sampling stations by means of seismic reflection profiling techniques. The methods require an acoustic source and a receiver to provide a time-scaled acoustic section of the geological structure beneath the sea-bed which is convertible to a true-scale horizontal section of the strata. For shallow penetration applications, appropriate to aggregate prospecting, the acoustic sources are commonly the pinger, a powerful echo-sounder, which provides good resolution (±10 cm) in sediments a few metres thick, and the boomer, an electro-mechanical device which generates a shock wave from which reflections to depths exceeding 30 m in drift sequences can be reliably recognised. Fine detail can be discerned in the uppermost 5 m, with differentiation of layers down to 0.3 m in thickness.

Most surveys for sand and gravel include geophysical investigations coupled with grid sampling using a sea-bed drill such as a vibro-corer, but the major shortcoming of sea-bed sampling is the loss of the fine sand, silt and clay fractions from the samples. In these circumstances bulk sampling with a dredger may provide better guidance to the commercial potential of the shallow part of a deposit, but it can give little indication of the quality of the deposits as a whole or of the likely volume of the resource.

A valuable illustrated review of techniques applicable to exploration in the shallow seas is given by McQuillin and Ardus (1977).

3.7 Presentation of field and test results

The following procedures, suitably adapted, can apply whether the investigation is wide-scale (regional) or site-specific in approach.

3.7.1 Cross sections and maps. These should show all geological data needed for estimating the volume and quality of the prospect, and form the basis of an extraction plan, if any. Maps and sections should be drawn to true scale (without vertical exaggeration) and should cross-refer.

In particular, the nature and recorded thickness of overburden and of weathered material should be shown at all data points on both maps and sections: the latter should intersect as many borehole sites as possible, to avoid projecting information onto lines of section.

The results of physical tests on samples should be correlated with the geological interpretation (preferably by plotting them on the maps and sections) so that the incidence of substandard or interbedded waste materials can be readily assessed. For more detailed guidance and for the calculation of volumes of useful and other materials must be made to borehole logs and laboratory data sheets.

3.7.2 Estimation of aggregate reserves and resources; definitions of terms. The large literature on the classification of mineral resources is testimony to the complexity of the subject, yet the establishment of limiting physical criteria, which are themselves largely conditioned by economic constraints, is necessary before estimation and comparison of the quality and amount of potentially useful prospects can proceed. American usage has been widely accepted and a review of terminology (Bureau of Mines and Geological Survey, 1976) defines a mineral resource as a concentration of naturally-occurring materials in such form that economic extraction is currently or potentially feasible. Material classified as a reserve is

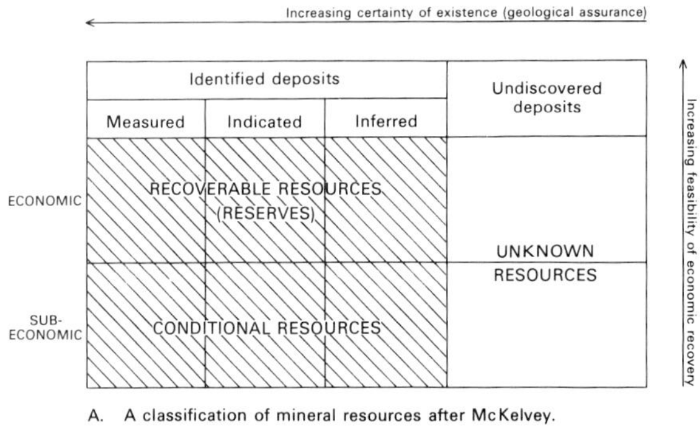

A. A classification of mineral resources after McKelvey.

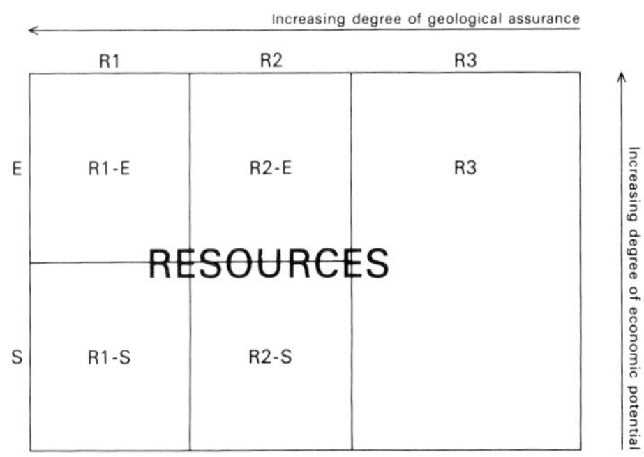

B. A classification of mineral resources, simplified from the United Nations system.

FIG. 3.8. Classification of mineral resources, applicable to aggregates. A. After McKelvey, 1972. B. After United Nations Experts Group, 1979.

that part of an identified resource producible at a profit at the time of classification. A diagrammatic presentation of this classification known colloquially as 'the McKelvey diagram' after its originator (McKelvey, 1972) is in Fig. 3.8,A. It shows how total resources can be categorised in terms of economic feasibility of extraction (economic constraints) and the degree of geological assurance (the level of available knowledge of the physical characteristics).

A simpler system has recently been proposed for international application by an Experts Group on Definitions and Terminology for Mineral Resources in the United Nations (United Nations, 1979).

3.7.3 Basic resource categories. These, categorised as R-1, R-2 and R-3, are distinguished in order of decreasing geological assurance. Each of these categories can be further subdivided into those considered to be exploitable under the prevailing socio-economic conditions (sub-category E) and other resources (sub-category S). These categories are set out in Fig. 3.8,B. Whilst there may be an inclination to equate the proposed categories with more familiar terms (R-1 = demonstrated, explored; R-2 = inferred, estimated; R-3 = potential, speculative) in the long term, the letter–number combinations should gain acceptance where for reasons of language and established often rather complex usage, the existing subjective descriptive terms will not. It is hoped that increasing acceptance of alpha-numeric terms will encourage clearer understanding of resource evaluations between all users of such data.

3.7.4 Methods for estimating amounts. Whether the investigation of a source of aggregates has been designed as a detailed appraisal, for example, to extend the known reserves for an existing quarrying operation for which the limiting criteria are well established, or it is part of an inventory of resources on a regional or national scale, the requirement is for numerical estimates of the quantity and quality of useful material in the prospect, upon which forward planning can be reliably based. Very commonly, graphical methods have been used in calculating the volume of occurrences, based on geological mapping and drill-hole data. If a deposit is largely regular in shape, true-scale cross-sections may be drawn parallel to each other, at regular intervals and the sectional areas of useful material measured. The volume is given by the product of the mean of these areas and the distance between the end sections. Alternatively, the volume of material between adjacent cross-sections within the prospect can be calculated and summed. Isopachytes may be drawn on a plan, joining points of equal thickness as demonstrated by borehole or other records (Fig. 3.9). The area between each successive pair of isopachytes is obtained by planimetry and multiplied by the average thickness value represented by them. The sum of the results is the volume of the deposit. Rather more complex calculations based on the construction of triangles with thickness data-points at their corners, or of polygons representing the areas of influence of data points, may be undertaken by reference to the relevant texts.

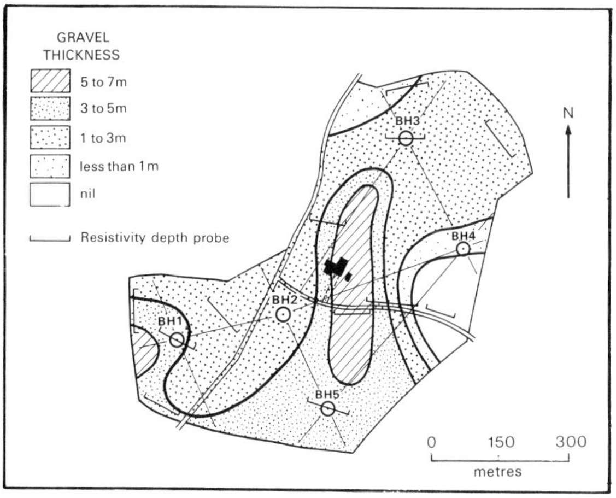

FIG. 3.9. Isopachytes drawn from borehole and intervening resistivity depth probe data for a sand and gravel prospect.

3.7.5 Reliability of estimates. The shortcomings of these largely graphical methods is that the reliability of the estimates cannot be readily demonstrated. The use of simple (or classical) statistical procedures can remedy this (Fig. 3.10), and the calculation of confidence limits using the 'students-t' distribution (Davis, 1973) is being increasingly adopted for sand and gravel surveys in the UK. A procedure is set out in each of the Mineral Assessment Reports of the Institute of Geological Sciences dealing with sand and gravel appraisal, which is generally applicable. The reliability of the estimates of the volume of a source of aggregate is normally a direct reflection of the degree of geological assurance available to the investigator, and this in turn enables him to classify a prospect according to the terminology demonstrated in Fig. 3.8.

In practical terms, the approach to the assessment of aggregates adopted by local and central government planners, prospectors for construction materials and trade associations is necessarily generalised and wide-scale; they are mainly concerned with the appraisal of resources (McLellan and Bryant, 1975; Thurrell, 1971, 1981). On the other hand, members of the extractive minerals industry must be involved with relatively much more intensive investigations of socio-economic as well as the geological and geotechnical aspects of prospects in order to categorise them as commercially viable.

Even when the physical and chemical characteristics of the aggregate source itself have been shown to be acceptable, the influence of operational, marketing, environmental and land-use planning factors, which can be expected to affect each prospect in possibly wide different ways, must be considered before a particular quarrying prospect can be regarded as a reserve. Changing economic circumstances may alter

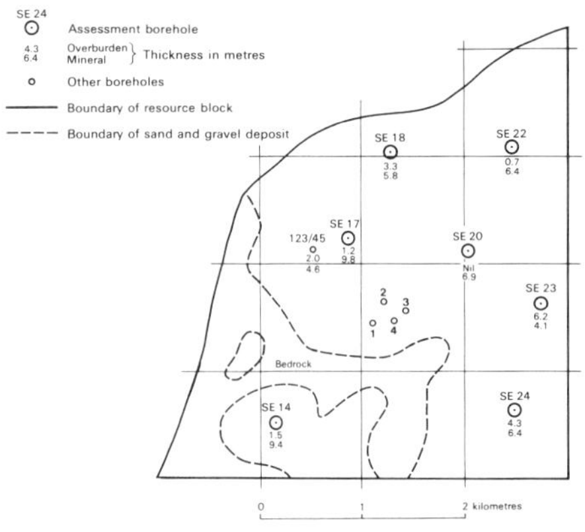

Block calculation

Scale: 1:25 000
Block: Fictitious

<u>Area</u>
Block: 11.08 km^2
Mineral: 8.32 km^2

<u>Mean thickness</u>
Overburden: 2.5 m
Mineral: 6.5 m

<u>Volume</u>
Overburden: 21 million m^3
Mineral: 54 million m^3

Confidence limits of the estimate of mineral volume at the 95 per cent probability level: $\pm$ 20 per cent
That is, the volume of mineral (with 95 per cent probability): 54 $\pm$ 11 million m^3

<u>Thickness estimate</u> (measurements in metres)
l_o = overburden thickness l_m = mineral thickness

Sample point	Weighting w	Overburden l_o	wl_o	Mineral l_m	wl_m	Remarks
SE 14	1	1.5	1.5	9.4	9.4	Assessment boreholes
SE 18	1	3.3	3.3	5.8	5.8	
SE 20	1	nil	-	6.9	6.9	
SE 22	1	0.7	0.7	6.4	6.4	
SE 23	1	6.2	6.2	4.1	4.1	
SE 24	1	4.3	4.3	6.4	6.4	
SE 17	½	1.2	1.6	9.8	7.2	Hydrogeology record
123/45	½	2.0		4.6		
1	¼	2.7	2.6	7.3	5.8	Close group of four boreholes (commercial)
2	¼	4.5		3.2		
3	¼	0.4		6.8		
4	¼	2.8		5.9		
Totals	Σw = 8	Σwl_o = 20.2		Σwl_m = 52.0		
Means		$\overline{wl_o}$ = 2.5		$\overline{wl_m}$ = 6.5		

<u>Calculation of confidence limits</u>

wl_m	$\lvert(wl_m - \overline{wl_m})\rvert$	$(wl_m - \overline{wl_m})^2$
9.4	2.9	8.41
5.8	0.7	0.49
6.9	0.4	0.16
6.4	0.1	0.01
4.1	2.4	5.76
6.4	0.1	0.01
7.2	0.7	0.49
5.8	0.7	0.49

$\Sigma(wl_m - \overline{wl_m})^2 = 15.82$

$n = 8$

$t = 2.365$

L_V is calculated as

$1.05\,(t/\overline{wl_m})\,\sqrt{[\Sigma(wl_m - \overline{wl_m})^2/n(n-1)]} \times 100$

$= 1.05 \times (2.365/6.5)\,\sqrt{[15.82/(8 \times 7)]} \times 100$

$= 20.3$

$\approx$20 per cent.

FIG. 3.10. Calculation of sand and gravel resources in a hypothetical map area (resource block) with limits quoted at the 95% probability level. (Source, Mineral Assessment Reports, e.g. Dunkley, 1979).

the status of a previously classified prospect so that it may become more or less attractive for exploitation.

It should be borne in mind also that the proportion of the potentially suitable aggregate material in a prospect that could be extracted in practice is likely to be much less than the gross estimated volume in the ground. The workable volume may often be limited by a variety of operational and environmental factors. In normal practice overburden must be dug to clear a quarry face by a distance equal to the depth of overburden, and must be formed to a stable slope, which may be as shallow as 1 in 5 in clay. Within the mineral formation access ramps may sterilise some material, and safety considerations will dictate a stepped profile to the working face. In unconsolidated formations of sand or gravel a safe face height is unlikely to exceed 10 metres and in rock formations, practice in Britain is that benches should not generally exceed 20 metres at 15 degrees inclination, although locally such factors as geological structure, blasting technique and safety, planning or legal constraints may dictate other limitations. In deep rock excavations the overall sidewall angle should not exceed 60 degrees at 65 metre depth, or 40 degrees at 300 metre depth (Roberts, Hoek and Fish, 1972). In developed regions of the world a variety of environmental constraints may prevail, and it is the concern of the land-use and development planning interests to resolve conflicting issues.

3.8 Exploration and development of aggregate prospects

A suggested procedure for examining and testing potential aggregate sources is given in Table 3.1. The effective management of a project requires the appointment of a co-ordinating geologist or engineer who will define the objectives of the investigations, and provide the field and laboratory personnel with precise terms of reference for each stage through to the production control programme (Morray-Jones, 1968; Lovell, 1968).

It is advisable to prepare reports at the conclusion of each of the four stages (A, B, C and D of Table 3.1) and a final report should summarise them. The co-ordinator can by then be confident in commencing the concrete mix design or other use for the aggregate and set up trials *in situ*.

TABLE 3.1 *A systematic procedure for the examination and testing of potential aggregate sources*

A Reconnaissance of the available materials and establishment of their performance in likely applications

1 Establish the geological basis for the investigation by means of a desk study to collate all relevant data and undertake a walk-over field survey (see 3.2 and 3.4.2), including any special geological investigations (3.3) to establish existence of potential sources.

2 Undertake field sampling of likely sources; investigate the characteristics of any existing nearby aggregate sources including the range of end-uses of the products, whether or not they are similar to those for which the present appraisal is being conducted.

3 Carry out laboratory examinations and tests on samples obtained during phases 1 and 2 above; assess the performance of existing locally available aggregates in use: for example the performance of aggregate *within* concrete as well as of the concrete itself.

B Examination of potential aggregate sources—site investigation

1 Carry out full-scale trenching, pitting and/or drilling (see 3.5.2 and 3.5.3) and sampling of the potentially useful occurrences identified from investigations so far, including material from existing local aggregate workings, whether or not it is intended for use.

2 Carry out laboratory investigations and tests on the samples (see Section 7) to determine their physical characteristics and potential suitability; relate this information to the unknown geology of the rock mass(es) under investigation and obtain an indication of the size and distribution of the potential mineral reserve.

3 In the light of all the new information assess the extent to which selective extraction and special processing techniques may be required to ensure a bulk supply of aggregate of sufficient quality. Review the need for particular processing plant designs.

C Aggregate production trials

1 Plan and supervise materials extraction and aggregate production trials including representative sampling of end products, selective sampling of both accepted and rejected material.

2 Carry out further laboratory tests on samples collected from the extraction and production trials, review and finalise extraction methods, production procedures and target qualities for the processed aggregates for approval purposes.

D Production control programme

1 Establish, approve and implement sampling testing regimes for production control Producer and specification compliance Engineer purposes.

2 Devise, approve and maintain data records, reporting proformas and control charts to monitor aggregate quality and variability.

References

ALLUM, J. A. E. 1966. Photogeology and Regional Mapping. Pergamon Press.

ARCHER, A. A. 1970. Standardisation of the size classification of naturally occurring aggregates. *Geotechnique,* **20,** 103–107.

AVERY, T. E. 1968. Interpretation of aerial photography. 2nd Ed. Burgess, Minneapolis, USA.

BARRINGER, A. R. 1971. Airborne exploration. *Ming. Mag. London,* **124,** 182–189.

BEAUMONT, T. E. 1979. Remote sensing for the location and mapping of engineering construction materials in developing countries. *Q. J. eng. Geol. London,* **12,** 147–158.

BEAVEN, P. J. 1976. Terrain evaluation. Section 8 in Manual of applied geology for engineers. H.M.S.O., London, 171–195.

BS 812:1975. Methods for sampling and testing mineral aggregates, sands and fillers. British Standards Institution, London.

BS 4019:1974. Rotary Core Drilling Equipment. British Standards Institution, London.

BS 5930:1981. Code of Practice for site investigations (formerly CP 2001) British Standards Institution, London 147 pp.

BUREAU OF MINES AND GEOLOGICAL SURVEY, 1948. Mineral Resources of the United States (Washington, DC; Public Affairs Press), 14–17.

BUREAU OF MINES AND GEOLOGICAL SURVEY, 1976. Principles of the mineral resources classification system of the US Bureau of Mines and US Geological Survey. Bulletin No. 1450-A, USGS, 1–5.

BURNETT, A. D. 1975. Engineering geology and site investigation Pt. 2: Field Studies. *Ground Eng.* **4,** pp 29–32.

CANNELL, B & CROFTS, R. G. 1984. The sand and gravel resources of the country around Henley-in-Arden, Warwickshire. *Miner. Assess. Rep. Br. Geol. Surv., No. 142,* 169 pp.

CASSINIS, R. 1977. Use of remote sensing from space platforms for regional geological evaluation and for planning ground exploration. *Geoph. Prosp.* **25,** 636–657.

CHAPLOW, R. C. 1975. Engineering geology and site investigation Pt. 1. Introduction. *Ground Eng.* **8,** (3) pp 34—38.

CLARKE, M. R. 1983. Geophysical evaluation of scattered occurrences of sand and gravel in drift deposits *in* Prospecting and Evaluation of Non-Metallic Rocks and Minerals (Edit. K. Atkinson and R. Brassington), Instn of Geol, London.

CRATCHLEY, C. R. 1976. Geophysical methods. Section 6 in

Manual of applied geology for engineers. H.M.S.O., London, 110–146.

CULLEY, R. W. 1973. Use of airborne resistivity surveys for gravel location. *Canadian Mining Metall. Bull.*, **66**, No. 731, 1-5.

DAVIS, J. C. 1973. Statistics and data analysis in geology, Kansas Geol. Surv. Wiley.

DOORNKAMP, J. C., BRUNSDEN, D., JONES, D. K. C., COOKE, R. U. & BUSH, P. R. 1979. Rapid geomorphological assessments for engineering. *Q. J. eng. Geol. London*, **12**, 189–204.

DUMBLETON, M. J. & WEST, G. 1970. Air photograph interpretation for road engineers in Britain. Dept. of the Environment, TRRL Rep. LR 369. Crowthorne.

DUMBLETON, M. J. & WEST, G. 1976. Preliminary sources of information for site investigation in Britain. Dept. of the Environment, TRRL Report LR 403 (Revised edition), Crowthorne (Transport and Road Research Laboratory).

DUNKLEY, P. N. 1979. The sand and gravel resources of the country around Maidenhead and Marlow. *Miner. Assess. Rep. Inst. Geol. Sci., No. 42,* 56 pp.

DUNN, J. R. & CUTCLIFFE, W. E. 1971. Selecting aggregate deposits: a geologic view. *Rock Prod.,* March 1971, 75–79.

EASTAFF, D. J., BEGGS, C. J. & MCELHINNEY, M. D. 1978. Middle East-geotechnical data collection. *Q. Jl Engng Geol.* **11,** pp 51–63.

FOOKES, P. G. & SWEENEY, M. 1976. Stabilisation and control of local rock falls and degrading rock slopes. *Q. Jl Engng Geol.* **9,** 37–55.

FOX, R. A. 1979. The role of the geologist in quarrying. *Quarry Mgmt Prod.* **6,** 239–245.

GEOLOGICAL SOCIETY, 1970. Engineering Group Working Party Report on the logging of rock cores for engineering purposes. *Q. J. eng. Geol. London,* **3,** 1–24.

GEOLOGICAL SOCIETY, ENGINEERING GROUP WORKING PARTY REPORT 1972: The preparation of maps and plans in terms of engineering geology. *Q. J. eng. Geol. London*, **5**(4), 293–382.

GLOSSOP, R. 1968. The rise of geotechnology and its influence on engineering practice. *Geotechnique. London,* **18** (2), 107–150.

GORDON, P., HEYES, A. J., MARKER, B. R. & WOOD, R. J. 1982. Geological basis of land-use planning: a Dept. of the Environment philosophy for commissioning research. *Trans. Instn. Ming. Metall.* (*Sect. B*: *Appl. earth sci.*)*,* **91,** 152–156.

HESS, H. D., 1971. Marine Sand and Gravel Mining Industry of the United Kingdom. NOAA Tech. Rept ERL 213-MMTC 1, US Dept Commerce, Washington, DC.

HILL MCDONALD, 1972. Sand and gravel exploration methods. *Proc. Forum Geol. Miner. Florida Dept. Nat. Resour. Spec. Publ.* **17,** 6 pp. 6 figs.

HIMUS, G. W. & SWEETING, G. S. 1968. The elements of field geology. London 2nd Ed. (University Tutorial Press).

INSTITUTE OF GEOLOGICAL SCIENCES, 1977. Quaternary map of the United Kingdom. (Two 1:625 000-scale sheets, compiled by Edmonds, E. A., 1977, mainly from work of IGS).

INSTITUTE OF GEOLOGICAL SCIENCES, 1981. Annual Report for 1980 and 1981. (London: Institute of Geological Sciences).

LAHEE, F. H. 1952. Field geology. McGraw Hill, 5th Edn.

LATHAM, L. H. 1965. Aerial photographs in field geology. Holt, Rinehart, Winstal.

LOVELL, S. M. 1968. The customer and the quarry operator. *Quarry Managers' J. London*, **52,** 25–29; discussion 30–36.

MANUAL OF APPLIED GEOLOGY FOR ENGINEERS (Military Engineering, Vol. XV) H.M.S.O. 1976, 378 pp.

MCKELVEY, V. E. 1972. Mineral resource estimates and public policy. *American Science,* **60,** 32–40, Connecticut.

MCLELLAN, A. G. & BRYANT, C. R. 1975. The Methodology of Inventory—A practical technique for assessing provincial aggregate resources. *Canadian Mining and Metallurgical Bulletin,* **68,** 102–108.

MCQUILLAN, R. & ARDUS, D. A. 1977. Exploring the geology of the shelf seas. Graham and Trotman.

MORRAY-JONES, R. H. 1968. Specification and quality control of quarry products. *Quarry Managers' J. London.* **52,** 15–24; discussion 30–36.

NORMAN, J. W. 1968. The air photograph requirements of geologists. *The Photogrammetric Record* **4,** 133–149.

NORMAN, J. W. 1969. Photogeology in site surveys. *The Consulting Engineer,* **33,** (Sept.) 74075.

PATOUREAUX, Y. 1979. An example of the value of infra-red multispectral remote sensing in mapping. *Q. J. eng. Geol. London*, **12,** 181–188.

PIPER, D. P. & ROGERS, P. J. 1980. Procedure for the assessment of the conglomerate resources of the Sherwood Sandstone Group. *Miner. Assess. Rep. Inst. geol. Sci., No. 56*, 14 pp.

ROBERTS, D., HOEK, E. & FISH, B. G. 1972. The concept of the mammoth quarry. *Quarry Managers' J. London*, Vol. **57,** No. 7, 229–238.

ROGERS, P. J., PIPER, D. P. & CHARSLEY, T. J. 1982. The conglomerate resources of the country around Cheadle, Staffordshire. Description of 1:25000 sheet SK 04. *Miner. Assess. Rep. Inst. geol. Sci. No.* **57,** 32 pp.

ROSE, E. P. F. 1976. Geological maps and other information sources. Section 7 in Manual of applied geology for engineers. H.M.S.O., London.

SABINE, P. A. 1977. The geological background to highway engineering. *Highway Engineer,* **24,** 23–27.

SAND AND GRAVEL ASSOCIATION, 1967. Pit and Quarry Textbook. London, 239 pp.

SIMPSON, P. 1979. Marine geophysical surveys. Hydrography (CLS/CMS Autumn 1979) pp 25–35.

THURRELL, R. G. 1971. The assessment of mineral resources with particular reference to sand and gravel. *Quarry Managers' J. London*, **55,** 19–25.

THURRELL, R. G. 1981. The identification of bulk mineral resources; the contribution of the Institute of Geological Sciences. *Quarry Manage. Prod. London* **8,** 181–193.

UNITED NATIONS, 1979. International classification of mineral resources. *Mining Mag.* June 1979, 533–536.

VANN, G. 1964. Location and evaluation of sand and gravel deposits by geophysical methods (Symposium on Opencast Mining, Quarrying and Alluvial Mining, Paper 1). Institution of Mining and Metallurgy, London.

VEENSTRA, H. J. 1969. Gravels of the southern North Sea. *Marine Geol.*, **7,** pp 449–464.

VERNEY, Sir R. R. (Chairman) 1976. "Aggregates: the way ahead". Report of the Advisory Committee on Aggregates (HMSO, for Dept. of the Environment, London), 118 pp.

WILLMAN, H. G. 1942. Geology and mineral resources of the Marseilles, Ottawa and Streater quadrangels. *Bull. Illinois State Geol. Surv.*, **66,** 343–344.

ZALASIEWICZ, J. A., MATHERS, S. J. & Cornwall, J. D. (1985). The application of ground conductivity measurements to geological mapping. *Q. J. eng. Geol. London*, **18,** 139–148.

Appendix A3

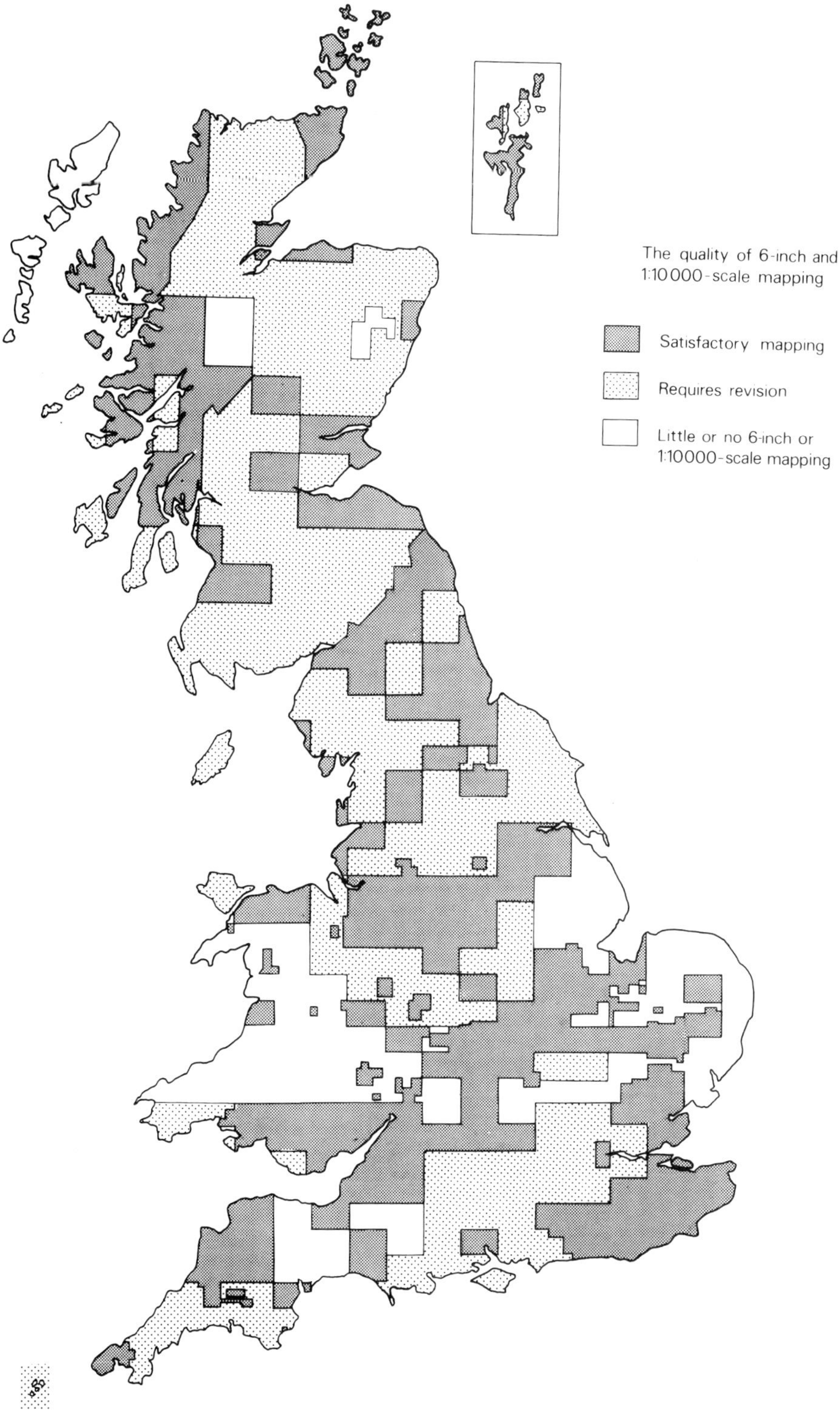

FIG. A3.1. The quality of 1:10 000 and six-inch geological mapping in Great Britain (British Geological Survey).

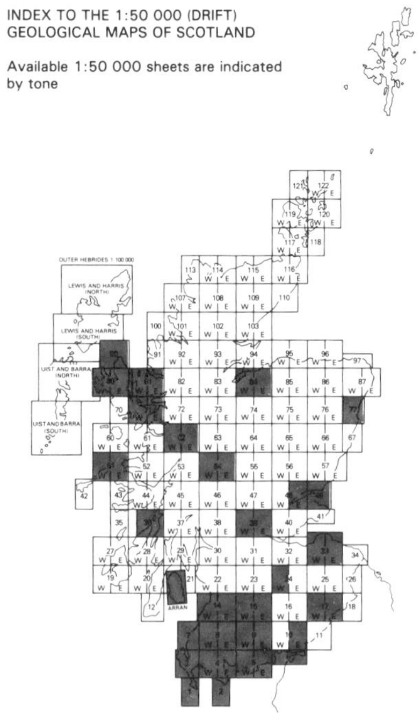

FIG. A3.2(i)

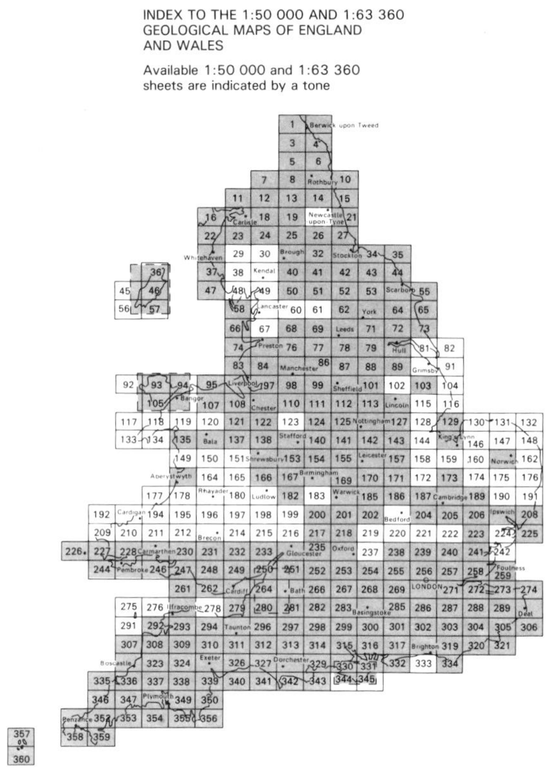

FIG. A3.2(iii)

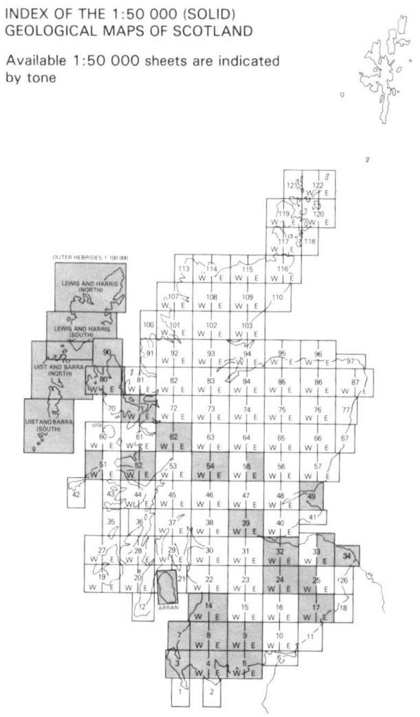

FIG. A3.2(ii)

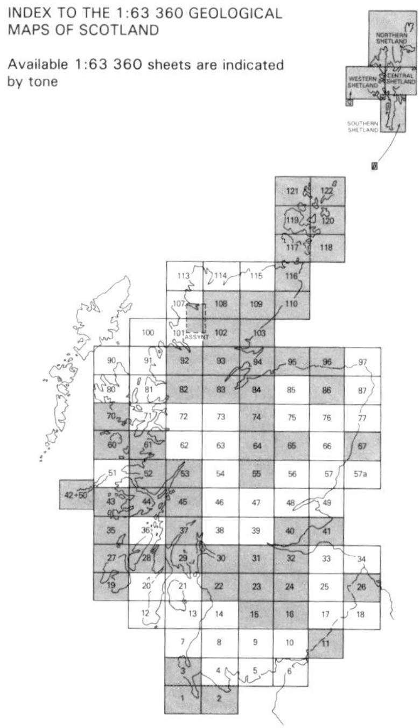

FIG. A3.2(iv)

FIG. A3.2. Availability of 1:50 000 and 1:63 360 scale geological maps of Great Britain (British Geological Survey).

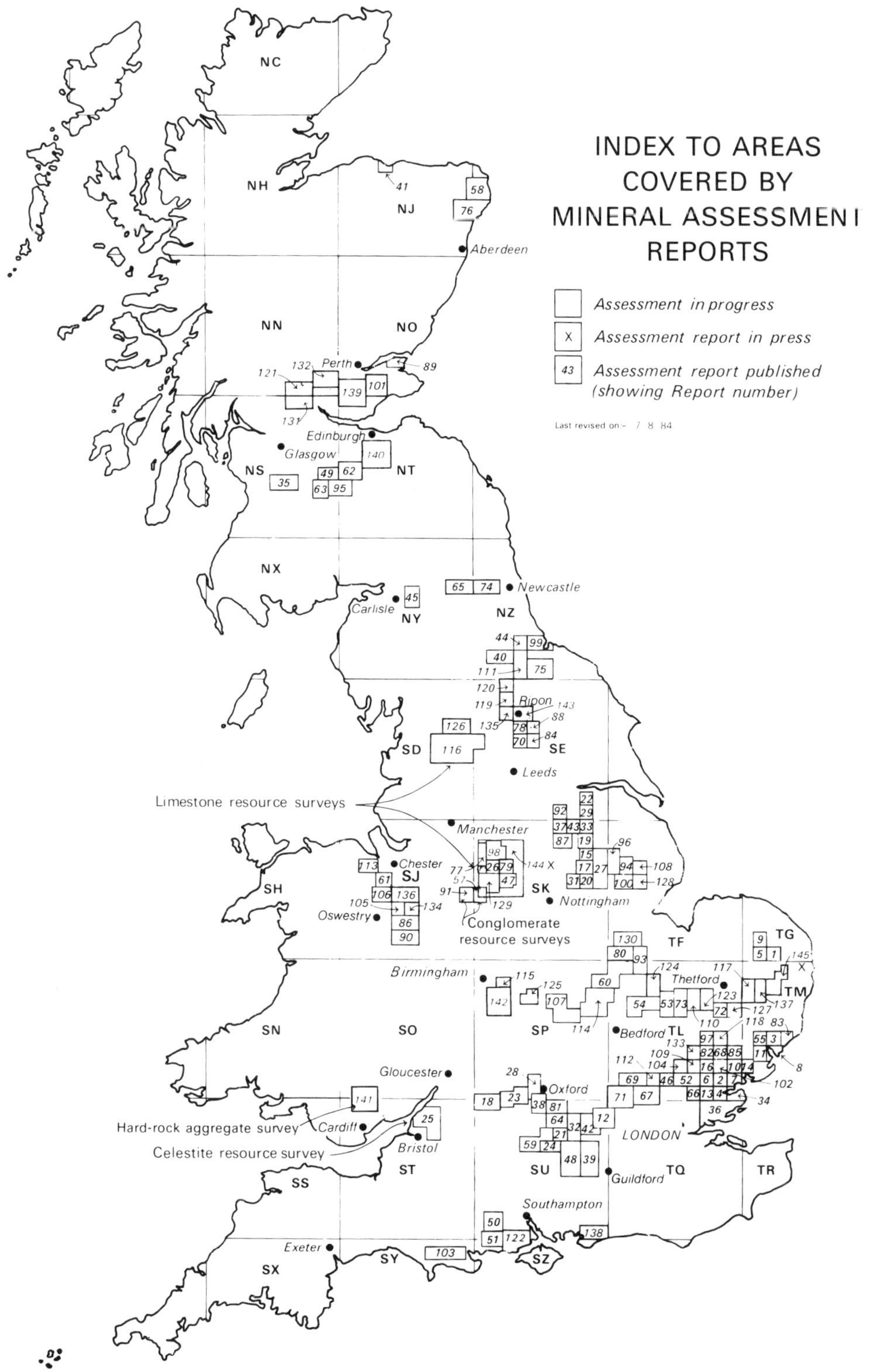

FIG. A3.3. Index to the location and progress of bulk mineral resource surveys in Great Britain. The numbers denote areas covered by reports on the assessment of British sand and gravel resources published in the Report Series of the Institute of Geological Sciences (Nos 1 to 12) and in the Mineral Assessment Reports of the Institute (No 13 onward). Some reports (as indicated) deal with resources of limestone, conglomerate, hard rock (sandstones) and celesite. (Updated from Institute of Geological Sciences 1981).

Chapter 4. Extraction

4.1 Introduction

Prior to 1967 sand and gravel deposits provided more than 50% of the aggregate produced in the United Kingdom, for geographical factors, ease of extraction, and lower production costs favour these deposits. As shallow sand and gravel deposits become exhausted and planning constraints sterilise replacement reserves the emphasis has been for the aggregate industry to turn increasingly to hard rock sources. The swing can be illustrated by comparing production for 1965 and 1973, the latter year representing the peak production before the economic recession which was experienced in the British construction industry (Table 4.1):

TABLE 4.1. *Comparison of aggregate production in two contrasting years in the United Kingdom (HMSO 1982)*

	Million Tonnes	
	1965	1973
Limestone	58.0	105.9
Sandstone	6.1	16.7
Igneous Rocks (includes metamorphics)	19.7	49.0
	83.8	171.6
Land sand & gravel	95.2	126.4
Marine sand & gravel	7.2	16.0
	102.4	142.4

The geological occurrence of aggregates is detailed in Chapter 2, together with the types of deposits and their geological origin influencing the consolidation characteristics which in turn govern the equipment that will be used to win the materials.

Sands and gravels, which are formed in the first place by the disintegrating action of frost and rain and by the dissolving action of acid and alkaline ground-water on the various massive rocks, can be divided into four different types of deposits (a) fluvial (b) glacial (c) residual and (d) marine and lake deposits (Industrial Minerals, 1971). Crushed rocks produced from the indurated deposits found on the earths surface can be obtained from the three main rock categories (a) limestone (b) sandstone and (c) igneous and metamorphic.

The selection of open-pit excavating equipment depends on the competence and "diggability" of the ground to be worked, many factors such as impact strength, bulk density etc. influencing this diggability (Atkinson, 1971). Deposits of compact rock must normally be drilled and blasted before the mineral can be excavated for processing. The equipment in use in the extraction of aggregates has a common application in most opencast mining of minerals and in the earth-moving industry. Whilst there are a number of general considerations which must be taken into account when selecting machinery it has been suggested that a combination of practical knowledge of equipment and mathematical calculation will give the best results (Shand, 1970). Many factors will influence the materials handling system and final choice of equipment.

In recent years marine deposits have contributed increasingly to the needs of the aggregate industry in the United Kingdom. They are worked from the sea bed where extensive glacial and fluviatile deposits accumulated prior to the rise in sea level at the end of the last ice age.

4.2 Overburden

In the quarrying of aggregates the overburden is the cover material which has to be removed to expose the payable rock to be quarried, and its thickness can vary from a few millimetres to several metres depending on the geology and climatic conditions.

Weathering, which is a process of alteration of rocks occurring under the direct influence of the hydrosphere and atmosphere, is an important factor in considering the amount of overburden which needs to be removed to expose the payable rock for aggregate production. Rock material tends to deteriorate in quality due to the effects of weathering and/or alteration and these changes can be detected by systematic measurement (Geological Society, 1977).

Although British rocks are not generally subject to decomposition and disintegration due to weathering and alteration to any great extent, the relationship between weathered and unweathered rocks can be an important factor when considering igneous and metamorphic rocks. In particular, fossil weathering overlain by fresh younger material may be encountered and is often found in the volcanic sequences being quarried in Northern Ireland. Other overburden types may be superficial unconsolidated cappings to hard rocks, as for example glacial deposits which are often encountered in many areas in the UK, and a variety of bedded strata with little or no economic value as aggregates.

Hard rock deposits may thus be covered by a variety of different types of overburden which will require stripping, and finding the most suitable method of removal will depend on whether the overburden is soft and easily handled or is consolidated and requires blasting and ripping. In some cases scrapers can be utilised to strip overburden, unit costs being reduced and economy achieved because of the large-scale nature of the operation. Excavating machines used in removing overburden in hard rock quarries are commonly face shovels and draglines, with a recent trend towards the use of hydraulically operated machines, the production rate anticipated governing which bucket size is used. Drag-lines tend to have a longer digging cycle time than a shovel, but this is compensated by the drag-lines ability to carry a larger bucket (Dover, 1964). It should also be noted that the amount of blasting required with draglines is more critical than with shovels, since the lump size must not be so great that the bucket is thrown over on striking a lump, but at the same time overblasting may result in the benching, from which the drag-line must work, becoming unstable.

When variable overburden thickness occurs, and accurate spotting is required a face-shovel is often used, as for example at Moons Hill Quarry, in Somerset (Mine and Quarry, 1973). In that particular excavation the rock aggregate of pyroxene-andesite occurs beneath up to 13 metres of soft rock and tuff conglomerate overburden. Removal of the latter by the face-shovel is arranged by placing into lorries and dump trucks for disposal at waste tips. Some occasional drilling and shotfiring is required for the harder rock overburden, and once a year contractors are employed to remove softer overburden with tractors and scrapers.

With some sand and gravel deposits, the overburden can be expected to occur at a regular thickness, up to 3 metres in those low level fluvial deposits associated with river systems of an area. Glacial deposits on the other hand often have characteristically variable overburden, the geological conditions associated with the formation of these deposits (Chapter 2) producing rapid fluctuation and mixing between clay and sand and gravel. Clay overburden in areas where glacial sand and gravels are being worked can reach 10 metres in thickness, and sometimes two or three layers of workable aggregate deposit will have interbedded boulder clays associated with them. Some sand and gravel deposits are located in areas where the overburden occurs partly or wholly below water table. In these circumstances the overburden can often be removed by first lowering the water table by pumping and then using a drag-line or scraper to excavate the material.

In recent years the removal of overburden has become a carefully controlled operation, with the topsoil and subsoil being transferred only when weather and ground conditions are suitable. Successful agricultural and other after-uses of the restored land requires special consideration, and, in particular, very careful planning and execution of all soil movements (Verney, 1976). The degree of control or the movement of the overburden depends on the proposed restoration, the nature of the after-use of the land dictating which method of overburden excavation to be utilised. Dry excavations in sand and gravel can often be restored with the equipment which was used to win the aggregate.

In both hard rock and sand and gravel excavation the technique known as "cut and fill" is often used to avoid additional cost of double handling of overburden (Fig. 4.1). This technique has the advantage of giving the shortest possible distance for the movement of overburden between cut and spoil bank (Dover, 1964). For sites where the aggregate to be quarried lies above the water table, single bucket hydraulic and drag-line operated machines can achieve restoration of worked out areas with a minimum of cost if they have sufficient reach to dump the spoil without intermediate rehandling. More recently there has been a trend towards the use of scrapers for stripping overburden, a complete area for a years production of hard rock and sand and gravel often being stripped in one operation. A wide variety of earth-moving equipment is available for overburden removal, and although there is often a case for using the same machinery for excavating the overburden as for the rock and sand and gravel, geological conditions usually dictate which equipment is most suitable.

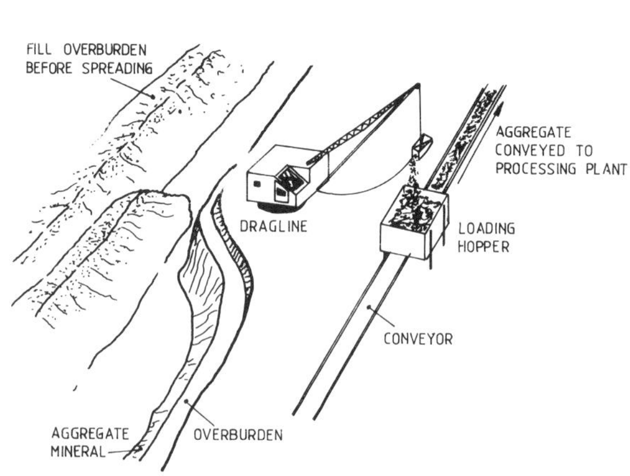

FIG. 4.1. Diagram showing "cut and fill" operation (after Atkinson 1971).

Estimates of direct operating costs for different machines of similar capacity often show the comparable costs per unit of output, and where the overburden is harder and rockier the case for using a face-shovel in preference to a drag-line becomes apparent.

4.2.1. Removal. The types of plant available for overburden removal and transportation are illustrated in Appendix A4.1 and can be summarised as follows:–

4.2.1.1. Crawler tractor (A4.1.1). Sometimes referred to as "track-type tractors" these machines have limited application in sand and gravel operation for overburden removal. In some dry quarries, the overburden can be pushed over the working face to the bottom level where restoration at the lower level can take place (SAGA, 1967). Difficulties can arise when using this technique when the working face becomes contaminated by the overburden. It can only be used where separation of the topsoil and subsoil is not required. In hard rock quarries the bulldozer has considerable application, the tracked ripper variety being widely used as a machine for breaking up weathered overburden rock, which, together with a towed scraper, can remove that overburden quite easily.

4.2.1.2 Crawler loader (A4.1.2). Sometimes referred to as "dozer-shovels" they are generally used in hard-rock excavation these machines having the advantage of operating well in conditions where tough excavating and backfilling work may be required.

4.2.1.3 Four-wheel loader (A4.1.3). The use of the rubber-tyred wheel loader has found wide acceptance in the hard rock and sand and gravel industry as a versatile machine in stripping overburden. In hard rock quarries prior breakage of the overburden is necessary either by blasting or ripping, the loader handling the broken rock with little or no hard digging (Sievert, 1973). Soft sediment overburden capping hard rock and sand and gravel are easily handled by the loader, some "articulated" varieties adding to the manoeuvrability of the machines. The hydraulically operated loader bucket enables these machines to be operated in conjunction with dumpers which can transport the materials to areas requiring restoration.

4.2.1.4 Face-shovel rope excavator (A4.1.4). Where hard compact overburden is being encountered in both hard rock and sand and gravel quarries the face-shovel rope excavator can be the most practical machine to excavate overburden, it being used in conjunction with lorries and dump-trucks to remove the material, some drilling and blasting being required for the harder rock (Mine and Quarry, 1973). In recent years the more robust and larger hydraulic versions of the face-shovel have been found to be practical and are replacing the rope machines to a significant extent.

4.2.1.5 Scrapers (A4.1.13). Motorised scrapers have been widely accepted for overburden stripping, particularly where there is a considerable depth of overburden to move. The scrapers can operate as a complete rubber-tyred variety or the rubber-tyred scraper unit can be towed by a tracked tractor (or bulldozer) if ground conditions are unsatisfactory.

As self-loading units, the scrapers can move large quantities of overburden which can be spread at the receiving end thereby facilitating restoration and reducing costs.

4.2.1.6 Drag-line excavator (A4.1.7). As a widely accepted method of overburden removal the drag-line excavator can be used in hard material in addition to its common application in soft rock overburdens. With hard rock excavations the rock must be well prepared for loading by using blasting or ripping techniques, and although likely to be expensive the drag-line may have considerable advantages in facilitating the requirements for restoration (Lakin, 1971). A common method of overburden removal, the working drag-line can stand on the overburden surface and can dispose of the overburden material in various ways as follows:

(a) Casting into worked-out pit. This is one of the simplest methods of strip-mining, and in sand and gravel workings a variation is to use barges if the excavation is wet (SAGA, 1967). With the aid of bottom-discharge barges the overburden can be removed from the area of excavation and disposed of in a part of the lake where it does not contaminate the mineral being won.

(b) Loading into dumpers (A4.1.6). If the working programme necessitates the transport of overburden to other parts of the site, material can be conveyed by dumper. Various types of dumpers are available, the size of the vehicle depending on the size of the drag-line and the length of the haul.

(c) Loading on to belt conveyor. Only on rare occasions is the belt conveyor used for transporting overburden because of the difficulties involved in moving the conveyor line as stripping progresses and in handling material containing clay. However, in certain circumstances where deep irregular overburden needs to be moved over long distances the method can be advantageous.

4.3 Hard rock excavations

The more common hard rock sources of aggregates use extraction processes which are closely linked to the type and extent of the deposit being worked, the nature of the rock influencing the technical operations in winning, loading and haulage of the materials (Flachsenberg, 1964). Although the production operations at the face of hard rock quarries comprise three separate activities—drilling, blasting and loading out of the stone, undertaken in that order—it is becoming increasingly difficult to discuss any one of those in complete isolation (Grimshaw & Watt, 1971). Drilling and blasting are complementary, their purpose being to break out the stone effectively from the solid face, and also to control the shape of the rock pile together with the degree of fragmentation to enable the loading equipment to work with maximum efficiency.

4.3.1 Drilling. There are a number of drills available on the market and in Great Britain there are two main groups that are used namely (Teichmann, 1960):

Large diameter holes (127 mm to 229 mm)	Medium diameter holes (63 mm to 102 mm)
(a) Churn drills	(a) Percussive rotary drills
(b) Rotary drills	(b) Rotary drills
(c) Percussive rotary drills	

A smaller third group which will drill small diameter holes of 50 mm and less embrace the wagon drill and the small diameter deep hole. Hole diameter is important because it determines the choice of explosive types that can be used, the lower the sensitivity of an explosive composition the greater the critical diameter of the explosive for reliable initiation and detonation throughout a column (Grimshaw & Watts, 1971). Thus a typical ammonium nitrate/fuel oil mixture closely poured may have a critical diameter of around 76 mm which means that it should be used in holes of greater diameter than this to get optimum results.

Modern drilling techniques tend to favour inclined drilling techniques, it being stated that compared with a vertical hole, twice the explosive energy is available for fragmentation from explosive charge detonated at the bottom in a 45° inclined hole (Kochanowsky, 1961) (Fig. 4.2). Faces 25° to 30° off-vertical are suggested as giving the most satisfactory results in practice (Pass, 1970). Some of the benefits of inclined drills can be listed as follows (Kochanowsky, 1961):

(i) Safer operation for operatives and machines.
(ii) Higher benches may be blasted and number of haulage levels reduced.
(iii) Toe and back breakage can be eliminated.
(iv) Better fragmentation will occur,
 (a) Because of better utilisation of energy and reduced rock resistance against blasting at the bottom of the bench.
 (b) Because of the reduced rock volume around the stemmed part of the borehole at the top of the bench, where there are usually large boulders.
(v) Cheaper blasting operations will be obtained because less footage on drill and/or less explosive per ton of rock are required.
(vi) Higher efficiency and cost reduction in loading haulage, and crushing operations through better fragmentation.
(vii) Damaging effect of vibrations in the surrounding area is decreased.
(viii) The throw distance will be greater.
(ix) In the same bench height, the angle hole is longer. Therefore larger tonnage can be blasted per foot of bench face length with the same burden.

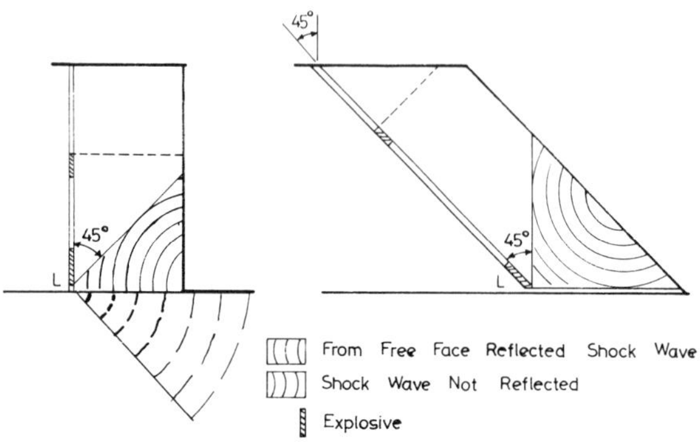

FIG. 4.2. Shock wave propogation through rock generated by detonation of explosive charges located at point *L* (after Kochanowsky 1961).

4.3.2 Blasting. All phases of quarrying are affected directly by the results achieved from primary blasts, and it is therefore important that their design is compatible with the widely varying natural conditions that exist in quarrying areas in general, and the changing geological conditions that occur with time, even within one quarry (Greenland, 1973). Satisfactory fragmentation of rock is essential to producing small material which will present easy loading conditions reducing wear and tear on machinery, less blockages at the primary crusher, and therefore increased output (Lakin, 1971). There are a number of widely used terms that are applicable in blasting, these being shown diagrammatically in Fig. 4.3. The main factors in the breaking of rock by blasting can be summarised as follows:

(i) Depth of shot holes.
(ii) Burden of shot holes.
(iii) Spacing of shot holes.
(iv) Charge in shot holes.

The efficiency of explosives as tools for rock breakages lies in their ability to deliver almost instantaneously large amounts of energy to a limited portion of rock (Greenland, 1973). When a charge of explosives is detonated in a shot hole, the surrounding stone is subjected to this release of energy in the form of gases at high termperatures and pressures which produce a system of intense stress waves in the stone (Grimshaw, 1963). The mechanism of the transfer of this energy to the rock and the resulting breakage are complex and not yet completely understood. The important factors in producing the breakage process can be grouped into three categories (a) explosive factors (b) charge loading factors and (c) rock factors (Greenland, 1973).

Explosive factors include density, velocity of detonation, detonation impedance, denotation pressure, gas volume and available energy. Charge loading factors are hole diameter, stemming, decoupling, type of initiation and point of initiation.

Rock factors affecting the fragmentation process include density, propogation velocity characteristic, impedance, energy absorption, compressive strength, tensile strength, variability and structure.

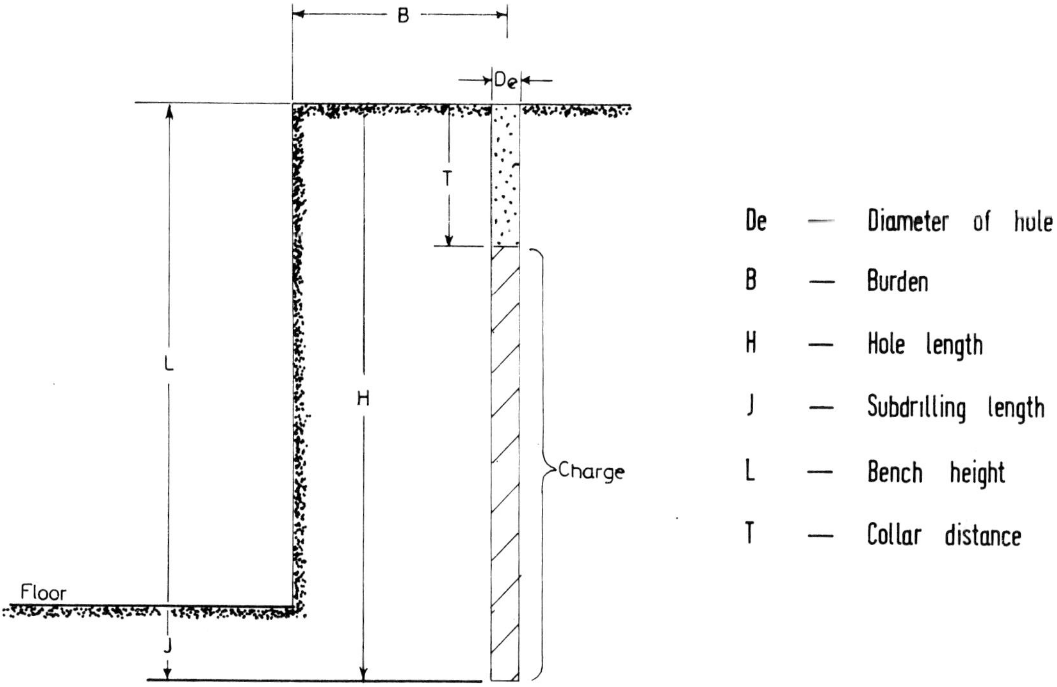

FIG. 4.3. Bench cross-section view showing blast terminology (after Pugliese 1973).

The structural features of a material usually have the greatest influence on blast effects (Ash, 1963). Of primary importance is compression jointing existing within all rocks (igneous, sedimentary and metamorphic) and composed of planes along which there is no resistance to separation. Igneous rocks may also have tension jointing formed during the cooling process. A joint cannot, by its very nature, transmit tensile strength, the surface of joint walls reflecting stress waves and impeding their progress through the rock mass, especially when the joint is filled with clay or similar material. When considering the blast geometry for hole depth, hole burden, spacing and number of holes, one of the primary factors controlling the results obtained is bench height (Fig. 4.3). The shot hole related to the bench height should be so located that the burden (the distance of the shot hole from the nearest free face) should always be less than the depth of the hole.

When rock in a blast is completely fragmented but not displaced, the critical burden has been obtained (Greenland, 1973). Although there are many formula relationships for approximating the burden for various explosive/rock combinations, complexities of the breakage mechanisms are involved, and it is common practice to conduct field trials to establish the most practical burden for specific blasting conditions. Collar and stemming are terms that are frequently used to mean filling part of the blast hole with inert material to trap the explosive gases, the energy travelling through solid rock much more rapidly than through the loose stemming material. The confinement allows the gases to fracture the solid rock before rock movement and before compaction causes ejection of the stemming.

In the spacing of shot holes the main factor of importance is the diameter of cartridge used, the larger the explosive shot then the greater the radius of its disruptive action. Generally, the spacing of shot holes should be approximately the burden on them.

TABLE 4.2. *Typical burden and spacings for different borehole diameters (NECL Undated)*

Minimum hole diameter		Burden		Spacings	
mm	in	m	ft	m	ft
25	1.00	1	3.0	1	3.0
31	1.23	1.4	4.5	1.4	4.5
35	1.37	1.5	5.0	1.5	5.0
44	1.75	1.8	6.0	1.8	6.0
57	2.25	2.4	8.0	2.4	8.0
76	3.00	2.7	9.0	2.7	9.0
82	3.25	3.0	10.0	3.0	10.0
102	4.00	3.6	12.0	3.6	12.0
168	6.62	6.1	20.0	6.1	20.0
229	9.00	7.6	25.0	7.6	25.0

Although many factors, particularly jointing, can effect the blast hole pattern a general guide can be obtained from Table 4.2 (NECL, undated).

However successful the fragmentation of a primary blast has been there will always remain some boulders too large to be handled in succeeding processes (Pass, 1970). Several options are open for reducing oversize boulders, breakage by explosives including "pop" drilling and "plaster" shooting resulting in heavy labour utilisation and high explosive costs respectively. "Drop-balling" requires adequate space on the working face, whilst pneumatic impact breaking has been suggested as being able to provide a solution to some of these problems (Pass, 1970).

4.3.3 Ripping. The technique involves breaking *in situ* rock by means of a single tooth mounted at the rear of a powerful tractor (A4.1.14). The major advantages of ripping in some open pit applications have been the considerable economies in ground preparation costs compared with drilling and blasting (Atkinson, 1970). In recent years the advent of more powerful crawler tractors and improvements in metallurgy of the materials has enabled ripping to enter into the realm of rock breaking previously reserved for drilling and blasting. Some of the applications where ripping can now show the greatest production and economic advantages are (Shand, 1970):

(a) Ripping and dozing into stock piles.
(b) Ripping and subsequent scraper loading.
(c) Ripping and dozer scraping.

Only certain types of rock can be subjected to ripping depending on their physical characteristics. The factors involved are:

(i) Fractures, faults, and planes of weakness.
(ii) Weathering.
(iii) Brittleness.
(iv) Crystalline nature and grain size.
(v) Stratification and lamination of beds.

The actual selection of the ripping tractor will depend on the rippability of the rock, and in some cases an indication of this can be obtained by seismic wave velocity techniques. These well known geophysical methods have an advantage of being quantitative and make use of the principle that seismic (sound) waves travel through sub-surface materials at different speeds, dependent upon hardness, degree of brokenness and the orientation, openness, filling, continuity and surface texture of the planes of the breaks (Atkinson, 1970). Experience has shown that there are various levels of seismic velocity at which different rock types are likely to be workable (Pass, 1970).

Although seismic wave velocity techniques can give a quick and inexpensive assessment of the physical characteristics of the rock, it is not always reliable and confirmation by core drilling and testing is always recommended followed by on-site pilot tests (Shand, 1970). Detailed assessment of rock characteristics such as "fracture spacing" and "point load strength" made on borehole cores can provide a practical use as a means of assessing ease of excavation (Fookes *et al.* 1971): an example from Dartmoor rocks shown in

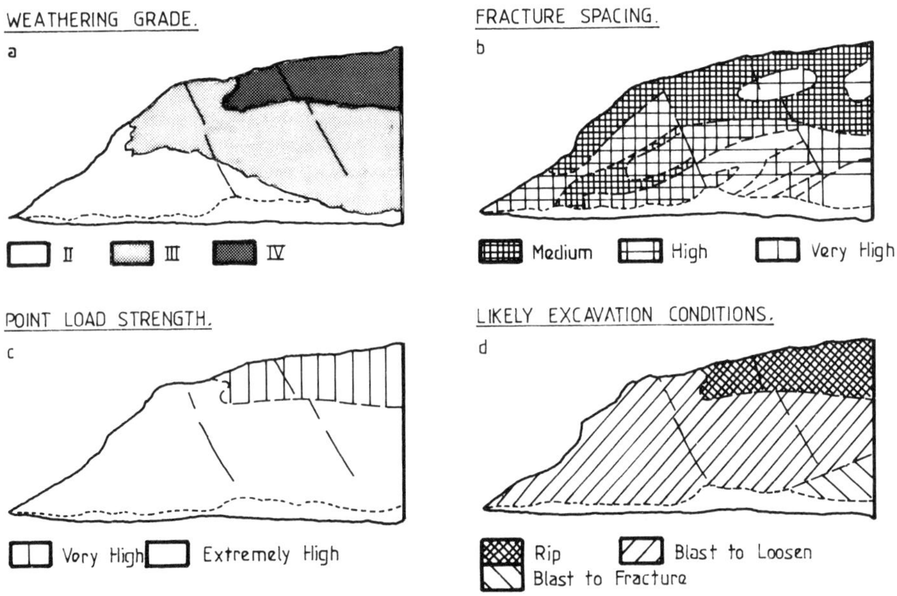

FIG. 4.4. Geotechnical properties and engineering appraisal of the quarry face in Cherts and limestones. Meldon Quarries, Devon (Fookes *et al.* 1971). (a) Weathering zones defined in terms of grade. (b) Discontinuity spacing variation. (c) Strength zones determined by point load strength tests. (d) Engineering appraisal in terms of likely excavation conditions.

Fig. 4.4 indicates how these authors were able to relate an engineering appraisal to likely excavation conditions.

Rock conditions which are not conducive to ripping are (Atkinson, 1970):

(i) Massive, monolithic structures with a very low degree of brokeness.
(ii) Fine grained materials with strong, non-porous cement agents.
(iii) Non-crystalline, non-brittle materials.
(iv) Plastic materials, e.g. wet clays.

Ripper selection will depend on the prevailing rock condition. The key to economic production is a combination of (a) selecting the correct tractor size for the rock and (b) matching the production required to the machine size—the larger the tractor size the higher the hourly operator's costs together with (c) selection of the ripper beam and ripper shanks suited to the duty (Atkinson, 1970). A disadvantage of ripping, as opposed to blasting for primary breaking is that it can produce excessive material of 40 mm size and below. Since one of the first stages in a crushing process is the removal of dirt by screening or "scalpings", over a 40 mm to 50 mm screen, ripping can lead to a large proportion of good quality rock being lost with the rejected "scalpings".

4.3.4 Loading. After primary fragmentation has been achieved selection of the correct equipment for loading and transport of the fragments to the processing plant depends on a number of factors, not least being the desired level of production to be achieved from the plant. A face-shovel excavator (A4.1.4) capable of handling hard, dense, abrasive rocks is one of the most popular types of loading machinery in hard rock quarries (Atkinson, 1971, and Table 4.3). It has the ability to "spot" accurately for loading into dump trucks, rail wagons, and loading hoppers together with a robust construction and simple action. Face-shovels can be rope operated or can incorporate hydraulic mechanisms.

The drag-line excavator (A4.1.7) has limited application as a hard rock loading machine. It has a less positive digging action than a face-shovel, and the small sizes are not suitable for loading dense, badly fragmented rocks (Table 4.3). Also the cycle time is longer than that of an equivalent face-shovel and it has poor "spotting" ability. In some modern hard rock quarries the wheel loader (tractor shovel) has become a popular machine for loading, and although its bucket size has been a limiting factor in the past, this has been overcome in the latest products which have bucket capacities up to 27 m^3. As a general guide when a wheel loader replaces a face shovel, the former should have double the bucket capacity of the latter for an equal output. The wheel loader is popular

TABLE 4.3. *Summary table of advantages and disadvantages of drag-line excavators and face excavators (crawler mounted)* (after Atkinson 1971 & Huddart & Runge 1978)

	Advantages	Disadvantages
Drag-line excavator (crawler mounted)	1. Greater reach and dumping radius enables deeper excavations to be made; can be used in wet pit operations.	1. Less positive digging action than the shovel, particularly when chopping down.
	2. More versitile than the face excavator—more easily manoeuvered.	2. Cycle time is longer than face excavator.
	3. Localised pitches & rolls in the mineral deposit can be easily adjusted for.	3. Poor 'spotting' ability when loaded.
	4. Can be located on any competent working horizon within the overburden.	
Face shovel excavator (crawler mounted)	1. A greater capacity than a drag-line (30% more) allowing it to be more productive	1. Higher capital investment than drag-line.
	2. Total & positive control of the dipper is always possible.	2. Because of its high bearing pressures the shovel is limited to the mineral beds it can stand on (it has a larger mass than a drag-line).
	3. Blockier material have less effect on the shovels ability to load efficiently.	3. Shovel has more wearing points (e.g. hydraulic levelling jacks) and incurs greater maintenance costs than drag-lines in hot dusty conditions
	4. Reduced operator fatigue compared with wheel loaders.	4. Support equipment may be needed for beach clean-up etc.
	5. Lower ground bearing pressures than wheel loaders and not so badly affected by bad floor conditions.	

TABLE 4.4. *Summary of table of advantages and disadvantages of crawler and rubber-tyred mobile equipment* (after Atkinson 1971 & Huddart & Rudge 1978)

	Advantages	Disadvantages
Crawler mounted mobile equipment	1. Strong digging capability as a front-end loader and as a tractor scraper. 2. Relatively low ground bearing pressure: and excellent tractive effort in bad conditions. 3. Good performance on steel inclines. 4. High degree of manoeuvrability.	1. Travel speeds are slow—can only be used where distances between digging and discharge points are short. 2. Maintenance costs in abrasive ground material are relatively high.
Rubber-tyred mobile equipment	1. They are very versitile, with excellent mobility and can travel quickly to any part of the pit or quarry. 2. Maintenance costs in easy to medium conditions are relatively low. 3. Travel speeds are relatively high. They can dig and transport excavated material over larger distances than crawler mounted machines. 4. Can operate on moderate grades without difficulty.	1. Tyres costs are high where angular and abrasive rocks are encountered and in wet floor conditions. 2. Less maneouvrable than crawler mounted units, unless articulated. 3. Ground bearing pressures are relatively high. The machines are not efficient in bad conditions where rolling resistence is high. 4. High operating costs.

because of its versatility; in addition to clearing up and grading, it can also be used for blending from two different loading points. However it lacks the ruggedness of a face-shovel rope excavator and in the early days of quarrying wheel loaders were restricted to handling soils and weaker, less consolidated rocks (Table 4.4). Crawler loaders are sometimes used in hard rock quarrying, but track wear is usually rapid on rocky surfaces. In recent years the development of larger hydraulic wheel loaders with bigger buckets, together with the increased hydraulic power has made it possible to use rubber-tyred machines more widely in hard rock quarries. Although they are short life machines compared with excavators, modern trends confirm that larger machines with greater capacity for handling rock will be developed in the future (Atkinson, 1971), because the advantages of low capital cost is one of the important considerations when using the equipment.

More recently the manoeuvrability of rubber-tyred shovels has been much enhanced by articulation as a means of steering instead of front wheel steering, and virtually all of the larger shovels on the market are now of the articulated variety.

Another important consideration with rubber-tyred shovels is the disadvantage of tyre wear, but this can be alleviated to some extent by the use of protective chains. Other protective methods are achieved by using tyres which are manufactured with steel-tracked treads, these being usually only available for the larger models. More recently a tyre development which has become popular are "slick", or smooth tyres, which achieve greater flotation by spreading the tyre load over more rubber in contact with the ground, and on hard dry surfaces they produce better traction than lug tyres (Sievert, 1973).

4.4 Sand and gravel extraction

Deposits of sand and gravel, usually unconsolidated drift but sometimes, cemented conglomerates, can be extracted by means of normal earth moving machinery (Archer, 1972). For convenience, they can be divided into either dry pits or wet pits in which the water table is so near the surface that some or all of the deposit is under water. In some wet pits artificial dewatering or partial dewatering may allow the site to be more easily excavated, but the economies of this method depend on the permeability of the sand and gravel deposit, its thickness and the use of the ground-water aquifer in water supply. Although some types of equipment such as drag-line excavators can be used successfully in both wet and dry workings, in most cases the different types of operation use different extraction machines as listed below (SAGA, 1967).

4.4.1 Wet operations

4.4.1.1. Suction dredgers (Figs 4.5 & 4.6). In loosely consolidated deposits suction dredgers have proved to be a widely acceptable method of winning sand and gravel. They can work in association with a pressure jet and a cutter head if the deposit is hard and needs breaking up. Pump dredgers are particularly suitable for deep deposits of free-running material, considerable success and economy of running costs being achieved with some of the thick Folkestone sand deposits in Southern England.

4.4.1.2 Floating cranes and floating grab dredgers (A4.1.8). Both types of equipment have advantages when working deposits of great depth, and deposits containing large stones which would interfere with the operation of a pump. The grab dredger type can avoid the use of a jib, several types are available, which are particularly useful for excavating deep deposits where the material is free flowing. High outputs can be achieved, the sand and gravel usually being delivered into barges or on to floating conveyors.

4.4.1.3 Ladder dredgers (A4.1.9). Commonly used in maintenance dredgers working in harbours and estuaries, the latter dredgers are usually mounted on a pontoon, and may have application in sand and gravel excavation where the material is hard to dig by other means.

4.4.1.4 Slack-line excavators (A4.1.10). Combining the excavation and transportation of gravel in one operation, the method has considerable application in large alluvial or glacial deposits. It is very cheap to operate, and changing the position of the towers allows deposits to be worked systematically. Although not commonly used in the British Isles, it has frequent application on the Continent.

4.4.2 Dry operations

4.4.2.1 Bulldozers (A4.1.1). Where the processing plant is located at a low level the bulldozer can excavate and transport very lage quantities of sand and gravel over short distances with the assistance of gravity. A rooter (or ripper) attachment to the rear of the machine can be used to break up hard material.

4.4.2.2 Tractors and scrapers (A4.1.13). These can be successfully employed if conditions are suitable, the method allows for the transport of the material to a convenient point where it can be discharged over a hopper to feed the processing plant.

4.4.2.3 Crawler shovels (A4.1.2). These are mounted on crawler tracks and have the advantage that they are robust and can work in the more clayey deposits where ground conditions may be slippery, particularly in areas of high rainfall.

4.4.2.4 Four-wheel loader (A4.1.3). The rubber-tyred nature of these machines enables them to be used successfully in dry workings which are excavated in shallow benches. Where large outputs are not required these loaders can also be used to convey the material to the processing plant.

4.4.2.5 Monitors (A4.1.11). Not greatly used in the UK except in china clay workings, jets of water at high pressure being applied to break up and flush down a high sand and gravel face, the resulting slurry being fed into a shute leading to the processing plant

4.4.3 Wet and dry operations

4.4.3.1 Excavators (A4.1.7 & A4.1.4). One of the most versatile types of machines used in extraction, the excavator is usually fitted with a "drag-line" although forward shovels (Rope excavator) may be used. Limitations to the drag-line are the depths to which it can work, a deep excavation demanding a correspondingly long jib. Dry excavations can work up to 20 metres of material depending on the size of the excavator, but under wet conditions this is usually reduced to 12 metres, the skill of the Operator being an important factor.

4.4.3.2 Multi-bucket excavators (A4.1.12). Usually in deeper deposits of well mixed materials, they have been widely used on the Continent. These machines provide a continuous flow of raw material and can therefore yield the high output sometimes required. Their high capital cost can only be contemplated with very large operations.

4.5 Marine aggregates

Sand and gravels are being recovered from the sea bed on a large scale in the United Kingdom, the United States (Archer, 1972) and Japan, and smaller scale operations occur in many other countries with continental shelves.

Dredging off the coast of the United Kingdom has taken place for well over half a century and dates back to the mid 1920's when sand was recovered from the shallow protective waters of the Bristol Channel (Hess, 1971). The major growth occurred over the past decade (Cottell, 1978), and the marine aggregate industry now supplies about 13% of all United Kingdom sand and gravel production. In the United States of America marine sand and gravel production represents about 2 to 3 per cent of the total production of that material, whilst in Japan recent statistics indicates that it is a major source of sand, at 19 per cent of total production (Washington Academy of Sciences, 1975).

The origins of the marine deposits are as varied as the sand and gravels that occur on land, but only recently have Geologists turned their attention to studying the off-shore superficial materials around the British coast. The sand and gravels show considerable differences in composition, grain size, and roundness, the components usually consisting of fragments derived from rocks in the immediate surroundings (Chapman and Roeder, 1969). In some cases the deposits are re-worked river sediments, or fossil beach gravels of the type frequently found in the Thames estuary; in other areas the sea bed is capped by extensive glacial deposits which can be unconsolidated sand and gravels or indurated boulder clays. All these materials may be covered by sand waves, up to 20 metres high, constantly on the move as a result of the seasonal tides and currents (Veenstra, 1969).

Marine dredged sand and gravel has been used for many years and no serious objections have been raised as to their suitability; they differ from land materials in the presence and type of impurities, salt and shell being peculiar to these deposits.

Although flint and quartzite are the common aggregate types in the southern North Sea and the English Channel, granites, sandstones, or limestones may form the major ingredients of the deposits in other parts of the British Continental Shelf. The petrological nature of marine aggregates depends on the source i.e. the dredging grounds (Chapman & Roeder, 1969).

4.5.1 Extraction. Materials on the sea bed exploited for aggregates are all unconsolidated

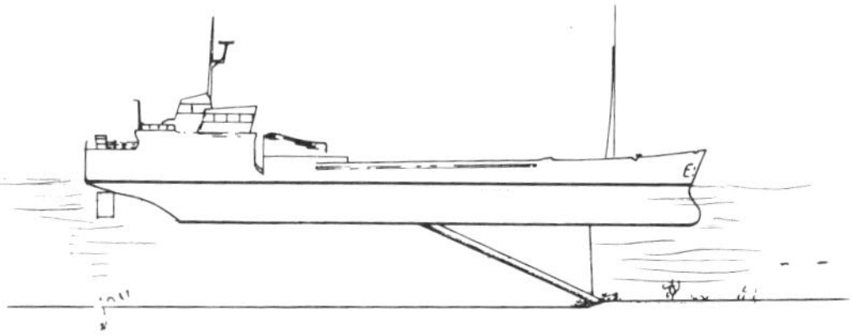

FIG. 4.5. Suction hopper dredger with forward leading pipe (at anchor).

sedimentary deposits which can be worked by dredging. The type of dredge is determined by the softness of the deposit, the size of the area to be dredged, the thickness of overburden, the compactness of the deposit, the depth of water and the weather conditions (Flemming, 1972). Except for a few old barge-mounted grabs, the UK sea dredging fleet is made up of suction hopper dredges. Some operate with forward leading pipe (Fig. 4.5), but most of the newer vessels trail the pipe (Fig. 4.6).

By using hydraulic pumps on the ship to create suction in a pipe leading from a dredge head on the sea floor, suction hopper dredgers are able to operate successfully in up to 36 metres of water. The volume of water lifted in the operation is generally ten times the volume of solid, and the cargo hold which is initially filled with water gradually becomes filled by the solids as the water is displaced. Some unwanted fine material may be washed overboard during the dredging operation, but sea-dredged aggregates must normally be processed, the complexity of the operations depending on the composition of the "as dredged" material (e.g. the clay and shell content) (Archer, 1972). When so designed it is possible for some dredgers to utilise screens to select the particle size of material collected, but dredging time, which is costly, will often dictate the need for this selectivity.

The size of the ship depends on several factors, the trend toward larger vessels being supported largely by changing conditions and new industry requirements. As steaming distances get longer and reserves in deeper water need to be exploited, more efficient dredging equipment and practices must be employed (Hess, 1971). Typical dredgers have increased in size over the years, and whereas the average gross tonnage operating in the UK was 800 tonnes in 1968, the gross capacity of the dredgers built in the 1970's has ranged from 2500 tonnes to 10 000 tonnes (Webb, 1978).

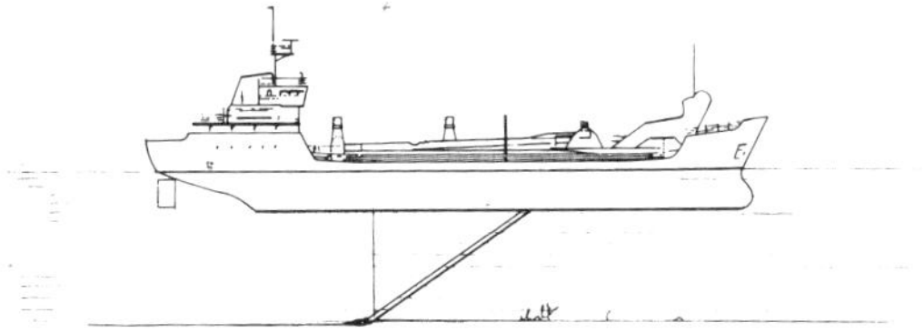

FIG. 4.6. Suction hopper dredger with trailing pipe.

Major dredging activities off South-East England have resulted in the steady deterioration of the quality of the deposits, partly as a result of the return of sand to the sea, and partly because the best material has been used up (Verney, 1976). The future sources of gravel and coarse sand will need to be obtained from water depths in excess of 36 metres (120 feet), the limit of present suction dredgers.

Currently the dredged material is discharged at wharves serving the main United Kingdom demand centres such as London, Southampton, Bristol and Liverpool, but materials are also shipped to continental markets in France, Belgium, Holland and Germany. Unloading techniques used by the dredgers at the wharves are varied, a number of different types of methods having been tried, only a few having been found acceptable in terms of efficiency and economy (Hess, 1971). Although deck mounted grabs, or clamshells, are a common means of unloading, the more recent dredgers have facilities for pumping ashore and by self-discharging with elevators and/or belt conveyors. Self-discharge by scraper buckets coupled with over-the-side conveyor belt booms has provided more flexibility at different wharves, the more modern vessels tending to adopt this technique thus speeding up the turn-round cycle (Webb, 1978).

4.6 Discussion

In the extraction of aggregates the variety of equipment available to facilitate the efficiency of the operation is extensive. The choice of equipment will depend on the climatic conditions, the production rate necessary, and not least the life of the operation. Thus a machine which may be suitable for large scale heavy duty operations may not have the manoeuvrability that is required for a smaller operation. Future operations on a large scale may tend to use bigger more capital intensive machines to enable greater production to be achieved, but each excavation site needs to be considered on its merits. Mobile earth-moving equipment has found greater application in the aggregate extraction industry in recent years, and will continue to play a major role with both crawler-mounted and rubber-tyred varieties.

Marine operations of the future will continue to make their contribution to the supply of aggregates with a dominant trend towards larger vessels capable of dredging from deeper water, thereby overcoming problems of limited materials of decreasing quality in shallow water, and the restricted domestic markets available to operations in home waters.

References

ARCHER, A. A. 1972. Economics of off-shore exploration and production of solid minerals. Inter-regional seminar on the development of the mineral resources of the continental shelf, Port of Spain, Trinidad. United Nations, New York, 75–93.

ASH, R. L. 1963. The mechanics of rock breakage. Parts I–IV. *Pit & Quarry* (USA) 98–131.

ATKINSON, P. 1970. Ground Preparation by Ripping in Open Pit Mining. *Min. Mag. London* **22** (6), 458–469.

ATKINSON, T. 1971. Selection of open pit excavating and loading equipment. *Trans. Instn Ming. Metall London* **80** A101–A129. A129.

CHAPMAN, G. P. & ROEDER, A. R. 1969. Sea-dredged sands and gravels. *Quarry Managers J. London.* **53** (7), 251–263.

COTTELL, D. S. 1978. Some international opportunities for marine aggregates. *Bull. Instn Ming Metall London.* **87,** 84–88.

DOVER, T. M. 1964. Stripping of Rock overburden symp. on opencast min. Qu. and Al Min. London. *Instn Ming. Metall.* 1–14.

FOOKES, P. G., DEARMAN, W. R. & FRANKLIN, J. A. 1971. Engineering aspects of rock weathering *Q. J. eng. Geol. London.* (**4**). 139–185.

FLASCHENBERG, P. 1964. Loading and haulage in quarries. *Instn Ming & Metall Symp. London.*

FLEMMING, N. C. 1972. Methods and technology of off-shore production of solid minerals. Inter-regional seminar on the development of the mineral resources of the continental shelf, Port of Spain, Trinidad, United Nations, New York, 63–74.

GEOLOGICAL SOCIETY ENGINEERING GROUP REPORT 1977. The description of rock masses for engineering purposes. Rep. Geol. Soc. Engng Group Working Party. *Q. J. eng. Geol. London* **10,** 355–388.

GREENLAND, B. J. 1973. Primary blasting practice, the factors affecting performance and results. *Quarry Manage. Prod. London.* **12,** 421–429.

GRIMSHAW, G. B. 1963. Primary blasting efficiency in large and medium sized limestone quarries in Derbyshire. *Quarry Managers J. London.* **12,** 487–491.

GRIMSHAW, G. B. & WATT, R. 1971. The current scene in quarry blasting. *Quarry Managers J. London.* **4,** 119–131.

HER MAJESTYS's STATIONERY OFFICE. 1982. Production of Aggregates in Gt. Britain 1–19.

HESS, H. D. 1971. Marine sand and gravel mining industry of the United Kingdom. NOAA Technical Report CRL 213-MMTC 1.

HUDDART, H. & RUNGE, B. E. 1979. Pit design optimised for drag-line productivity *Trans. Instn. Ming. Metall. (sect. A. Min. Industry)* **88,** A6–A11.

INDUSTRIAL MINERALS 1971. *An introduction to aggregates. London.* **47,** 9–17.

KOCHANOWSKY, B. J. 1961. Inclined drilling and blasting, *Mining Congress Journal*, U.S.A. 1–8.

LAKIN, A. E. 1971. Excavation Techniques. *Quarry Managers J. London.* **55,** 5, 155–165.

MINE & QUARRY 1973. *How Moons Quarry faces are worked.* **2** (2), 9–15.

NOBEL'S EXPLOSIVES COMPANY LIMITED. Undated—Explosives in Quarrying—I.C.I. Publ.

PASS, R. D. 1970. Planning for production. *Quarry Managers J. London* **54** (5) 163–176.

PUGLIESE, J. M. 1973. A comparison of calculated patterns with plans used in quarrying limestone and dolomite with geological considerations. *Pit & Quarry* 85–88.

SAND & GRAVEL ASSOCIATION OF GREAT BRITAIN 1967. Pit and Quarry Text Book Macdonald and Co. (Publishers) Ltd. London. 1–239.

SHAND, A. N. 1970. The basic principles of equipment selection for surface mining. *S. Africa Inst. of Min. & Metall. Johannesburg Symposium.* 235–249.

SIEVERT, R. A. 1973. Load and carrying for quarries. *Cem. Lime Gravel.* **48** (5) 91–93.

TEICHMANN, G. A. 1960. Recent developments in blasting techniques. *Quarry Managers J. London.* **44** (7) 251–258.

VEENSTRA, H. J. 1969. Gravels of the southern North Sea, *Mar. Geol.* **7,** 449–464.

VERNEY, SIR R. B. (Chairman) 1976). "Aggregates: the way ahead". Report of the Advisory Committee on Aggregates. H.M.S.O. for Dept. of the Environment.

WASHINGTON ACADEMY OF SCIENCES 1975. Foreigh Continental Shelf Developments. Mining in the outer continental shelf and in the deep ocean 1–176.

WEBB, P. 1978. Legal and economic aspects of dredging marine aggregates in the UK, International Seminar, Orelans, France, BRGM 7 49–60.

General References

CHUGH, Y. P. 1980. Surface mining of minerals by draglines in the U.S.A., *Trans. Instn Ming. Metall. London (Section A: Min. Industry)* **89,** Oct.

CHURCH, H. K. 1984. Excavation Handbook, McGraw-Hill Book Company, New York.

CRUISKSHANK, M. J. 1964. Mining offshore alluvials. *Symp. on Opencast Min. quarrying and alluvial ming. Instn Ming. Metall. London.*

KUTTER, H. K. 1969. Consideration on blasting in jointed rock, *Quarry Managers J. London.* **53,** No. 4, 146–152.

Appendix A4.1 Diagrams of Extraction Equipment in use in 1984

A4.1.1. *Crawler Tractor* (Bulldozer)

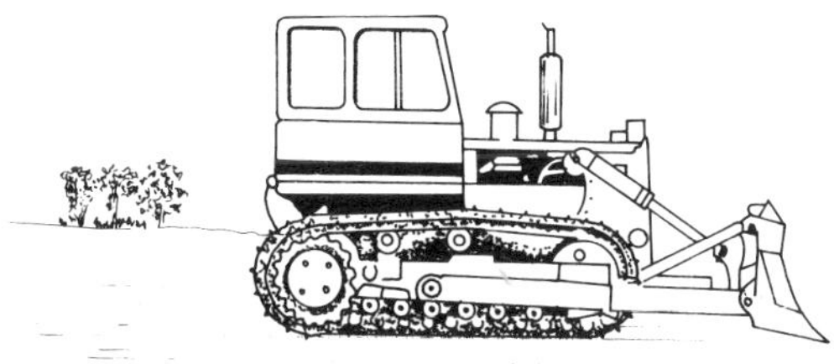

Consists of power unit fitted with a crawler or caterpillar track enabling the machine to move over most types of surface. The front of the machine has a bow-shaped blade fixed at right angles, the two outer edges of the blade being blocked. Particularly applicable to dozing and ripping.

A4.1.2. *Crawler Loader*

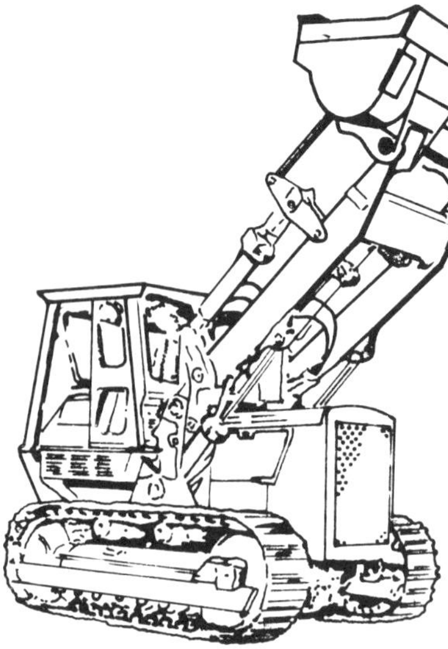

These models are particularly applicable to tough excavating and backfilling, the power unit being mounted on crawler or caterpillar tracks with automatic bucket controls for loading ability.

A4.1.3. *Four-Wheel Loader*

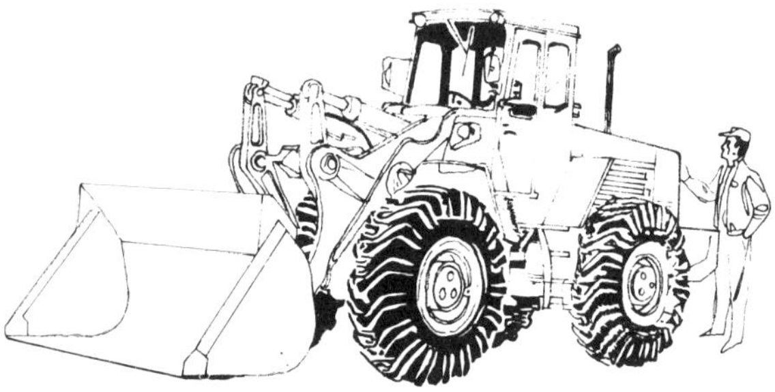

Mounted on rubber-tyres and normally with front-wheel steering, greater manoeuvrability has been achieved with these shovels in recent years by use of articulation, most of the larger models today being of this type.

A4.1.4. *Face-Shovel Rope Excavator*

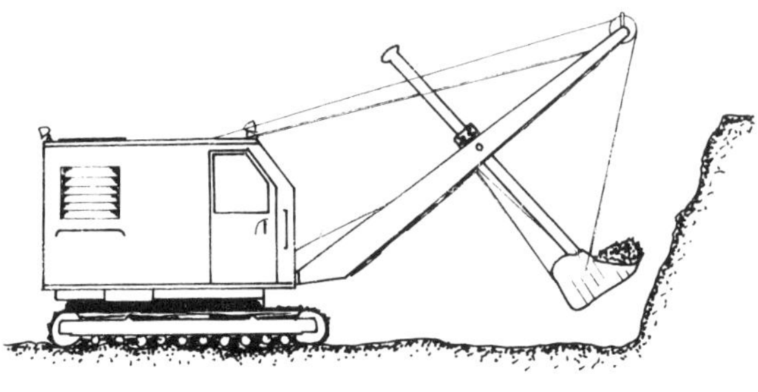

Consisting of a boom with a saddle attachment at about half its length where the bucket arm is attached, the boom and bucket are controlled by cables (ropes) from the cab. The base of the machine is normally on crawler tracks, but may sometimes use tyres in the smaller varieties.

A4.1.5. *Back-Hoe Hydraulic Excavator*

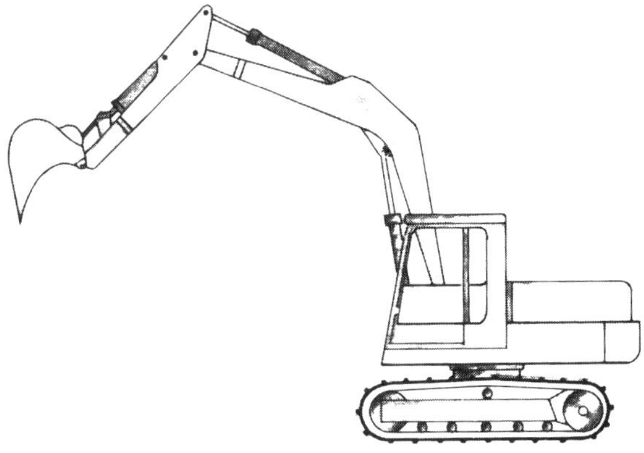

Usually mounted on driven tracks but sometimes on wheels, the hydraulically rotating boom with full circle slew gives considerable flexibility as a digger loader.

A4.1.6. *Dump-truck*

The primary requisites in a modern dump truck for surface mining work are simplicity of control. The overall height of the truck should be such that it makes a good target for the shovel dipper, the length being at a minimum and with a short wheel base with ability to turn round in cramped situations. A wide range of machines are available varying in shapes and sizes to include rear-dumps, side dumps and bottom dumps, together with rigid and articulated types.

A4.1.7 *Drag-line Excavator*

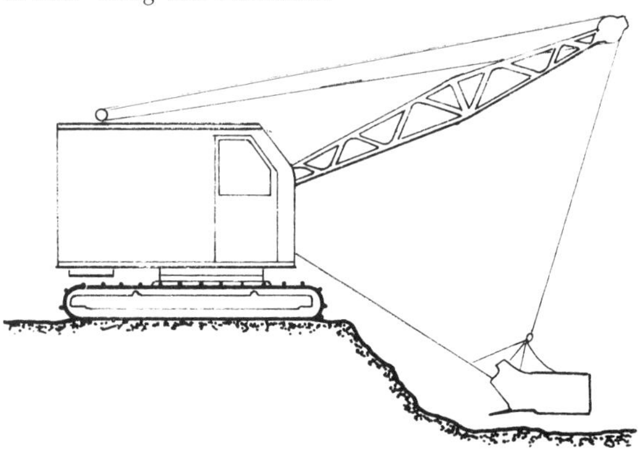

Used mainly for loading directly into hauling equipment the drag-line is usually mounted on crawler tracks, the long side tracks reducing bearing pressure and increasing stability. The machine is usually situated on a bank and is able to strip material to a considerable depth below the bank depending on boom length and digging slope angle. Because of the increased radius of the boom the machine is capable of excavating over a larger area of ground than any other static plant equipment.

A4.1.8. *Floating Cranes and Floating Grab Dredgers*

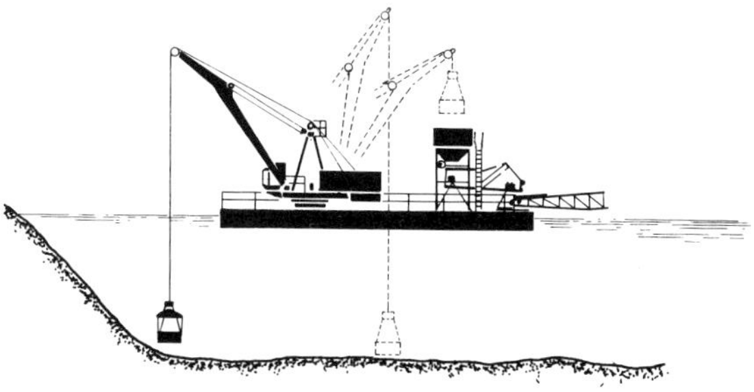

Usually mounted on one end of a rectangular pontoon for loading into barges, floating cranes and grabs are capable of very high speeds of operation. Grabs can deliver to a surge bin with a grid to separate oversize material picked up by the grab the undersize material being discharged to conveyors and thence to the processing plant.

A4.1.9. *Ladder Dredgers*

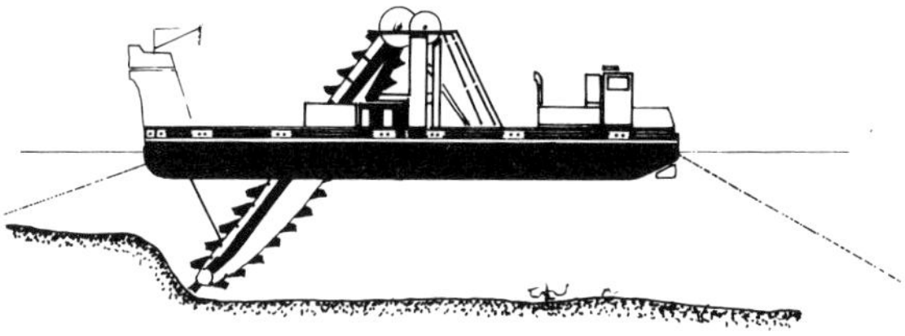

Although considered to have limited use in sand and gravel dredging from below water they do have the advantage of being able to dig hard material.

A4.1.10. *Slack-Line Excavators*

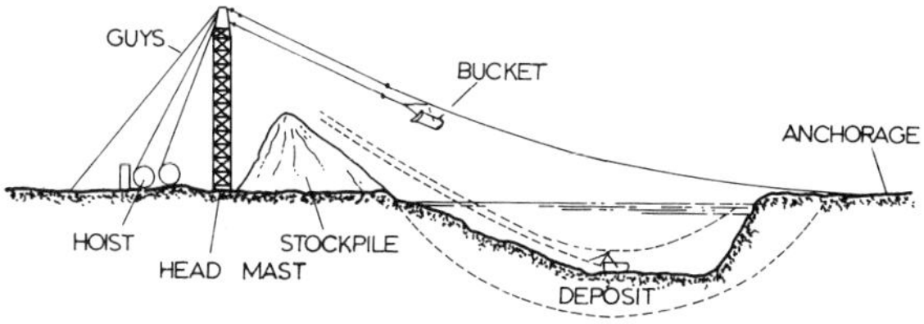

Combining the excavation and transportation of the mineral in one operation, the method is most successful in free-running material.

A4.1.11. *Monitors*

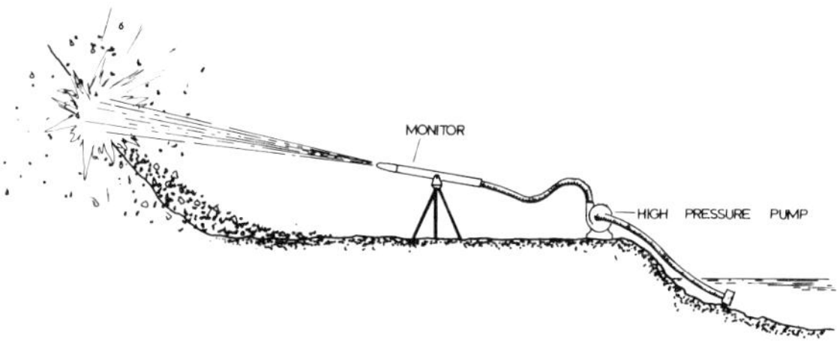

Water nozzles constructed with a curved inlet and a swivel bearing in such a way that the reaction thrust of the nozzle is taken by the pipe-line anchor. This enables the monitor operator to swivel the jet and aim it at the base of the bank to be undercut. Its use requires a plentiful supply of water, the water jets used to break up sediments and wash it to a sump and treatment plant.

A4.1.12. *Multi-bucket Excavators*

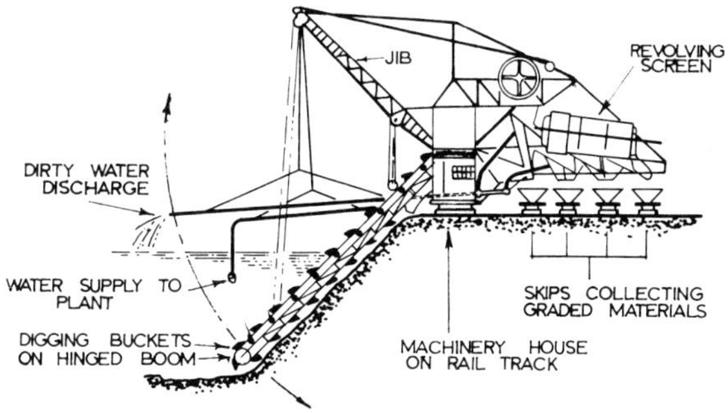

The two most important types of continuous excavators are the bucket-chain excavators and bucket-wheel excavators. Mounted on crawler undercarriage, the structure has mobility. The material dug by the buckets is transferred to a conveyor belt which carries the material away. The figure shows a bucket-chain excavator.

A4.1.13. *Scrapers*

Usually associated with bulk excavation of earth-works, it consists of a bowl mounted on wheels with pneumatic tyres and may be towed by means of a tractor, or contain its own power unit. To facilitate the movement of the material into the bowl a bucket elevator is sometimes incorporated into the machine.

A4.1.14. *Ripper*

This equipment has to be towed by a tractor or a power unit, and consists of a drawbed and a frame mounted on steel wheels. The "tines" enable the ground through which it is travelling to be broken up, the equipment having wide application in friable rock and unconsolidated materials.

Chapter 5. Processing

5.1 Introduction

Materials raised in sand and gravel extraction and in hard rock quarrying, by methods described in the previous chapter, are generally only sold in an "as dug" condition for fill but, even for this application, it is not unusual to pass the material through a single crusher or screen to ensure a uniform maximum size. For most other uses, including concrete aggregate and various types of roadstone, size reduction and particle sizing operations are necessary to bring the naturally occurring or crudely broken material within the limits imposed by end use specification.

Production of aggregates usually involves the use of heavy, capital intensive machinery. However in some rural areas abroad aggregate may still be made by breaking rock manually with hammers, although such methods are viable only where labour is very cheap, demand is for relatively small quantities and fairly coarse material can be used.

Mechanical size reduction is mainly carried out by crushing and grinding machines based on the principle either of nipping the rock (e.g. gyratory and jaw crushers) or of direct impact (e.g. hammer mills), while particle sizing of coarse material is based on the use of screens. Sizing of fine material is normally based on the exploitation of differences in particle velocity in a stream of water.

Separating equipment, of the type commonly found in mineral beneficiation plants, (e.g. shaking tables, flotation cells) is seldom used in aggregate treatment, although occasional use is made of the cheaper processes. Similarly the economics of aggregate production rarely justify the use of such fine grinding equipment as ball mills or rod mills.

5.2 Crushing

5.2.1 General. The crushing action in size reduction equipment results from stresses applied to rock particles by moving parts of the machine. Inevitably the action is crude in its effect, allows for only a limited amount of control and is difficult to analyse mathematically.

The object of crushing in aggregate production is size reduction to a specified size range, with a minimum production of finer material. Machines can only be adjusted to produce material finer than a given size and it is impossible to control the minimum particle size. Although some machines produce more fines than others and to some extent there may be variations in product shape, the differences tend to be relatively small. However, the crushing characteristics are reproduceable and from testwork, for a given machine and rock type, it is possible to predict the yield of crushed material into various size fractions.

Energy requirements in crushing are relatively large as the rock is broken essentially by compressive forces and machines are always of robust construction. The relationship between the energy used and the amount of breakage achieved is complex. To some extent it is proportional to the new surface area produced in a given rock type but further details can be obtained from the specialist literature (Lowrieson, 1974) for a more thorough treatment of the subject.

5.2.2 Nipping machines. Jaw, Gyratory and Cone crushers which are of great importance in aggregate production, achieve size reduction mainly by compressing particles between relatively slow moving inclined surfaces. The feed material is received from above where the working surfaces are furthest apart and is nipped into successively smaller fragments as it descends to the narrowest zone between the crushing surfaces before it is discharged by gravity.

The nipping action is conferred by a reciprocating motion of relatively small amplitude. The maximum size of feed accepted by the machine is controlled by the maximum distance between crushing surfaces, while the maximum size of the product depends on the minimum distance between crushing surfaces in the discharge zone. In practice crushing surfaces are arranged in the form of rectangular plates in jaw crushers or as truncated conical shells in gyratory and cone crushers.

5.2.2.1 Jaw crushers. Most jaw crushers consist of one fixed crushing face and an inclined mobile face which moves alternatively towards or away from the fixed face by a small throw. Because the loading varies greatly during the operating cycle, it is usual to incorporate a heavy flywheel in the design. Jaw crushers are classified by the method of pivoting the mobile jaw. If the jaw is pivoted near the top it is known as a Blake crusher and is characterised by having a fixed opening for feeding. If pivoted near the base it is known as a Dodge crusher and has a fixed discharge area.

Dodge crushers produce a closely sized product

because of the constant width of the discharge opening, but they suffer from the disadvantage that the minimum leverage is applied to the largest particles, which confines their application to light duties and the treatment of relatively small fragments. Blake crushers, which exert maximum leverage on the largest material, are widely used and suitable designs have treated material up to 3 m in diameter.

Important variations in the basic design of Blake crushers are concerned with different methods of applying the rotary motion of the drive shaft to the moving jaw plate. In one design the moving jaw is hung on an eccentric top pivot and the plate moves with a gyratory action in the vertical plane. Such machines are known as overhead eccentric or single toggle types. The gyratory action results in a downward rubbing action in addition to the usual nipping action which effectively increases the capacity of the crusher. In double toggle machines the moving jaw is hung from a non-eccentric pivot and is driven through a separate eccentric and double toggle system which results in a simple lateral jaw motion (Fig. 5.1).

The double toggle machine is relatively expensive and is used for tough abrasive materials such as quartzites, otherwise single toggle machines with their higher capacity are preferred although they tend to experience greater wear on their crushing surfaces. In terms of usage there are probably four times as many single as double toggle machines. As a rough working rule some manufacturers hold that double toggle machines are necessary only when the crushing strength of the rock exceeds 250 MN/m^2, a value which in general is exceeded in only the tougher igneous and metamorphic rock and in some greywackes and quartzites.

Major variables in jaw crusher operation are the angle of the jaws, rate of jaw movement, the displacement or throw and the distance between the jaws at the discharge end which controls the product size.

5.2.2.2 Gyratory crushers (Fig. 5.2). The essential features of a gyratory crusher include an upward pointing solid rotating cone set within an inverted cone of various angles. The inner cone rotates eccentrically so that it approaches and recedes from a given point on the outer surface during each revolution. Material fed to the crusher from above becomes lodged between the two surfaces where it is broken by being alternately compressed and released during rotation of the inner surface. The action is similar to that of a jaw crusher and the machine can be regarded as consisting of an infinite number of jaw crushers of infinitely small width. The eccentricity of the inner cone is arranged so that the apex maintains a fixed position while the maximum displacement occurs in the lower or discharge area. This action helps to prevent blocking by fine material and, as with the Blake crusher, ensures that maximum leverage is applied to the coarser material. Hence the gyratory crusher is suitable for handling coarse, tough and abrasive material; machines are made to handle up to 5000 tonnes/hour and to take feeds of up to 2 m diameter. In one variety of gyratory crushers, known as "parallel-pinch" machines the apex does not maintain its position but rotates in an eccentric circle so that the same amount of pinch is exerted at all points on the face of the crushing head.

In comparison with the jaw crusher the gyratory has several advantages. Because of its rotary motion, it makes more economic use of power and requires a less massive construction for a given throughput. The rotary motion also enables wet and slightly clayey materials to be more effectively handled. However, jaw crushers have the advantage that they can handle occasional oversize fragments, whereas a gyratory designed to crush such material would result in a

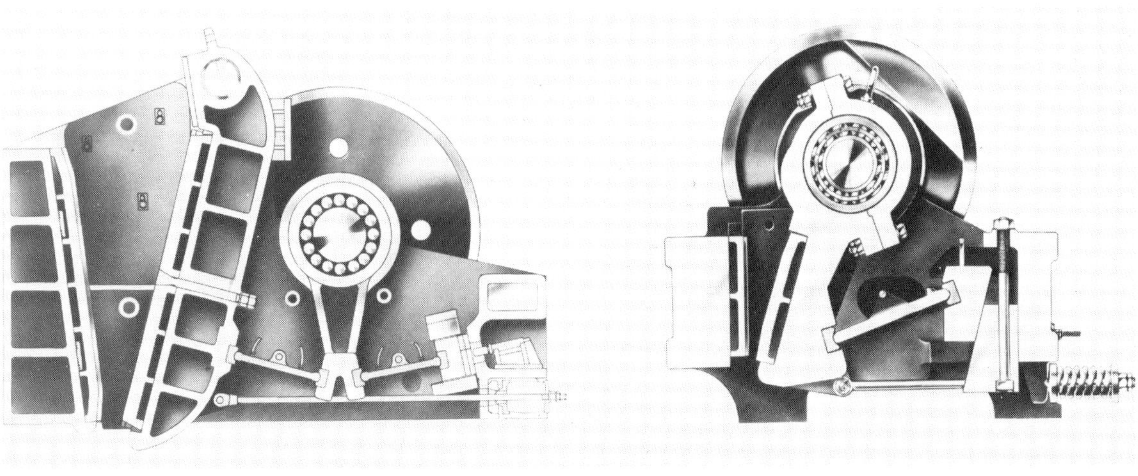

FIG. 5.1. Sections through double toggle (left) and single toggle (right) jaw crushers. (Courtesy W. H. Baxter Ltd).

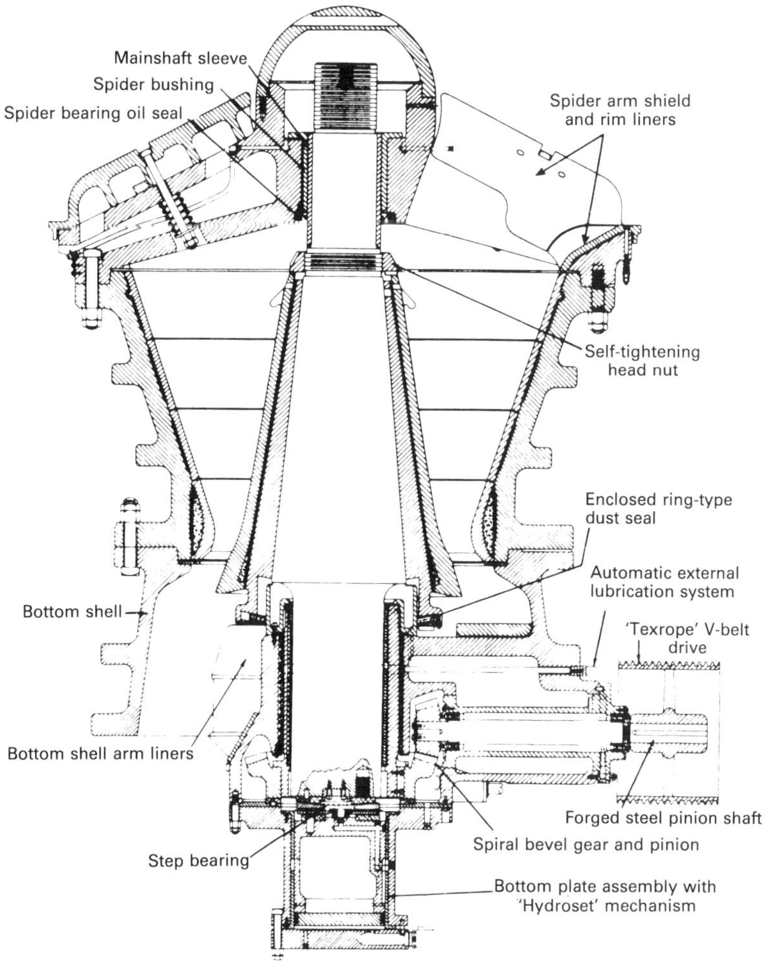

FIG. 5.2. Section through a gyratory crusher. (Courtesy Allis-Chalmers Ltd).

machine of excess capacity. Hence gyratories, although approximately three times more expensive than jaw crushers capable of receiving the same sized feed, are justified where their higher throughput can be exploited by operating at near peak conditions. Jaw crushers offer more flexibility and are often preferred for intermittent operations.

5.2.2.3 Cone crushers (Fig. 5.3). Cone crushers are similar to gyratories except that the outer crushing surface is in the form of a cone with an upwards pointing apex, which results in the two crushing surfaces being nearly parallel. The design allows the crushed material to spread out as it works its way downwards and prevents close packing and blockages in the discharge area. Crushing surfaces of different geometry may be used depending on the particular duty.

The two surfaces, held together by springs, can separate under load to allow any tough oversize material to pass through without interrupting the working. These characteristics together with a relatively high speed of rotation make the cone crusher ideally suitable for achieving high throughputs with feeds of intermediate particle size where a limited size reduction is required. In practice cone crushers are usually used to treat the products of gyratory or jaw crushers, i.e. for secondary crushing. They suffer from one significant disadvantage in that they tend to produce more flaky particles than other crushers.

5.2.3 Impacting machines (Fig. 5.4). Machines of the rotary hammer type are widely used in aggregate production. They are characterised by a fast moving rotor with beaters or hammers attached which transfers some of its kinetic energy to rock fragments by striking

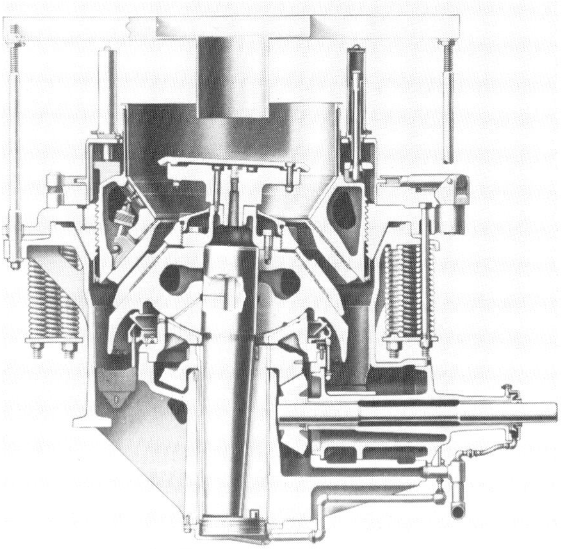

FIG. 5.3. Section through a cone crusher. (Courtesy Rexnord Process Machinery).

them. Internal stresses may then cause a particle to shatter immediately or it may be flung, together with broken fragments, against an impact plate, where further shattering can occur.

Machines vary in the number and shape of the hammers and whether these are fixed or pivoted. Some machines discharge through gratings, others have an open discharge and some have double rotors. The breaking plates also vary in design and although usually stationary some have been designed to rotate at various speeds with or against the rotor.

Swing hammer machines are more flexible in that the hammers can avoid any excessively large or tough rock fragment, although the hammers themselves are less effective in achieving breakage. Double-rotor machines are more effective in treating softer materials which might otherwise cause blockages.

The maximum product size is controlled by the distance between the hammers and breaking plates. In some machines the first set of breakers near the feed zone are set furthest away from the hammer to produce a coarse fragmenting zone, which is followed by a finer zone where the gap between breakers and hammers is reduced.

In general the advantages of impacting machines include a lower energy requirement than most other crushers for a given duty, an ability to cope with a high proportion of clay or shale, a marked tendency to produce more cube like particles, the capacity to achieve a large reduction ratio and a limited space requirement. The major disadvantage is that tough abrasive fragments cause intense wear on the moving parts and for this reason impact mills are mainly used for treatment of limestone which is not abrasive. However they are sometimes used on gravel and other rocks, particularly where their tendency to produce minimum amounts of elongated particles is of value in meeting aggregate specifications with difficult materials.

5.3 Sizing

5.3.1 General. Most aggregate sizing operations are carried out by means of a screening operation, which grades particles according to the minimum cross section presented to a wire mesh, a hole in a plate, or gap between parallel bars through which undersized material can fall. Because screens pass particles on the basis of their approximate diameter, the weight of given material falling through a single opening varies according to the cube of the aperture size. At the same time the number of openings per unit of screen surface varies inversely with the square of the aperture size. Consequently the rate of passage of material through a screen surface is directly proportional to the size of the apertures and for this reason coarse screens have very large capacities, while the capacities of successively finer screens eventually becomes so limited that industrial screening is impractical.

Fine sizing is carried out be exploiting differences in particle velocity in water. For many practical purposes

FIG. 5.4. Section through an impact crusher. (Courtesy Frederick Parker Ltd).

the settling rate is approximately proportional to the square of particle radius (Stokes Law). This principle, known as classification, can form the basis of a sensitive sizing method.

In aggregate sizing, screening is not normally carried out at sizes below 3 mm, and classification is usually only applied to materials which are finer than 5 mm.

5.3.2 Industrial screening An effective industrial screening process involves the presentation of particles to screen apertures in such a way that material which is fine enough to pass through does so without involving an unduly extensive screen area or excessive period of time. The major factors contributing to screen efficiency involve transporting material onto and across the screen deck, stratifying the material so that the fines are selectively presented to the apertures and preventing blockages of the mesh.

In simple static screens, transport is arranged by inclining the surface sufficiently for material to be moved by gravity. In slightly inclined cylindrical screens, transport is effected by gravity and the rotary motion, while in other inclined screens it is provided by the application of vibration together with gravity. In horizontal or near horizontal screens transport is effected by a directed vibration only.

It is possible to present fine material to screen openings by repeatedly turning the particles over so that eventually each will meet with an aperture. To a greater or lesser extent, screens which utilise, or partly utilise, gravity for transport exploit this method. However, by means of a suitable vibration it is possible to stratify the feed by causing the larger particles to work their way to the top of the bed leaving finer particles in contact with the screen opening, where they have a maximum chance of passing through without the apertures being obstructed by oversize.

There are two conditions which lead to blocking of screens, pegging and blinding. Pegging involves the wedging of single particles in the apertures and becomes prevalent when a high proportion of "critical" material, that is with diameters between 70% and 110% of the mesh size, are present in the feed. The tendency to peg increases with the thickness of the screen but can be decreased to some extent by the use of flexible wire or rubber screens, or by the application of a suitable vibration. Blinding is caused by an accumulation of adhering fine particles which block the screen openings. It is usually caused by excess moisture or clay particles in the material and is normally avoided either by prior washing and desliming or by screening under a water spray. The use of a water spray on a suitable designed screen is generally beneficial in that it helps to lubricate the particles and prevent adhesion. However, totally submerged screen surfaces would be inefficient as the buoyancy effect of the water decreases the downward force between particles and the screen deck. As with pegging, blinding is less likely to occur with flexible wire or rubber screens. It can also be diminished by using polished stainless steel surfaces or electrically heated screens although these are rarely economic in aggregate production.

Screen apertures are normally square or circular or, to a lesser extent, slotted in shape. With slotted apertures the slots may be of various lengths and either in line with the flow of material or transverse. Normally it is preferable to use square apertures, particularly for screening gravel, but round apertures sometimes result in less pegging when used with angular material and are of value in handling flaky particles. Short slotted holes are mainly used for finer material and for handling particles which are otherwise difficult to screen. However, in most cases selection of the shape of the aperture is of less importance than selection of the right hole size (Fig. 5.5). A new screen type, made up from wave formed wires connected and supported at suitable longitudinal intervals, has proved successful in avoiding blinding with material finer than 6 mm. The extra flexibility of the wave formed wires allows them to shake off or cut through adhering material.

In recent years there has been a growing tendency to replace steel screens with rubber or other polymer surfaces. Various types are available with or without steel or fabric reinforcement and in one popular variety the screen mat consists of a top layer of soft, abrasion resistant rubber, bonded onto a harder lower layer to provide support. The main advantage of polymer screens lies in their resistance to wear. As a rough guide it is estimated that although rubber screens are approximately three times as expensive as steel screens they last eight to ten times as long. Another major advantage is a much lower level of operating noise, a factor of increasing importance in view of recent health and safety legislation.

Because of their flexibility, polymer screens are less prone to pegging, provided the screens are not too thick. To compensate for wear it is clearly desirable to make screens as thick as possible but as pegging and blinding increase with thickness, it is usual to restrict the screen thickness to the size of the apertures. Because polymer is weaker than steel, apertures in polymer screens must be surrounded by a wider zone of supporting material which reduces the effective screening area. Hence a disadvantage is that larger surface areas are required to provide the same duty as a steel screen. In general polymer screens are more popular in heavily used plants treating abrasive material where their higher cost can offset against savings in maintenance.

The following types of screens are commonly used in aggregate production:

5.3.2.1 Inclined vibrating screens (Fig. 5.6). Probably the most widely used screen type in aggregate production is the relatively cheap, simple machine based on the vibration produced by a single eccentric driving shaft. Such machines have a circular motion and rely on a gravity component to assist in

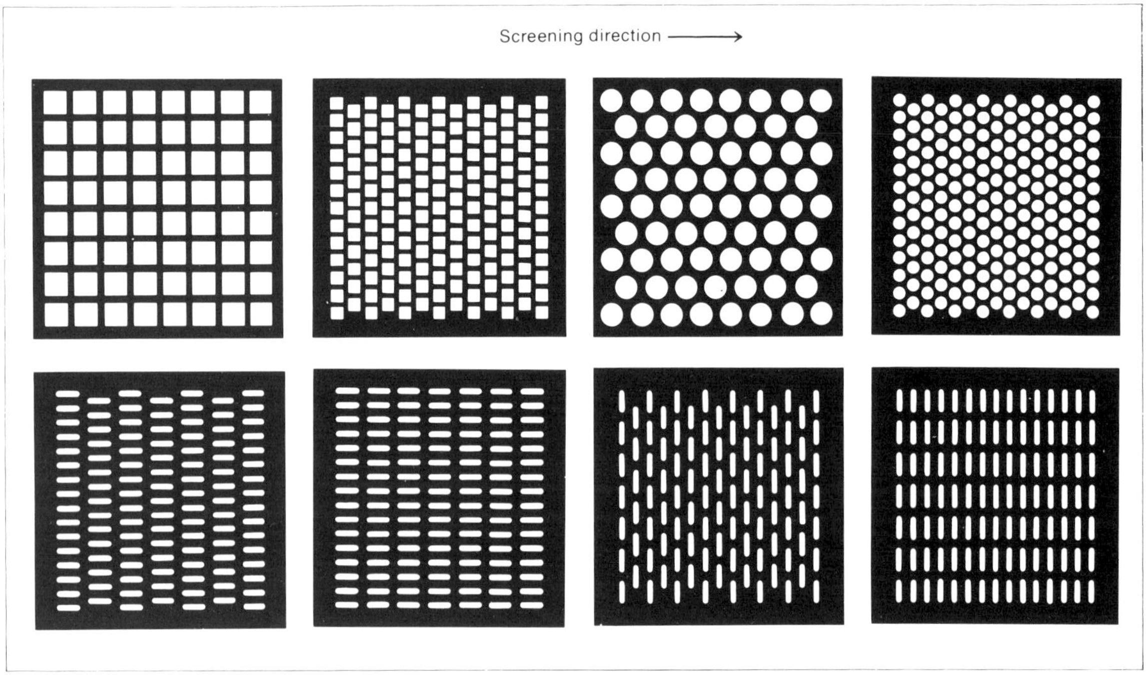

FIG. 5.5. Types of aperture in rubber screens. (Courtesy Trelleborg Rubber Ltd).

transport. Hence they are usually used at angles of up to 20 degrees if the direction of motion is the same as the transport direction and occasionally at 20 to 30 degrees if the motion is contrary to the material flow and a more vigorous action is required.

The maximum amplitude and frequency of vibration is limited by the stress which the machine can tolerate and by the fact that particles on leaving the deck, should not remain in the air for more than one cycle, while the minimum acceleration is that necessary to overcome the effects of gravity and friction on the particle. In practice amplitude and frequency are adapted to different aperture sizes, for example apertures of 100 mm require frequencies of about 800 revs/min while 6 mm apertures require about 1400 revs/min with correspondingly reduced amplitudes. The inclination is also critical as too much slope results in transport becoming too rapid for the undersize to pass the openings, while too little gives a reduced capacity. Hence there is a mechanical optimum screen motion, inclination and aperture size which limits the use of multiple parallel decks in a single machine. However, economic factors rarely justify the cost of a separate machine for each screening operation and double or triple decks are frequently used, the machines being optimised on the most critical size with excess capacity on the other sizes. In addition, surfaces are sometimes divided to provide finer apertures at the feed end of the screen and coarser apertures at the discharge end. Morgensen sizers have five decks arranged at progressively steeper angles as the mesh size decreases.

5.3.2.2 Horizontal vibrating screens (Fig. 5.7). Screens working at slopes of up to 5 degrees make use of a linear motion with horizontal and vertical components which stratifies material and transports it along the deck. The motion may be provided by two contra-rotating eccentric shafts or by a crank driven rod working against sloping leaf springs. In resonance screens, which also have a linear motion, the vibration stroke is limited by springs and buffers in such a way that the kinetic energy is largely retained in the oscillating system with consequent savings in energy.

Horizontal machines are more expensive than inclined machines but have a higher capacity and require less headroom. In aggregate production they are mainly used where the initial outlay can be off-set against savings in building costs and the need for elevators to feed material onto inclined screens. The relationship between vibration characteristics and particle size together with the use of divided and multiple decks involve similar factors to those considered for inclined screens.

5.3.2.3 Trommels (Fig. 5.8). Trommels are rotating, slightly inclined cylindrical screens made up from curved screen plates. Material enters the trommel at the higher end and is discharged as oversize at the lower end or as undersize through the walls.

As screening devices trommels have the advantage

FIG. 5.6. Double deck inclined vibrating screen (Courtesy Pegson Ltd).

of being simple, strong, cheap and free from vibration. However they are susceptible to pegging, have a poor capacity for their screen area and screen changing is difficult. In many applications they have been replaced by vibrating screens, although they are used to some extent in small mobile plants intended for intermittent use and they sometimes form an integral part of rotary washers used in gravel treatment.

5.3.2.4 Grizzlies (Fig. 5.9). Grizzlies are robust screens consisting of parallel bars of various cross section held together by widely spaced cross bars. They are used mainly to facilitate operation of the primary crushing or washing processes by removing unsuitable oversize or undersize material from the untreated quarry product. Static grizzlies are normally used to prevent the passage of occasional oversize material, they may be inclined to allow material to slide off the screen, or horizontal in which case they may also act as a anvil on which to break large fragments. Vibrating grizzlies are often used to feed coarse material into a crusher and at the same time to separate undersize which does not require treatment or which is contaminated with clay or shale to the extent that it is not suitable for good quality aggregate. Such material, often called scalpings, may then either be rejected, sold separately or subjected to extra treatment.

5.3.2.5 Piano wire screens. These screens consist of individually stretched parallel wires which effectively form very long apertures with flexible sides, ideal for the avoidance of pegging and blinding. They are extensively used for dry screening of building sand, where their function is to remove occasional oversize pebbles and lumps of clay.

5.3.3 Screening practice. The function of a screen is to protect a crusher or other machine from receiving oversize material which it cannot handle, to enable fine material to avoid a crusher set to give a coarser product, and to grade crushed material into specific size ranges.

In aggregate production a distinction is often made between screening procedures which allow unwanted material to be discarded as undersize (screening by rejection) and those which discard unwanted material as oversize (screening by selection).

The design of a screening plant, which has been said to be more of an art than science, depends on optimising several factors, both mechanical and economic, in terms of size distribution of material

FIG. 5.7. Single deck resonance screen. (Courtesy Hunt-West Ltd).

treated and markets to be served. Usually it is preferable to remove coarse material first and work down to the finer sizes as this avoids loading heavy particles onto fine screen surfaces. However, other systems are used. For example, sometimes the initial separation is at an intermediate size to allow division into bulk coarse and fine fractions for further screening. Hence generalisations are difficult to sustain and specialist knowledge and experience is usually needed in flowsheet design (5.7.1).

5.3.4 Classification (Figs 5.10, 5.11 and 5.12). Many designs of wet or dry classifiers are used in mineral processing, but those used in the treatment of fine aggregate in Great Britain usually incorporate a water filled tank which receives a slurry feed. Fine particles are carried upwards with the flow of water and are discharged over a weir, while coarser material settles to the bottom where it is either discharged downward through a valve in the base of the tank or elevated upwards by a series of buckets, or by a spiral or rake acting along an incline. The size at which separation occurs is a function of the height of the weir and the flow of water. A certain amount of drainage takes place as the coarser sand is moved up the incline and provision of drainage holes allows water to escape in the bucket flight.

Another type of classifier consists of a long rectangular tank with discharge points at regular intervals along the bottom, an overflow weir at one end and a low feed entry point at the other. The finest material follows a rising current of water from the entry point to the overflow while other material tends to settle along the bottom of the tank. The coarsest sand settles quickly while successively finer sizes are carried greater distances from the entry point. Thus the discharge points, which are automatically operated, give successively finer sized sand products.

Cyclones (Fig. 5.13) are now used extensively in sand treatment plants, particularly for the rejection of material finer than 75 microns and for dewatering. A cyclone is essentially a hollow inverted cone with axial discharge points and a tangential inlet near the top. Slurry entering the cyclone generates a vortex about the vertical axis, the resulting centrifugal force accelerates coarse particles to the wall of the cone where they encounter a downward flowing zone which

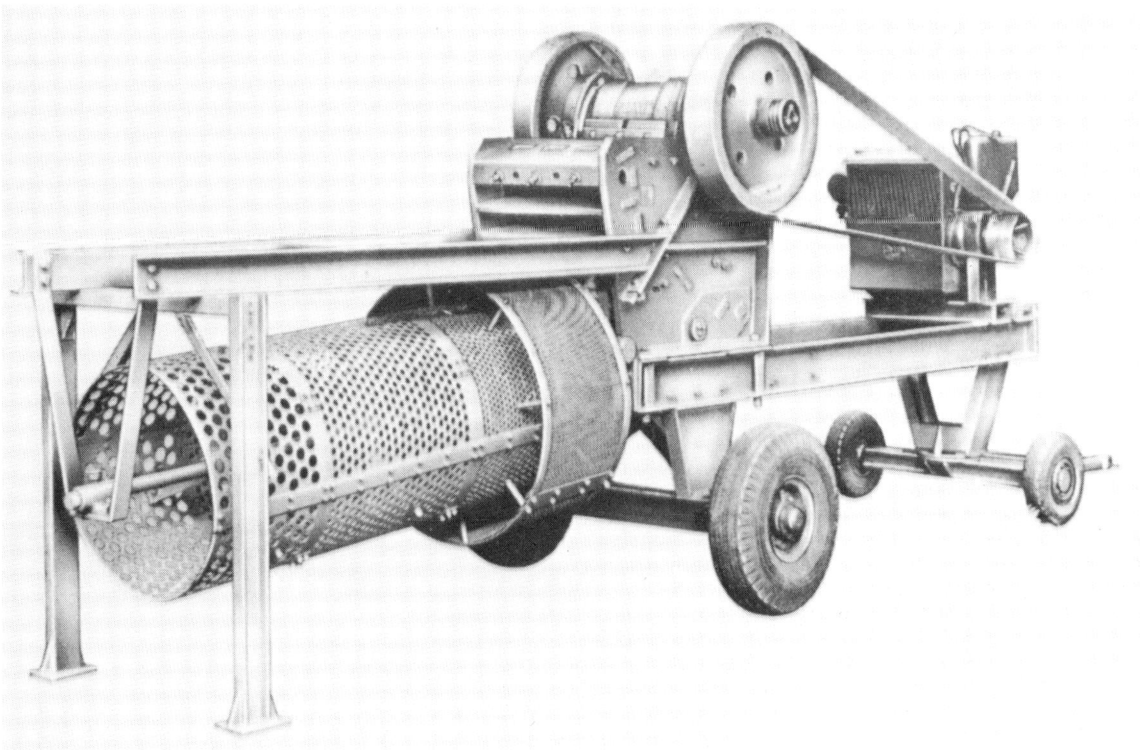

FIG. 5.8. Small mobile trommel and single toggle jaw crusher. (Courtesy Frederick Parker Ltd).

carries them to the discharge point in the inverted apex. At the centre of the cyclone the finer particles move into a zone of upwards flow which carries them through a short tube, or vortex finder, to the discharge point at the top of the unit. Compared with other classifiers, cyclones are simple, cheap, economical of space and free from moving parts, except for the necessary pump. Difficulties with internal abrasion have been largely overcome by the use of special rubber liners. The theoretical basis of their operation is somewhat complex, the size at which they separate being governed by such factors as feed rate, feed pressure, feed inlet diameter, solid–liquid ratio and diameter of outlet points. Hence each cyclone is optimised for its duty in a particular plant.

In desert areas where water is scarce, pneumatic methods of sand classification may have to be used. In principle these are comparable to hydraulic methods except that a current of air, rather than water, is used to carry away fine particles. However difficulties may occur in dispersing the dry material before separation and in clearing dust from the large volumes of air used in the process, environmental conditions apart.

5.4 Scrubbing

Sand and gravel deposits often contain clay to such an extent that they cannot be screened or classified without preliminary scrubbing treatment. Machines frequently used for this operation make provision for a deep bath of slurry within a long rotating cylinder fitted with lifter bars (Fig. 5.14). The presence of abundant water together with the self-scouring action of the tumbling pebbles breaks the clay to a fine slime which will easily pass through subsequent screens and classifiers.

There are several variations to the basic design. In some types the water flows in the opposite direction to the gravel. In others the discharge end of the washing cylinder is fitted with concentric cylindrical trommel-like screens enabling sized material to be produced directly.

Log washers may also be used for scrubbing. These consist of a long inclined trough containing two longitudinal contra-rotating shafts, attached to which are many paddles that mesh together in sequence (Fig. 5.15). Feed enters the lower end of the trough which

FIG. 5.9. Double deck vibrating grizzly. (Courtesy Pegson Ltd).

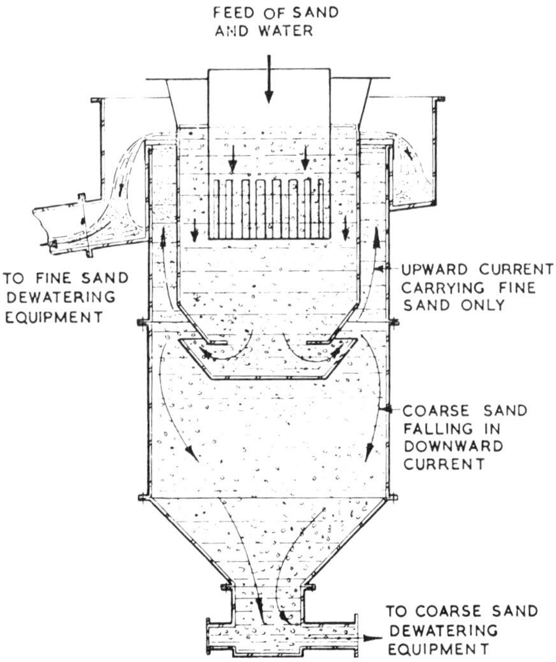

contains a pool maintained by a weir. The paddles then agitate the material and convey it to the discharge end of the trough against the incline and a flow of water and the slimes are discharged over the weir.

5.5 Dewatering

Wet screening, classification and scrubbing operations require relatively large quantities of water and deliver products in the form of fluid slurries. In order to recycle as much water as possible and to prepare easily handled sands for sale it is necessary to minimise the fluid content.

Classifiers often have a dewatering action in that the mechanism for the withdrawal of the coarse product is designed to allow time for drainage. For many purposes therefore a classifier may be an entirely adequate dewatering device which can deliver products containing only 25% water. Cyclones, as

FIG. 5.10. Diagram of a "Linatex" sand classifier. (Courtesy Wilkinson Rubber Linatex Ltd).

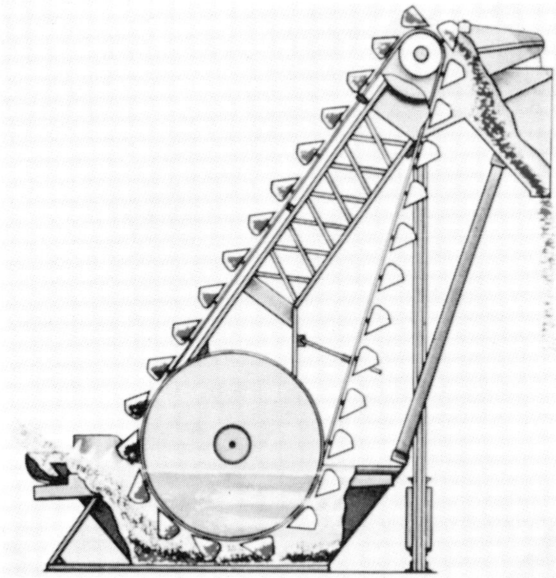

FIG. 5.11. Diagrammatic Section through a "Sandor" bucket classifier. (Courtesy Frederick Parker Ltd).

previously described, are also widely used for dewatering.

Screens offer a simple method of dewatering coarse material and the application of high frequency vibration has enabled this technique to be extended into fine size ranges. Dewatering screens operate at a shallow reversed slope and are driven by a linear vibration. The screen deck consists of slot apertures orientated perpendicular to the flow of the sand and is made from a plastic or rubber having a low coefficient of friction and high resistance to abrasion. Sand is transported across the deck as a thick cake and consequently only a small proportion of the sand feed comes into contact with the screen surface. The vibration causes the sand grains to become close packed in which condition the intergranular pore space, and hence the water content reaches a minimum. Products dewatered by this method can have water contents of as little as 13 to 15% and in this respect are superior to other dewatering devices. Since the screen openings are often larger than some of the sand grains, finer particles occasionally pass through the screen and may be recirculated.

In many plants effective dewatering is carried out by allowing the coarse products of classifiers or cyclones to stand in conical stockpiles for 24 hours. Pulp discharge is usually arranged into any of three stockpiles so that at any given time one stockpile is receiving wet material, one is draining, and drained material is being recovered from the third.

Dewatering of the finest waste slime containing clay or other fine particles is often the most difficult process to carry out conveniently. Usually it involves the use of large settling ponds of capacity sufficient to allow for long settling times under still water conditions. Eventually, however, the ponds fill with sediment which must be removed, or new ponds must be dug. In many cases the space requirements and handling difficulities associated with the treatment of slime preclude the working of sand and gravel deposits containing a high proportion of clay. Chemical flocculants are effective in increasing the settling rate of fine particles and although expensive they are now used to a limited extent in aggregate treatment.

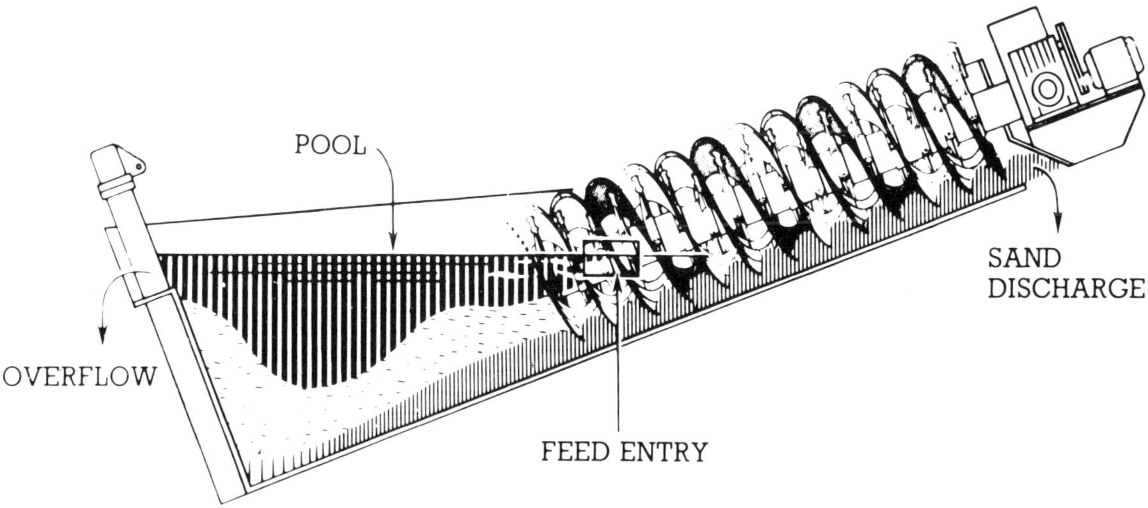

FIG. 5.12. Diagrammatic Section through a spiral classifier. (Courtesy Hunt-West Ltd).

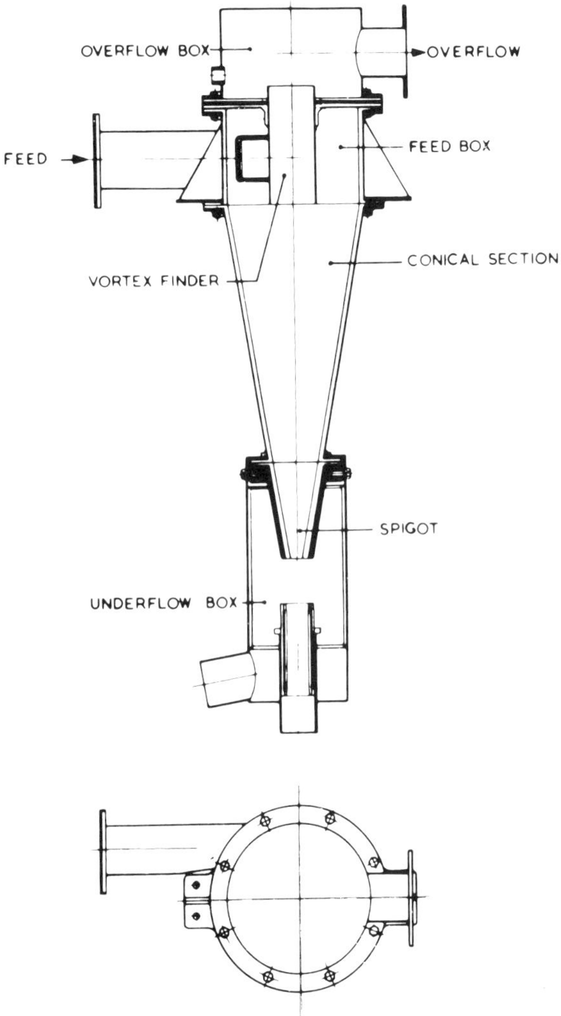

FIG. 5.13. Diagram of a "Linatex Separator" cyclone. (Courtesy Wilkinson Rubber Linatex Ltd).

5.6 Mineral separation

Machines for separating minerals, commonly used in ore beneficiation, are rarely employed in aggregate processing. The subject is too extensive to be treated in detail here. However, it is worth recording that jigs are occasionally used for separating coal from gravels, and a heavy media separator is used in at least one operation in the UK for removing dense particles of sedimentary iron ore from a gravel. The principle of hindered settling is sometimes employed to decrease the coal or lignite content of certain sands. Magnetic separation is used in some facilities to protect crushers from damage by large pieces of steel which may inadvertently become mixed with the quarry product.

5.7 Plant selection and design

There are some general principles applicable to the processing of different rock types which indicate the procedures most suited to particular materials. For example, in crushing, the use of impact machines is usually restricted to the treatment of limestone or plastic materials except in instances where other crushers fail to give a product of the required shape. Elongated or flaky aggregate particles are often produced from Lower Palaeozoic greywackes or other cleaved or laminated rocks and often there is no alternative to the use of impact mills even though severe abrasion problems may be encountered, particularly if the rock has a high quartz content. Otherwise primary crushing of abrasive rocks is carried out by double or single toggle jaw crushers depending on the strength of the material, or by gyratory crushers where larger continuous operations are involved. In secondary crushing, the availability of material with a uniform top size enables cone crushers with their characteristic high throughputs to be

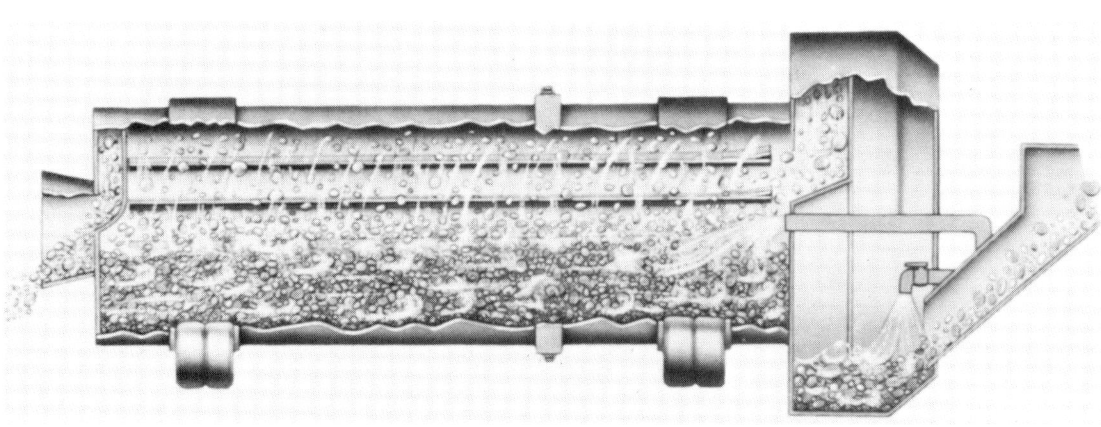

FIG. 5.14. Diagrammatic section through a "Scrubber-Mill". (Courtesy Frederick Parker Ltd).

FIG. 5.15. A log-washer. (Courtesy Hunt-West Ltd).

exploited, except where the production of flaky particles is a problem.

In all crushing operations greater control of particle size and sometimes shape is obtained by restricting the reduction ratio (i.e. the ratio of the average size of the feed to that of the product) attempted at each pass. Thus with more friable material, when the overproduction of fines becomes a problem, it is necessary to reduce the reduction ratio in a single machine, screen the product and recycle the coarser part of it. However, since there are obvious economic limits to crushing capacity and the size of recycling loads, similarly for diverting the coarser material to secondary or tertiary crushers, these factors have to be optimised against the production of unusable materials.

Apart from their use in conjunction with crushers and for the preparation of final size fractions, screens may be used to facilitate the rejection of weaker material from a product. Thus in a mixture of strong and weak rock, after crushing, the weaker material can be expected to concentrate to some extent in the finer fraction. Early rejection of finer material followed if necessary by successive stages involving further crushing and further rejection of fines should eventually give a coarse product containing little of the weaker component. However such methods carry a cost penalty in that a high proportion of the original material may have to be discarded to obtain the required product. Optimum screening can also be effective in reducing the proportion of flaky or elongated particles, but progress in this area is largely a matter of empirical test work ideally at the commissioning stage.

In the treatment of sands, flaky mica-like materials, or low density materials such as lignite, which have a low particle velocity in water compared with rounded quartz gains of the same size, can often be removed by using cyclones or classifiers. However as the particle velocity depends on size as well as shape, the methods can only be effective in sands with the appropriate size distribution of both quartz grains and undesirable flaky minerals. For optimum results it is preferable to screen the sand into close size fractions before classification, otherwise conventional mineral separation techniques such as tabling might be used, but the economics of the aggregates industry rarely

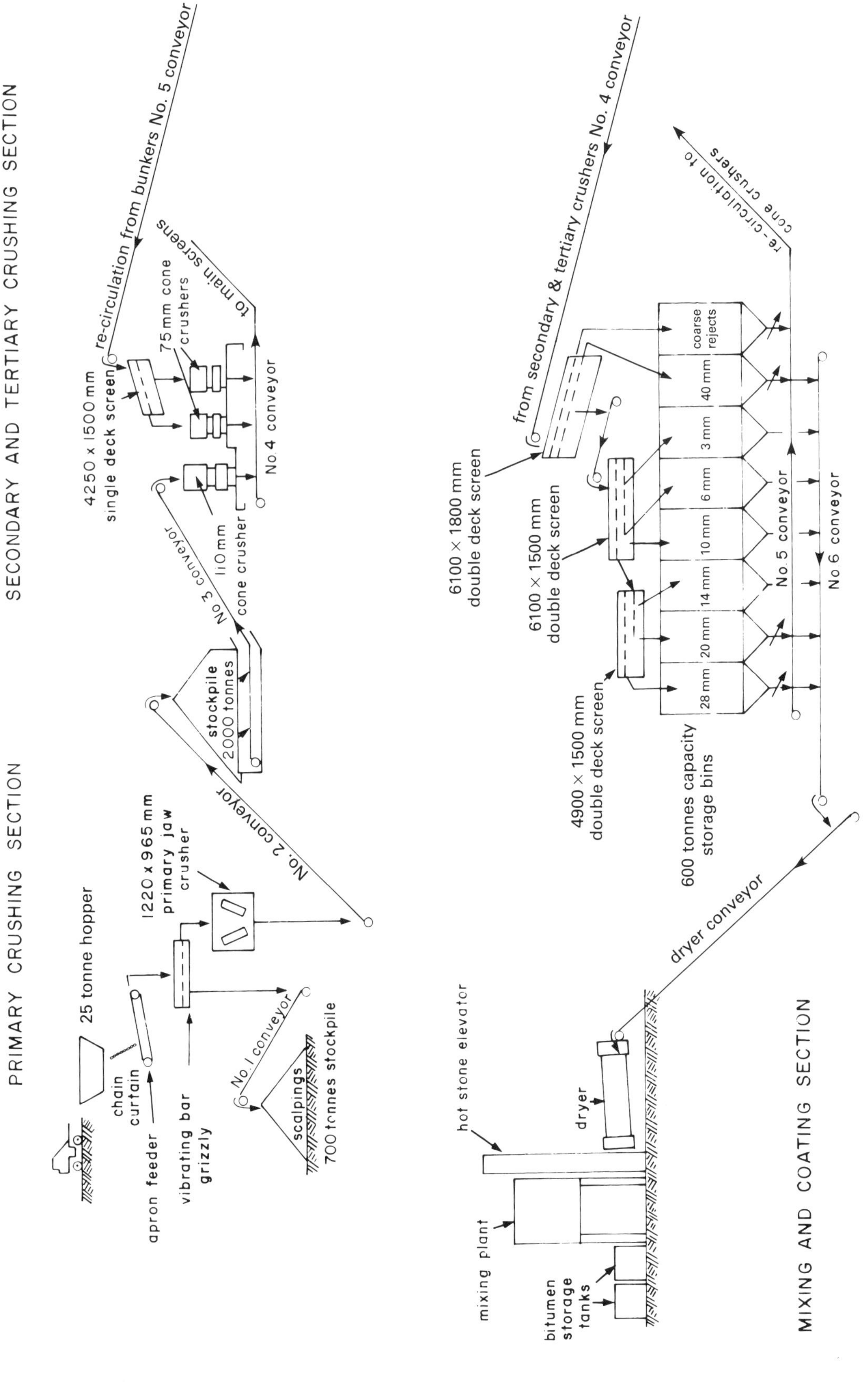

FIG. 5.16. Plant for the production of roadstone from metamorphic rock.

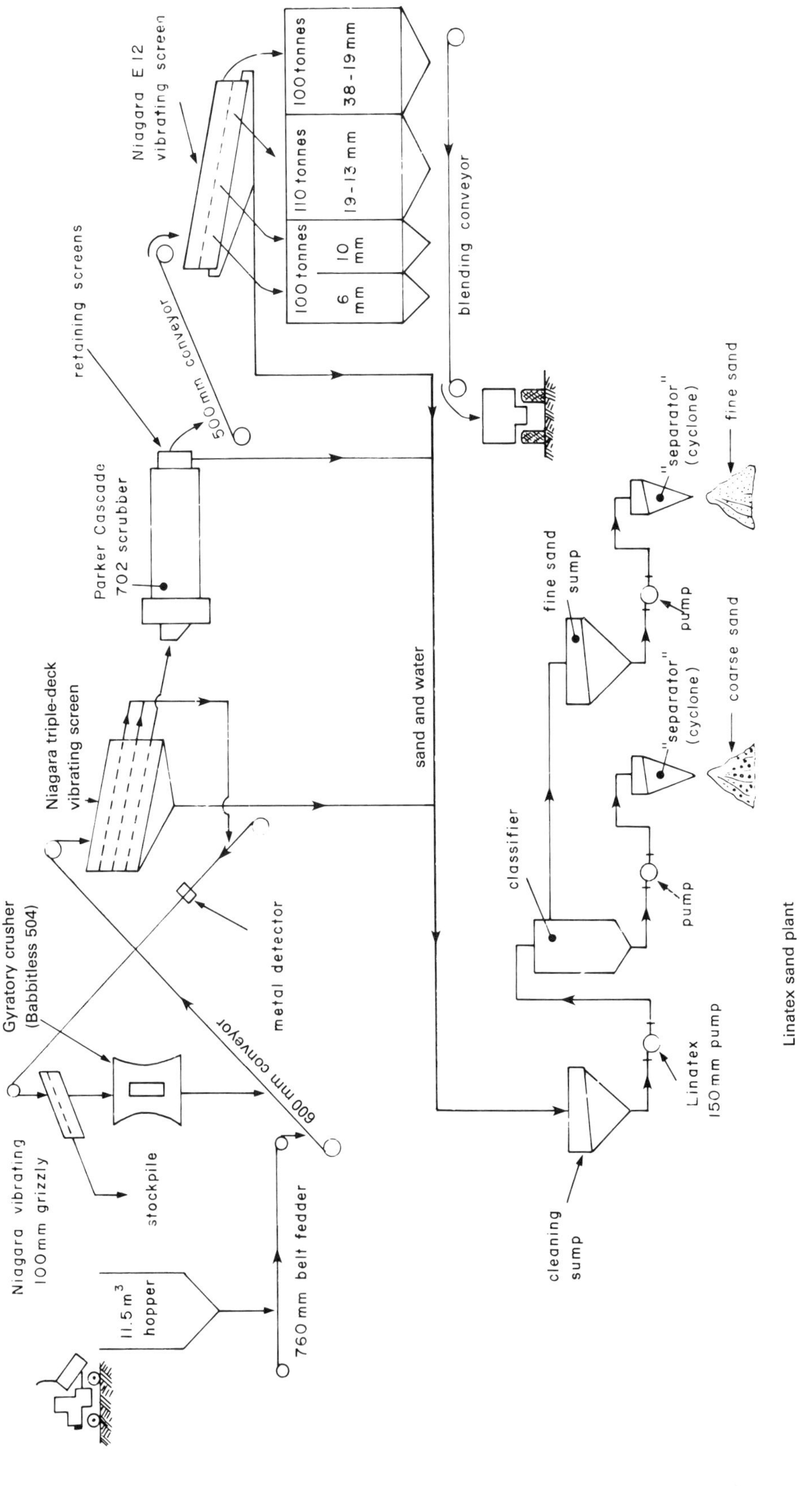

FIG. 5.17. Sand and gravel plant.

justify these type of treatment. In some desert areas the reduction of gypsum in sand can be effected by a scrubbing and desliming operation, which breaks up the more friable gypsum crystals and enables them to be discarded as a fine product in suspension.

The inability of many machines to treat damp or sticky material is a source of difficulty which can result in blinded screens and blocked crushers. Some moisture on the surface of large fragments can be tolerated, particularly when extensive areas of fresh dry surface are to be created by further crushing, and impact crushers can handle wet or sticky materials to a limited extent. Otherwise, the presence of excess moisture usually leaves no alternative to the use of a wet process. Clay particles, which tend to retain water and often become coated on other fragments, usually have to be removed by scrubbing and desliming operations and soluble salts must also be removed or at least substantially reduced by washing. Compared with dry processes, wet operations involve the additional expense of providing a suitable water supply, product dewatering equipment, slime settlement, water circulation and liquid effluent control.

The design of a processing plant should depend mainly on the nature of the raw material, the market to be served, the water resources available and the scale of the operation. Since it is unusual to find that these factors are the same in different quarries or pits, it follows that flowsheets are rarely identical except perhaps in very simple operations or when mobile plant is used, in which case optimisation may be sacrificed in the interests of mobility and utilisation of existing machinery.

Hence these comments are intended only as a guide to the principles involved in processing, and plant selection and design. Development of a permanent flowsheet should be based initially on laboratory test work and commercial optimisation, followed by trials during plant commissioning to realise the potential value from a relatively high capital investment. Although individual items of equipment may be fairly standard, the concept of a standard flowsheet has little relevance and attempts to design a complete plant by reproducing a system operating elsewhere usually results in an inefficient and, therefore, necessarily costly operation.

5.7.1 Flowsheets. From the great variety of flowsheets in existence, a crushed rock and a sand and gravel plant are briefly described to illustrate some of the principles involved.

Figure 5.16 illustrates a 700 tonnes per day plant designed to produce roadstone from a tough metamorphic rock. Crushing is effected by a double toggle primary jaw crusher followed by secondary and tertiary cone crushers which form the basis of a flexible plant suited to moderate output requirements.

The vibrating bar grizzly rejects a minus 32 mm scalping, subsequently used for sub-base, from the untreated quarry product. The jaw crusher, set to a minimum gap of 150 mm, delivers material via a stockpile to the largest of three cone crushers. Material then passes to a double deck inclined vibrating screen where a minus 40 mm product is passed for further screening and plus 50 mm rejects are returned for reduction in the two remaining cone crushers. However, before crushing, recycled material is further screened into two size fractions which are fed to separate cone crushers optimised to receive different sizes of feed, the discharge from both crushers being returned to the double deck screen. Minus 40 mm material is treated on two double deck horizontal screens, one of which has a lower deck divided into two mesh sizes. Provision is made to allow any of the plus 20 mm final size fractions to be recycled for further crushing, should there be an increase in demand for finer products thus providing important flexibility. As the plant produces roadstone, a bitumen coating section is included.

Figure 5.17 illustrates a 1000 tonnes per day plant designed to process a river terrace gravel which contains a high proportion of quartz and quartzite pebbles together with sand and a relatively small amount of clay or other fine adhering material. The products are for use in concrete, except for some fine sand which finds an outlet in asphalt and mortar.

The quarry produces material, which is finer than 130 mm. It is normally received in a fairly dry condition, through a hopper, to a triple deck inclined vibrating screen fitted with water sprays. The decks in descending order separate at 38 mm, 19 mm and 5 mm. Only a partial recovery of minus 5 mm material is made at this stage and particles of this size, which pass the finest screen together with wash water, travel to a separate sand plant. The plus 38 mm fraction is subsequently screened on a grizzly where the small proportion of plus 100 mm material is rejected and sold separately. Minus 100 mm plus 19 mm material is then crushed in a gyratory machine set to deliver a minus 16 mm nominal product which is recycled to the triple deck screen. Minus 19 mm plus 5 mm material from the screen, together with adhering fines, forms the feed to a scrubber where it is subjected to intense attrition under plentiful water, sufficient to break up any remaining undispersed particles. The discharge end of the scrubber is fitted with rotary screens to remove minus 5 mm material from the outflow and effectively dewater the oversize. The washings pass to the sand plant, while the clean minus 19 mm plus 5 mm fraction goes for final sizing to an inclined vibrating screen fitted with water sprays and a single deck divided into three separate mesh sections.

Wash water and fine particles from the screens and the scrubber collect in a sump at the sand plant. Subsequently material from the sump is pumped into a sand classifier where the coarse sand sinks to a bottom discharge and fine sand passes out with the overflow. Each fraction then travels to a cyclone (Linatex Separator) where it is dewatered and discharged with the underflow onto stockpiles. The overflow from the cyclones passes to settling points where the silt is allowed to settle. The clean water then returns via a reservoir to the plant.

General references

FARAHAR, R. M. 1976. Impact Crushing. *Quarry Manage. Prod. London* **3,** No. 2 35.

FREEBURY, J. 1978. *Practice of Screening Mine & Quarry* **7,** (6) 56, 58, 60, 62.

FIELDHOUSE, W. 1965. The application of horizontal screens in quarries. *Quarry Managers. J. London,* **49,** (8), 327–335.

JOHNSTON, K. J. R. 1964. The upgrading of fines by dry processing. *Quarry Managers J. London,* **48,** (7) 277–283.

LISSENDEN, A. W. J. 1975. Fixed plant—the influence of technical development. *Quarry Manage. Prod. London,* **2** (5), 121–126.

LOWRIESON, G. C. 1974. Crushing and Grinding. The size reduction of solid materials. Butterworths, London.

PRYOR, E. J. 1965. Mineral Processing, 3rd Edn. Elsevier Publishing Co. Ltd, Amsterdam–London–New York.

RASPASS, F. W. 1980. Developments in fine aggregate processing. *Quarry Manage. Prod. London* (8).

SAND AND GRAVEL ASSOCIATION 1967. SAGA Pit and Quarry Textbook. MacDonald and Co (Publishers) Ltd., London.

SHERGOLD, F. A. 1961. Results and conclusions from research on granulators. *Quarry Managers J. London* **45** (4) 133–139.

SIDEBOTTOM, J. B. & RASPASS, F. W. 1964. The upgrading of fines by wet processes. *Quarry Managers J. London* **48** (6) 227–239.

TAGGART, A. F. 1945. Handbook of Mineral Dressing. Ores and Industrial Minerals. John Wiley & Sons, Inc., N. York.

Chapter 6. Classification

6.1 Classification systems

The description and, in particular, the classification of aggregates in a manner appropriate to their use in the construction industry has long posed problems, not only of a scientific nature but also from practical and commercial points of view.

Naturally occurring rock materials can be classified in a variety of ways, the method chosen depending on the observations made and the use for which the classification is required. Age, colour, fossil content, grain size, mineralogy, mode of formation and compressive strength are but some of the many approaches that have been used. The most common method is that developed from the classical geological approach, which is based essentially on the mode of formation. Hence natural rock material is divided into three main classes: igneous, sedimentary and metamorphic. These groups are then subdivided, principally on the basis of their mineralogy, texture and chemical composition.

The numerous subdivisions possible in this fundamental geological system inevitably results in a nomenclature which is too cumbersome for general use in the construction industry. To obtain a simpler approach, a "Trade Group" classification, based essentially on the classical system, was introduced for use with aggregates into British Standard, BS 63 in 1913. Apart from minor modification this system has been retained in subsequent Standards, including the most recent (BS 812:1975), which is currently under revision, where it is designated as the "Petrological Group Classification".

As indicated in 6.4 below this "group system", which relates essentially to crushed rock, is unsatisfactory from many points of view and it is therefore suggested that it should be replaced by a system more appropriate to the needs of industry. In order to be of value for commercial purposes, any alternative system must be realistic and should be capable of being understood by vendor, purchaser, engineer and geologist alike. The system suggested is intended to meet these requirements.

6.2 Historical development of classification and descriptive systems

6.2.1 Local and traditional terms. Due to the complexity of the normal geological classification systems, local and traditional names were, and often still are, commonly used to describe the materials from which aggregates are commercially produced. In some instances the names are used to describe rocks which are vastly different from those suggested by the more usual scientific meaning of such terms. Thus, the term "granite" is commonly used to describe a "hard" rock and even some sedimentary rocks like greywackes and gritstones have been referred to by this name. Similarly, the term "whinstone" may be used to describe dolerites, greywackes or other 'compact' rocks, while "marble" is loosely used to describe many compact limestones and other rock types which have not been metamorphosed.

Table 6.1, which shows some typical traditional rock names together with their petrological names or types and the appropriate British Standard group classification, illustrates some of these anomalies.

TABLE 6.1. *Some traditional names used to describe rock materials*

Traditional name	Petrological name or type	B.S. group classification
Clinkstone	Phonolite	Porphyry and basalt
Cornstone	Impure limestone	Limestone
Elvan (Blue elvan)	Quartz porphyry	Porphyry and basalt
Flagstone	Micaceous sandstone	Gritstone
Freestone	Sandstone or limestone	Gritstone and limestone
Greenstone	General term for quarried materials but often igneous in origin	Basalt
Hassock	Sand	—
Hoggin	Sandy, clayey gravel	—
Hornstone	Chert and silicified tuffs	Flint
Pennant	Sandstone sub-greywacke	Gritstone
Rag (stone)	Calcareous sandstone	Limestone and gritstone
Thames ballast	Gravel	—
Toadstone	Altered dolerite and decomposed tuffs	Basalt
Trap (rock)	Igneous intrusive or lava	Basalt
Whin (stone)	Igneous intrusive	Basalt
Grey Whin	Moine siliceous granulite	Schist

6.2.2 British Standard Trade Group classification approach. In 1913 the British Standards Institution introduced a list of petrologically named rock types, assembled in groups known as "Trade Groups", in order to meet the need for a classification of roadstones suitable for engineering purposes. When first published in BS 63, the list was based on the following twelve trade groups:

Andesite	Gabbro	Limestone
Artificial	Granite	Porphyry
Basalt	Gritstone	Quartzite
Flint	Hornfels	Schist

The number of groups was later reduced to eleven by incorporating Andesite into the Basalt group (BS 812: 1943). Phemister *et al.* (1946) commented that the list "had been compiled with a view to associating in the same trade group, rocks which may be expected to behave similarly as roadstone". However, the range of properties of any one group (and even of any one rock type) of aggregate is so large as to make nonsense of any expectation that they will perform in a similar manner either in tests or in service.

Subsequently the British Standards Institution dropped the word "Trade", firstly suggesting that aggregate materials be given a "Group Classification" (BS 812:1967) and then more recently a "Petrological group classification" (BS 812: 1975). Apart from very minor variations within the groups, the system remains essentially unchanged and is really a "Quasi-petrological group classification" system. The full text of the relevant section in BS 812: 1975, i.e. 6.4, is given in Appendix A6.1, for ease of reference.

Two important statements are included in this extract from BS 812, under paragraphs (iii) and (vi). The first states that "the petrological group classification does not take account of suitability for any particular purposes" and the second requires that "the correct identification of a rock and its placing under the appropriate group shall be made by the Institute of Geological Sciences or by a qualified geologist". The significance of these statements in relation to the proposed system of classification and description is discussed in 6.4 to 6.9 below.

6.2.3 The American Society for Testing and Materials (ASTM) approach. A number of different methods of description and classification are in use in countries outside the United Kingdom. Probably the most comprehensive and best known is that of the American Society for Testing and Materials. ASTM publications include two important papers relating to aggregate materials which are largely based on the works of Mather (1950) and Mielenz (1966). The first of these, ASTM Standard C294–69 (1975) covers the standard nomenclature to be used in the description of mineral aggregates. The second, C295–79, deals with the recommendations for the petrographic examination of aggregates for use with concrete, but its contents are similarly applicable to aggregates for other construction purposes. Both aim to facilitate the description and classification of natural rock materials in a classical geological manner and will be referred to later in the sub-section of this chapter dealing with recommendations. The information required in a petrological description of concrete aggregate has been indicated by Mielenz (1978) and the manner in which the information obtained from such a petrographical examination should be presented is detailed in ASTM Standard C295–79. Extracts from both are given in Appendix A6.2.

6.2.4 International Society for Rock Mechanics (ISRM) approach. In 1967 the International Society for Rock Mechanics set up a Commission to consider the standardisation of laboratory and field tests on rock materials. In 1981 the Commission issued two categories of tests requiring standardisation, together with an agreed order of priority for the work within each category. 'Index tests' are those needed for the classification and characterisation of the rock material, while 'Design tests' are tests to measure qualitative material characteristics for design calculation purposes.

The suggested Index tests show that several test parameters, in addition to a petrographic description, are thought to be necessary in order to classify and characterise rock materials and that these include both laboratory and field tests. In particular they indicate that the system suggested for the petrographic description of rocks should be used, not only to obtain a geological classification of the material, but also to highlight any features which might be relevant to mechanical properties.

The drafting committee state "that to ensure correct classification the first step should be to ascertain the mineral composition and texture of the rock. Further investigation should include a fabric and mineral analysis in the case of strongly anisotropic rocks, the determination of the degree of alteration or weathering, grain size, micro-fracturing and porosity".

The list of Design tests includes those thought to be necessary in order to appraise the likely engineering performance of the materials. These are equivalent to the mechanical tests needed to assess aggregates as detailed in Chapter 7.

The ISRM approach to the classification of rock materials follows lines similar to those suggested by the ASTM, discussed in 6.2.3, both of which are described in greater detail in Appendices A6.2.1, A6.2.2 and A6.2.3. In neither case is it suggested that the classification process provides an end in itself, rather that it should be used as a basis from which the appropriate and likely mechanical characteristics of the materials can be assessed using standard tests.

6.2.5 Classifications based on engineering

criteria. A number of attempts have been made to describe and classify rock materials from both an engineering and a geological point of view. The majority are concerned with material in the *'in-situ'* condition and these methods of description have generally been associated with the subject field of rock mechanics. The historical development of this work, first documented in 1946 by Terzaghi, is well expounded in the Geological Society Engineering Group Working Party Report, "The description of rock masses for engineering purposes" (1977). There have been three main lines of approach, the first based on the petrography of the material, the second on rock parameters as derived from tests (usually related to strength characteristics) and the third on the field condition of the rock mass including details such as discontinuities and degree of weathering. The recommendations contained in the 1977 report, only marginally modified, are incorporated in the revised version of British Standard Code of Practice on "Site Investigation" BS 5930:81 (formerly BS CP 2001).

Fookes and Higginbottom (1975) proposed a system of classification and description of near-shore carbonate sediment rocks for engineering design purposes, using grain size and post-depositional induration as the main parameters of engineering significance. Clark and Walker (1977) have extended the Fookes and Higginbottom system to cover sedimentary materials ranging from total carbonate to total non-carbonate content, also incorporating variations in the particle size and degree of induration together with an approximation of the unconfined compression strength of the material. Burnett and Epps (1979) have also developed the original system to cover the carbonate suite rocks and soils.

The above methods are concerned with solely bulk rock materials. Ramsay *et al.* (1974) have prepared a classification of crushed rock aggregate based on mechanical factors which is independent of petrological boundaries. The factors used are those of particle shape and impact strength, which are quantified in the form of the flakiness index and aggregate impact value respectively. The system being based on strength and shape criteria, is limited in that it makes no provision for the assessment of other engineering properties of materials, such as the polishing potential and abrasion resistance, which can in some instances be critical in relation to their use and performance in-service.

Attention should also be drawn to some important work, developed by Leverett (1970) from Duncan (1969), which investigated the significance of the aggregate index properties: porosity, saturation moisture content and specific gravity, in relation to the engineering characteristics of the aggregate as measured by standard laboratory tests. Although these properties cannot be used to predict particular aggregate characteristics with any great accuracy, there is no doubt that some valuable general assessments can be made and the results obtained appear to justify further work in this area.

6.3 The description of weathered and altered aggregate source materials

Knight and Knight (1935) and many other workers have drawn attention to the importance of the degree of weathering and alteration on the potential performance of rock materials. As far as the aggregates currently in use in the UK are concerned, decomposition and disintegration due to weathering and alteration of the mineral components is not normally of importance. Nevertheless, evidence shows that these processes do exist and their possible effect should therefore be considered. In overseas practice, and particularly in unglaciated areas, weathering is often a vital factor in the selection of the source of aggregates.

Thus an accurate description of the state of weathering and alteration and its possible effect on mechanical properties should be included in the geologist's report on such materials. A more detailed discussion of the natural processes involved in weathering and alteration is given in Chapter 2 and a system to describe the degree of weathering and alteration is given in Table 2.1.

In practice it has been found that materials in grades 1A and 1B are generally satisfactory as aggregates, while those classed as Grade III or higher are usually unsuitable. Thus Grade II weathering forms the transition zone within which materials may or may not perform satisfactorily when used as aggregates. Within this grade then, it is essential to carry out tests to assess the soundness, or otherwise, of the material in relation to the proposes use (see Chapter 7).

Although there is obviously a need to sub-divide Grade II, further research is necessary before this may be satisfactorily achieved. Some of the problems associated with the use of weathered aggregates in the production of concrete have been indicated by Fookes (1980).

In southern Africa certain materials are particularly susceptible to environmental influences. Weinert (1980) has proposed a classification of rocks for use in road construction in this region which divides naturally occurring materials into nine groups. The groups are intended to be based essentially in the presence or absence of the mineral quartz, although the mode of formation is also included to some extent. The presence or otherwise of quartz is used to infer the mode of weathering, the durability and the possible suitability of the material for certain road construction purposes. The nine groups employed, together with their best known members, are:

Acid crystalline rocks:	granite (but also syenite)
Basic crystalline rocks:	dolerite
High-silica rocks:	quartzite (but also hornfels)
Arenaceous rocks:	sandstone (but also mica schist)

Argillaceous rocks:	shale (but also phyllite)
Carbonate rocks:	dolomite
Diamictites:	tillite
Metalliferous rocks:	ironstone
Pedogenic materials:	calcrete and ferricrete

6.4 Some comments on methods used for the classification and description of mass rock and aggregates

6.4.1 Classification and description systems in general. The classification and description systems adopted by the ASTM (6.2.3) and the ISRM (6.2.4) fall essentially into two distinct parts:

(i) a classification of the material in classical geological form based on a full petrological analysis,

(ii) technological description directly related to the properties of the material as determined from physical and mechanical tests.

Of the two the ISRM is the more specific as regards the Index tests needed to classify and categorise the materials, but this is probably due to the fact that it is concerned with mass rock in general, rather than aggregate material in particular.

In order to classify materials by using proper names a petrological analysis is normally necessary. This introduces a relatively expensive analysis which may be justified for a first time classification, but which would generally be unnecessary for the majority of subsequent investigations (6.4.2). The classifications of Fookes and Higginbottom, Clark and Walker, and of Burnett and Epps (6.2.5) are developments of the classical system as applied to rocks of sedimentary origin, where the degree of induration may be used to infer a likely compressive strength.

The main problem in attempting to classify aggregates in terms of their characteristic engineering properties lies in the number and range of possible parameters involved.

Early work concerned with engineering property classification tended to assume that strength was by far the most important characteristic, but for some years it has been generally recognised that the properties required of an aggregate must be specifically related to its intended use and, therefore, its fitness for purpose. Consequently it is unrealistic to propound a single comprehensive aggregate classification system based on one or more selected mechanical properties. The classification proposed by Ramsay *et al.* (6.2.5) was quite specific in that it indicated the purposes for which the system could be used, thus clearly recognising the limitations of this kind of approach.

In its original form the Trade Group system attempted to combine a classical method of classification with inferred common engineering properties. Knight and Knight (1935) and many subsequent authors have pointed out the disadvantages of the Trade Group approach. Knight and Knight not only considered the groupings to be misleading but also often a disadvantage to the users of roadstone. They stated "that it is not necessary for the road engineer to be familiar with the very large number of rock types recognised by petrologists, but he can decide the suitability of a stone for a particular purpose when he knows its essential features, although he may not be able to give it a specific name. Thus the engineer requires to know what is the *approximate* class of the stone, what is its mineralogical composition, how fresh it is, whether it is weakened or strengthened by its geological history, and so forth".

They were of the opinion that a classification based only on mineralogical constitution, structure and geological history was more likely to be appropriate and suggested the general classification of roadstones based on petrology shown in Table 6.2, with the intension that any stone so classified would also be accorded supplementary information of the type noted above.

Lees (1968) proposed a similar petrological classification system to that of Knight and Knight in which an indication of the mineral grain slab is included, as is shown in Table 6.3. Lees also intended that this classification should be augmented by further information of the type indicated by Knight and Knight.

Inspection of the systems suggested by Knight and Knight (Table 6.2) and Lees (Table 6.3), indicates that although these authors were appreciative of the problems involved in attempting to classify roadstones mainly on their petrology, they nevertheless still retained the established classical approach as the basis for their classifications.

Although the classification proposed by Weinert (1980) is said to be based on the presence or absence of quartz it also introduced the mode of formation into the system in some cases. In consequence mixed terms are used for group names, giving a lack of consistency. For example, some sandstones which would come within the arenaceous group, are likely to contain as much if not more quartz as certain quartzites and hornfels which are grouped as high-silica. Similarly the mineral composition, including the quartz present, of some mica schists is likely to be very similar to that of certain acid crystalline rocks. The group names Diamictites and Pedogenic material are concerned almost exclusively with mode of formation and, in consequence, will cover a wide range of possible mineral content with a wide variety of probable engineering performance characteristics, when embracing materials available world-wide.

Knill (1978), referring to the BS 812 Trade Group Classification, states "this classification has several limitations when the engineering properties of the aggregate are concerned but obviously no simple

TABLE 6.2. *Suggested "Classification of Roadstones" according to petrological characters (after Knight and Knight 1935)*

ROCKS		
Igneous	*Sedimentary*	*Metamorphic*
Plutonic	*Calcareous*	
Granite	Limestone	Gneiss
Granodiorite	Domomite	Epidiorite
Diorite	Ironstone	Hornfels
Syenite		Hornstone
Gabbro		Mylonite
Norite		Phyllite
	Arenaceous	Slate
		Schist
	Arkose	Hornblende schist
	Conglomerate	
Hypabyssal	Greywacke	Marble
Aplite	Grit	Quartzite
Granulite	Sandstone	
Pegmatite	Quartzite	
Diabase	Breccia	
Dolerite	Flint	
Lamprophyre	Chert	
Teschenite		
Felsite		
Granophyre		
Keratophyre	*Argillaceous*	
Microgranite		
Porphyry	Shale	
Porphyrite	Mudstone	
Volcanic	*ARTIFICIAL AGGREGATES*	
Basalt		
Spilite		
Rhyolite		
Trachyte	Slag	
Andesite	Clinker	
Dacite		
Greenstone		
Tuff		

classification will be entirely satisfactory. A petrological examination, coupled with the appropriate physical tests, is the only reliable means of assessing the properties of a particular roadstone".

In the majority of the systems so far discussed one common factor emerges. This is that for engineering purposes the description of the material itself and of its engineering characteristics need to be treated as two separate entities. In view of the foregoing statements the present British Standard system of classification and description of aggregates will be examined in greater detail.

6.4.2 The British Standard System. BS 812:1975 included the following statement relating to the description of samples of aggregate immediately before dealing with the Petrological Group Classification *per se*:

"The description of the material shall be given under the following general headings.

(a) Petrological group classification, e.g. granite, limestone etc.
(b) Petrological name and description. The correct petrological name accompanied by a brief description of such properties as hardness, colour, grain size, imperfections, etc.
(c) Description of the bulk, the degree of cleanness, e.g. freedom from dust, and reference to any pieces not representative of the bulk.
(d) Description of particular shape and surface texture".

The description of the "group classification of aggregates" then follows (as given in Appendix A6.1).

These statements give no indication of how often petrological analyses and descriptions should be carried out, but could be taken to infer that this is required every time an aggregate assessment is needed. However, once the nature of the material has

TABLE 6.3. *Suggested "Classification of Roadstones" according to petrological characteristics (after Lees 1968).*

ROCKS		
Igneous	*Sedimentary*	*Metamorphic*
Ic Coarse Granite Syenite Diorite Gabbro	*Sc* Coarse Conglomerate Breccia Gravel Scree	*Mc* Coarse Gneiss
Im Medium Microgranite Microsyenite Microdiorite Dolerite	*Sm* Medium Sandstone Sand	*Mm* Medium Quartzite
If Fine Rhyolite Trachyte Andesite Basalt	*Sf* Fine Shale Mudstone Clay *Sca* Calcareous Limestone Dolomite	*Mf* Fine Schist Slate Hornfele
ARTIFICIAL MATERIALS		
Slag Calcined bauxite Calcined flint Crushed Brick Pulverized Fuel Ash Symopal		

been established (usually when the reserve is being investigated) further analyses are rarely likely to be needed. Such analyses may be necessary when a change in the character of the material is suspected, or when the mineral content may be deleterious in relation to its intended mode of use. Also, when a petrological examination is carried out, the accompanying petrological description should not "be brief", as is suggested but should refer in detail to those properties which may influence the use or performance of the material. Where the composition or texture of the material is likely to cause problems, or be of special benefit, then these points should be highlighted in the descriptive report. In view of the foregoing it cannot be emphasised too strongly that these and similar analyses should only be carried out by those conversant with the problems of using aggregates in construction. The use of such techniques is discussed in greater detail in 6.8.

No mention is made under the BS 812 general headings of the aggregate particle size, but later paragraphs imply that this should be included under heading (d) along with particle shape and surface texture. Further comment on the inclusion of shape and surface texture in descriptive information is given in 6.6.2.3 and 6.7.

The main criticisms of the BS 812 Group Classification are that:

(i) It is a quasi-petrological classification in that it groups together those rocks which do not necessarily have a clear petrological relationship, yet from these groupings an assumed similarity of engineering characteristics may be and often is inferred; although, as is noted in the current BS 812, the group names in fact give no reliable guide as to the engineering characteristics of the included rock aggregates.

(ii) The use of precise geological terms as group names is misleading and can lead to unnecessary contractual disputes.

(iii) The mixture of terms within groups as employed in the classification is confusing and thereby open to misinterpretation by supplier and consumer alike. Again this can lead to contractual problems.

(iv) The system is unnecessarily biased towards igneous and away from carbonate rocks which provide a high proportion of crushed rock output in the United Kingdom. Thus it gives 42 descriptions of igneous and metamorphic rock aggregates which provide about 25 per cent of the UK production of crushed rock aggregate, as compared with three descriptions for limestones and the like which account for about 65 per cent.

(v) The system omits to classify sands and gravels although the consumption of these materials is generally of the same order as that of crushed rock aggregates, being for example about 103×10^6 tonnes of sand and gravel as compared with about 104×10^6 tonnes of crushed rock aggregate in Great Britain during 1979. The bulk of this material was used by the construction industry.

Note. The terms sand and gravel, as generally used, are based on their particle size rather than on the mineral composition of the material. Unfortunately the division into size ranges is also commonly referred to as "classification" rather than as "size classification", (Open University Press, 1974). Component grain size is discussed in further detail in 6.5.

In view of the above comments there would seem to be little point in continuing to use the BS Group Classification for the description of aggregate materials. Given that the existing classification is unsatisfactory, then what if anything would be a suitable replacement? Any substitute must be simple and should include information relevant to the specification and engineering use of the particular aggregate material. A system used to describe and classify crustal materials from a general engineering viewpoint must be more complex than that needed for aggregates since, when considering rock masses and non-indurated deposits, all possible types of material have to be dealt with.

However, when the naturally occurring material is to be used as an aggregate then certain minimum qualities are required and this automatically limits the types of material which will be considered. Thus many argillaceous, non-indurated or otherwise excessively weak rocks would not normally be considered for use as a source of good quality coarse aggregate, though circumstances do exist where these materials need to be utilised as a low-grade product.

From the foregoing it follows that there is a need for a simple, rational, descriptive classification of aggregate materials, capable of being generally understood yet based on sound geological principles. Such a system is described below and which, the Working Party believe, is not only a possible replacement for the existing British Standard classification, but also provides the basis for an international system of classification.

6.5 Principles of the proposed classification system

In order to be of value for both commercial and contractual purposes a classification system must satisfy the following basic requirements:

(i) Be simple in concept and yet at the same time use only terms which have a precise meaning capable of application in contract specifications.

(ii) Use properties relevant to aggregates materials as the basis for grouping them into classes.

(iii) Have no bias except that which may reflect on the importance of the material as an aggregate.

(iv) Be capable of being further expanded, as

necessary, by the use of supplementary information relating to the particular material or its intended use.

To comply with these principles the proposed classification employs a simple grouping or "Class" system based on the predominant mineral content of the materials and which is independent of physical properties, except for those which can be inferred from mineral composition.

The Class names have a precise geological meaning and are pertinent to the main types of material likely to be produced commercially as aggregate. However, the use of scientifically correct petrological terminology as appropriate to the material is encouraged in a supportive role to the Class name.

It is suggested that the Class be supplemented with some further simple descriptions which will then enable some generalisations to be drawn as to the physical condition of the material, to the extent that any such generalisations are worthwhile without some actual physical testing being carried out. The descriptions to be used are dependent upon the Class allocated and may be concerned with the geological age, the colour, the component grain size and the extent of any foliation of the material. The age used should conform to an established geological time scale. Similarly the colour description used should conform to a standard pattern i.e. such as that devised by Munsell (Geological Society of America, 1963). The grain size notation used should preferably be aligned to a more comprehensive and acceptable system.

A number of different systems are in use, but it is suggested that the one proposed by Sherman (1980) be used, which is based upon the particle size limits currently accepted as standard values in BS 5930:1981 and BS 1377:1975 (Appendix A6.3). Accordingly the divisions between coarse/medium and medium/fine coincide with the boundaries between macro/mesosize and medium-meso/fine-meso sizes respectively. They correspond to about 2 mm nominal diameter and the point where individual grains cannot be identified with the unaided eye (which occurs at about 0.2 mm nominal diameter) respectively. These boundaries and their correlation to Sherman's system, also to the proposed CADAM system (see 6.6 below), are shown in Table 6.4. The presence of any foliation should also be described since this may have an influence on aggregate particle shape.

It must be emphasised firstly, that the proposed basic system for the classification and description of

TABLE 6.4. *The relationship between the terms proposed to describe the grain size for the CADAM classes and those proposed by Shearman to describe the grain size of rock in general (after Shearman 1980)*

Descriptive terms as applied to Engineering Soils (BS 1377)		Component Size (in mm)	Proposed Rock and Soil Descriptive Terms*		SIZE TERMS TO BE USED IN CADAM
BOULDERS		↑	Coarse	MEGASIZE	COARSE GRAINED
		200			
COBBLES			Fine		
		60			
GRAVEL	Coarse		Coarse	MACROSIZE	
		20			
	Medium		Medium		
		6			
	Fine		Fine		
		2			
SAND	Coarse		Coarse	MESOSIZE	MEDIUM GRAINED
		0.6			
	Medium		Medium		
		0.2			
	Fine		Fine		FINE GRAINED (Grains not visible by means of the unaided eye)
		0.06			
SILT	Coarse	0.06	Coarse	MICROSIZE	
		0.02			
	Medium		Medium		
		0.006			
	Fine		Fine		
		0.002			
CLAY		↓	CRYPTOSIZE		

* Note: Shearman's proposals used the term "... grade" rather than "... size" as given above. The change has been adopted in order to avoid any possible confusion with the term 'grade' as used in relation to metamorphism.

TABLE 6.5.

1	Form of Aggregate (a)	Crushed Rock					Gravel	Sand
							Natural, Crushed, Mixed	Natural, Crushed Rock, Mixed
2	Class	Carbonate Class (b)	Quartz Class (c)	Silicate Class (d)			Class of major constituents as for crushed rock aggregates, with petrol. name (if known). Description of minor constituents, e.g. mica, shell fragments, etc. (Mainly applicable to natural sands) (f)	
				Igneous	Sediment	Metamorphic		
		Miscellaneous Materials (e)						
		Petrological Name (if known)						
3(l)	Geological Ass (g)	Essential for Sedimentary Rock Materials		—	Essential	—	Geological Age of deposit, if other than Quaternary (h)	
	Colour (i)	Optional	Optional	Essential	Optional	Essential		
	Component (j) Grain Size	Optional	Optional	Essential	Optional	Essential	Description of major constituent(s) as for crushed rock aggregates	
	Foliation (k)	If applicable	If applicable	—	If applicable	Essential		
4	Other (m) Comment	Any further information likely to be of help in assessing the material e.g. for a Gravel–Glacial deposit.						

Notes:

(a) Indicates the nature of the aggregate material being considered. In the case of Gravel and/or Sand an indication of whether the material is Land-won or Marine-dredged should be included.

(b) Includes ALL materials composed predominantly of Calcium and/or Calcium Magnesium Carbonate, irrespective of origin.

(c) Comprises ALL materials in which Quartz or Free Silica is the dominant mineral, and includes materials such as Flint, Vein Quartz, Quartzites, many Sandstones and Gritstones, some Greywackes and the like.

(d) Comprises ALL materials in which the rock forming silicates are the dominant minerals. These are subdivided according to origin into Igneous, Metamorphic and Sedimentary in order to differentiate between those containing mainly high temperature silicates and those containing a high proportion of clay mineral silicates.

(e) Materials not covered by (a), (b) and (d) above are NOT allocated a Class but the correct scientific name for the material MUST be recorded.

(f) In the case of Gravel and particularly in the case of Sand, the presence of certain materials, even in small quantities, can be important. Note should be made of the occurrence of such materials.

(g) The Geological Age scale to be used in the U.K. is as follows:— Quaternary (to include Pleistocene and Holocene), Tertiary, Cretaceous, Jurassic, Triassic, Permian, Carboniferous, Devonian, Silurian, Cambrian and Precambrian.
Where alternative or different age terms are applicable in countries outside the U.K. then these should be used as is appropriate.

(h) The majority of Sand and Gravel deposits worked in the U.K. are of Quaternary age and unconsolidated. The older the deposit the more likely it is to be consolidated.

(i) Colour descriptions should be to a common standard. Reference is made to the Munsell Colour Chart system, (Geological Society of America, 1963), which is useful for both field and laboratory assessments. Should a more accurate colour description be required then this should be carried out under laboratory conditions and it is suggested that the Lovibond–Schofield Tintometer or similar method be used (Hosking and Ritson, 1968). Normally however, such a detailed description would not be included in the CADAM.

(j) A standard Component Grain Size notation should be used. Grain greater than about 2 mm nominal diameter are termed Coarse; those which are less than 2 mm, but which can still be distinguished with the unaided eye, are termed Medium; and those which cannot be distinguished with the unaided eye are termed Fine. The component grain size is applied to the matrix.

(k) Foliation and fissility are particularly important with respect to certain Metamorphic and certain Sedimentary rock materials. Evidence of its occurrence should be recorded as is appropriate.

(l) Entries to be made with respect to Geological Age/Colour/Component Grain Size/Foliation for the different Classes are indicated in the above table as Essential, Optional or Not Applicable.

(m) Additional comments should only be added when that information is likely to be of help in assessing the material and is not to be included in any subsequent more detailed description of the material.

aggregate material makes no attempt to infer engineering characteristics relating to the material when in aggregate form and, secondly, that it is deliberately intended to allow flexibility of use. It aims by so doing to give general guidance to those producers and users who may not be familiar with petrological nomenclature but who nevertheless have a basic understanding of rock materials, and at the same time to provide a system whereby all aggregates offered for sale can reasonably be described within its terms as a minimum requirement.

6.6 The classification and description of aggregate materials (CADAM) system

6.6.1 Introduction

In order to provide a full description of an aggregate, three types of information need to be provided:

(i) Basic data relating to the nature of the aggregate material in a form which can be understood at virtually all levels within the industry.

(ii) Specialist information concerned with the physical, mechanical and chemical characteristics of the aggregate.

(iii) Specialist information concerned with the petrological characteristics of the aggregate.

The first of these is covered by the proposed Classification and Description of Aggregate Materials or CADAM system. The latter two come within the provinces of the professional materials engineer and geologist respectively and are dealt with in more detail in 6.7 and 6.8.

CADAM is intended to provide the first type of information in the form of an elementary system of description based on simple yet sound geological concepts. A typical pro-forma which can be used in the assembly of the required data is shown in Table 6.6, from which it will be seen that three simple sets of data are needed (apart from the essential information required relating to the supplying source and type of sample provided, on which the assessment is based). These three sets of data are:

(a) The form of the aggregate.
(b) The Class (and petrological name if known).
(c) The Geological age/Colour/Grain size/Fissility.

The basic rules to be followed in assembling the data for CADAM are given in 6.6.2 below and 6.3.3 shows how the CADAM is used when dealing with crushed rock aggregates and sand and/or gravel aggregates respectively in practice. Typical examples of the application of the CADAM are given in Tables 6.7 to 6.16 inclusive based on standard form shown in Table 6.6. The assessment of the aggregate material in toto is made on "representative" samples of that

TABLE 6.6.

AGGREGATE FORM	Crushed Rock ☐	Gravel	Natural ☐ Crushed ☐ Mixed ☐	Sand	Natural ☐ Crushed ☐ Mixed ☐	Land-won ☐ Marine Dredged ☐
CLASS (or MISCELLANEOUS)	Carbonate Class ☐	Quartz Class ☐	Silicate Class: Igneous ☐	Sedimentary ☐	Metamorphic ☐	Miscellaneous Material. (Correct name to be given below) ☐
Petrological name (if known)						
GEOLOGICAL AGE/ COLOUR/ GRAIN SIZE/ FISSILITY						
Comment (if any)						

Compiled by: ______________________ Date: __________

____

CADAM — CLASSIFICATION and DESCRIPTION of AGGREGATE MATERIAL

LOCATION AND SAMPLE DETAILS	Quarry/Pit address: ______ Operator: ______ Sample: Type ______ Size ______ Preparation ______ Supplied by ______	Grid Ref. ______ Date of sampling ______	Date Rec'd ______ Sampling Cert. No. ______

material. It is extremely important therefore that full and accurate details of the sample itself should accompany the assessment sample: this is discussed in 6.6.4.

Once the basic description of the aggregate material has been established, further specialist descriptions relating to the physical, mechanical, chemical and petrological characteristics, as appropriate to the material and its intended use, can be added along the lines indicated in 6.7 and 6.8.

It is assumed that the CADAM will normally be compiled by a qualified and experienced geologist but it is also recognised that in some circumstances this may not be possible.

6.6.2 Basic rules for the CADAM. The notes which follow on the information required for the classification and descriptions are also presented in tabular form in Table 6.5.

6.6.2.1 The form of the aggregate. Construction aggregates are obtained either from mass rock sources by means of crushing, or from particulate deposits which may be crushed or used in the uncrushed state. Mass rock materials include crystalline igneous and metamorphic rocks together with indurated sedimentary rocks and particulate materials comprising natural sands and gravels, the latter commonly referred to as granular material. In the CADAM, aggregate materials are noted as being either of crushed "mass rock" or "sand/gravel" in origin.

6.6.2.2 The Class. Three main groups of minerals are found in naturally occurring materials used for construction aggregates, irrespective of their geological origin. They are the carbonates, the quartz and the silicates. The quantity of aggregates falling outside these three main groups is relatively small.

Thus, three Classes of material result:

A. *Carbonate Class*; comprises all aggregate materials in which calcium carbonate (calcite or aragonite) and calcium magnesium carbonate (dolomite) are the predominant minerals, irrespective of the geological origin. It will therefore include all limestone, dolostone, marble and the like.

B. *Quartz Class*; comprises all aggregate materials in which quartz or free silica is the predominant mineral, irrespective of geological origin. It will therefore include flint, quartzite, vein quartz, granulite, the majority of sandstones and gritstones, some greywackes and the like.

C. *Silicate Class*; comprises all aggregate materials in which rock forming silicate minerals such as feldspars, micas, amphiboles, pyroxenes and olivines are the predominant minerals. Thus it will include all igneous and metamorphic rocks other than those composed predominantly of carbonate or quartz and a few sedimentary rocks such as some argillites, greywackes and arkoses in which quartz or carbonate are not dominant. In the case of the Silicate Class a knowledge of the geological origin can prove to be of value and the Class is therefore divided into three sub-groups,
(i) Igneous
(ii) Metamorphic and
(iii) Sedimentary
This provides separate groupings for the sedimentary rock materials in which clay minerals are likely to be important, for the common igneous rocks, and for the metamorphic rocks in which quartz and carbonate are subordinate.

D. *Miscellaneous*; as already mentioned, a few materials used for aggregate do not fall into one or other of the above three Classes. Such materials should be given their appropriate geological name or term such as barite, hematite, ironstone and the like. They should not be grouped as a "miscellaneous class" and thus robbed of their proper identification.

No matter which Class has been allocated to the aggregate material the addition of the petrological name, though not essential, is encouraged providing it is scientifically accurate.

In the relatively exceptional case where the Class mineral contents are similar, the *single* most appropriate Class should be allocated and attention drawn to the nature of the material in the Additional Comment box, as indicated in Section 6.6.2.5.

6.6.2.3 The Geological Age/Colour/Grain size/Fissility. The Class can usefully be supplemented with further information, depending on the particular Class allocated, related to the geological age/colour/grain size/fissility of the material, as considered below.

(a) Geological Age
Wherever possible the geological age of sedimentary rocks should be given. Thus in the UK and similarly in many other countries, limestones and many sandstones are recognised by their geological succession, which also to some extent gives an indication of their physical condition e.g. Carboniferous limestone, Jurassic limestone, Cambrian quartzite and Carboniferous sandstone.
The age of igneous and metamorphic rocks may not be important. However in the case of sands and gravels, the age of the deposit as well as of its constituents is likely to be of value, particularly if the deposit is other than Quarternary.

(b) Colour and Grain Size
Igneous and metamorphic silicate rock material should be described in terms of the colour and component grain size, since this will give an indication of the petrographic nature of the material, particularly in the case of igneous rock. Thus a basalt would be described as dark grey fine grained, as distinct from a gabbro which would be dark grey coarse grained.
Though not essential to the description, the colour and grain size of Carbonate, Quartz and

Sedimentary Silicate Class materials may be of interest and can therefore be added if desired.

It is suggested that acceptable standard methods should be used to describe colour and component grain size, such as those of The Geological Society of America (1963) and Shearman (1980) respectively, as described in 6.5.

(c) Fissility

Some rocks exhibit a distinct fissility which is likely to influence the shape of the aggregate particles and their possible performance. Although this is particularly applicable to the Metamorphic Silicates it also applies to the other Classes. It is therefore important that the presence of any such fissility should be recorded.

The essential supplementary information required for each Class may be summarised as follows:

CLASS	ESSENTIAL SUPPLEMENTARY INFORMATION
Carbonate	Geological age
Quartz	Geological age of sedimentary rocks. A double age entry may be needed where the aggregate source form is sand and/or gravel (see 6.6.4)
Silicate	
(i) Igneous	Colour and component grain size
(ii) Metamorphic	Colour, component grain size and extent of fissility
(iii) Sedimentary	Geological age and extent of any fissility

In the case of miscellaneous materials, the supplementary information appropriate to that material is to be given. Thus in the case of Barite, the colour and cleavage fissility would be noted.

It should be noted that the surface texture and the shape of the aggregate particles provided for the assessment are NOT included in the CADAM since these are as much influenced by the processing procedures as by the geological characteristics of the parent material (but see 6.6.4 (ii) and 6.7).

6.6.2.4 Additional comment. It is suggested that where a form of tabulation is used for listing the CADAM data, a box should be provided into which any additional relevant information can be entered, as is shown in Table 6.6. Thus in the case of natural sands and gravels it may well be advantageous to know the type of the deposit since this can often give some guidance as to the likely shape and surface characteristics of the particles.

6.6.3 Use of the CADAM system for crushed rock and sand and/or gravel aggregates. Material in which the particle size is predominantly greater than about 2 mm in diameter is commonly described as gravel. Where the particle size is less than about 2 mm and substantially greater than about 75 μm then the term 'sand' is generally used. It should be noted however that at present the various codes and classifications are not consistent in the dimension given for the lower boundary size for gravel (and hence for the upper for sand also) nor for the lower boundary size for sands which corresponds with the upper for silts (Dumbleton 1981). BS 812:1975 makes no recommendation whatsoever as to the appropriate boundary sizes. It is suggested that these limits should conform with those indicated in Table 6.4.

Sands and gravels commonly occur as a mixed deposit, gravel invariably containing some proportion of sand which may or may not be separated in the processing depending upon the use to which the material is to be put. The majority of sands and gravels are clastic sedimentary rocks having been transported and then subsequently deposited. Some also occur as the result of *in-situ* weathering or alteration of pre-existing rocks. They are usually obtained by marine dredging (or otherwise extracted from salt water or from land deposits which retain the characteristics of salt water environment) or else they are 'land won', which term includes includes all materials obtained from dry or fresh water workings. It should be noted that gravels may be marketed in a natural, crushed or mixed state and that sands may be natural, formed from crushed rock or gravel, or may also be a mixture. Sands produced by the crushing of suitable rocks and gravels are commonly described by the terms "crushed fines" or "crusher fines".

Crushed rock coarse aggregate and crushed rock sand (or fines) are treated in the same manner, the form of the aggregate, the class and the information supplementary to the class being given according to the rules detailed in 6.6.2.

Natural sand and/or gravel may need to be described in greater detail in order to differentiate between the characteristics of the deposit as a whole and those of the individual particles of which it is composed.

In the first instance the age of the deposit should be given providing this is other than Quarternary, after which the material is classified in terms of the major particle constituent. In some instances it may be desirable to record the percentages (be weight) of the different particles present in terms of their Class and the supplementary descriptions then added to each as is appropriate.

The main mineral constituent of natural sands is usually quartz, but other types, such as limestone sands, do exist. Hence the material should be classed according to the predominant constituent and a note made of any other constituents which may be present in significant amounts. Thus the presence of shell fragments in a quartz sand should be recorded. In the case of crushed rock and crushed gravel fines the nature of the material(s) should be described.

Ideally all commercial quarries ought to be capable of describing their products at least in terms of the CADAM before any material is offered for sale. However, the material may also be described by more precise petrological terms, such as granite, dolostone, marble, flint, provided that the terms used conform to proper scientific standards and have a precise meaning as noted previously.

Typical examples which show how the CADAM can be used to classify aggregates are given in Tables 6.7 to 6.16 inclusive. Abstracts of the information included on the CADAM proforma are given below. The underlying indicates information which is *essential* in describing the particular material being assessed. By referring to these examples and at the same time noting how the information contained is abstracted from Table 6.5, it can be seen that the compilation of a Classification and Description for an Aggregate Material is a relatively straightforward and simple procedure.

Table 6.7 gives the CADAM for a sedimentary limestone.

(1) The aggregate is of *Crushed Rock*
(2) The material is *Carbonate Class* (Shelly Limestone)
(3) The Geological Age is *Carboniferous* and the Colour is Very Light Grey

Table 6.8 gives the CADAM for a sedimentary quartzite.

(1) The aggregate is of *Crushed Rock*
(2) The material is *Quartz Class* (quartzite)
(3) The Geological Age is *Cambrian,* the Colour Pinkish-Grey and the Grain Size is Medium to Fine

Table 6.9 gives the CADAM for an igneous intrusive.

(1) The aggregate is of *Crushed Rock*
(2) The material is *Igneous Silicate Class* (Granite)
(3) The Colour is *Pinkish-Grey* and the Grain Size is *Coarse*

Table 6.10 gives the CADAM for metamorphic schist.

(1) The aggregate is of *Crushed Rock*
(2) The material is *Metamorphic Silicate Class* (Micaceous Schist)
(3) The Colour is *Grey,* the Grain Size is *Coarse to Medium,* and some Fissility is *Apparent* from the sample provided

Table 6.11 gives CADAM for a greywacke.

(1) The aggregate is of *Crushed Rock*
(2) The material is *Sedimentary Silicate Class* (Greywacke)
(3) The Geological Age is *Silurian,* the colour is Mid-Grey, the Grain Size is medium to Fine and Foliation is indicated by *Slight Fissility* evident in sample

Table 6.12 gives the CADAM for a river gravel.

(1) The aggregate is a *Land-Won Gravel,* both *Natural and Crushed*
(2) The material is *Quartz Class* being composed of approximately 60 per cent Flint and 40 per cent Limestone particles
(3) The Quartz Class material is *Dark Grey* in Colour. The Carbonate Class Material is *Jurassic* in Age and Light Grey in Colour

Table 6.13 gives the CADAM for a marine gravel.

(1) The aggregate is a *Natural Marine-Dredged Gravel*
(2) The material is *Quartz Class* being approximately 80 per cent Flint together with about 20 per cent of mixed Limestone, Granite and Gneiss particles
(3) The Carbonate Class material is *Carboniferous* in Age and Light-Grey in Colour
The Quartz Class material is *Grey* in Colour and of *Fine* Grain Size
The Igneous Silicate material is *Light Grey* in Colour and of *Coarse* Grain Size
The Metamorphic Silicate material is *Mid-Grey* in Colour, of *Coarse* Grain Size and is *Banded with no apparent Fissility*

Table 6.14 gives the CADAM for a dune sand.

(1) The aggregate is a *Natural, Land-Won Sand*
(2) The material is *Quartz Class.* (Quartz grains with occasional shell fragments)
(3) The Colour is Light Greyish Yellow.
(4) Comment that derived from *Coastal Environment*

Table 6.15 gives the CADAM for a crushed rock sand

(1) The aggregate is a *Crushed Rock Sand*
(2) The material is *Quartz Class* (Ferruginous Sandstone)
(3) The Geological Age of the deposit is *Triassic* and the Colour is Pale Red

Table 6.16 gives the CADAM for a miscellaneous aggregate–Barite

(1) The aggregate is of *Crushed Rock*
(2) The material is Barite ($BaSO_4$)
(3) The Colour is *Colourless to White* and Foliation note indicates that the aggregate particles show evidence of *Good Cleavage*

6.6.4 Assessment sample information. As was mentioned in 6.6.1 it is essential that specific information should be given relating to the material provided for making the assessment, since the value of the assessment is entirely dependent on the extent to which the sample represents the source material. This is supplementary to the CADAM data and should be listed separately under a heading such as "Location and sample details". The information is also likely to be of value in the provision of data relating to the physical and mechanical properties of the material in aggregate form, which is the subject matter of 6.7. In addition to giving the name, address, grid reference and operator of the quarry and the actual body supplying the assessment sample, the following details relating to the sample itself should be provided.

(i) The sampling location i.e. whether obtained from the quarry face, borehole, quarry stock-

TABLE 6.7.

AGGREGATE FORM	Crushed Rock	~~Gravel~~	~~Natural~~ / ~~Crushed~~ / ~~Mixed~~	~~Sand~~	~~Natural~~ / ~~Crushed~~ / -~~Mixed~~	~~Land-won~~ / ~~Marine-Dredged~~
CLASS (or MISCELLANEOUS)	Carbonate Class	~~Quartz Class~~	~~Silicate Class~~: ~~Igneous~~ / ~~Sedimentary~~ / ~~Metamorphic~~			~~Miscellaneous Material~~. (Correct name to be given below).
Petrological name (if known)	Shelly Limestone					
GEOLOGICAL AGE/ COLOUR/ GRAIN SIZE/ FISSILITY	Carboniferous Very Light Grey					
Comment (if any)						

Compiled by: C.R.Usher, Wellmeant Laboratories Date: 3/1/83

CADAM – CLASSIFICATION and DESCRIPTION of AGGREGATE MATERIAL

LOCATION AND SAMPLE DETAILS	Quarry/Pit address: Leane Quarry, Crodform, Temlock, Oakshire Operator: Wavering Stone Ltd Sample: Type Quarry Stockpile Size 75mm down, crusher run Preparation Supplied by Operator	Grid Ref. SE 219 536 Date of sampling 23/12/82	Date Rec'd 28/12/1982 Sampling Cert. No. Qs/1693

TABLE 6.8.

AGGREGATE FORM	Crushed Rock	~~Gravel~~	~~Natural~~ / ~~Crushed~~ / ~~Mixed~~	~~Sand~~	~~Natural~~ / ~~Crushed~~ / ~~Mixed~~	~~Land-won~~ / ~~Marine-Dredged~~
CLASS (or MISCELLANEOUS)	~~Carbonate Class~~	Quartz Class	~~Silicate Class~~: ~~Igneous~~ / ~~Sedimentary~~ / ~~Metamorphic~~			~~Miscellaneous Material~~. (Correct name to be given below).
Petrological name (if known)	Quartzite					
GEOLOGICAL AGE/ COLOUR/ GRAIN SIZE/ FISSILITY	Cambrian Pinkish-Grey, Medium to Fine					
Comment (if any)						

Compiled by: A.G.Reegate, Wellmeant Laboratories Date: 31/1/1983

CADAM – CLASSIFICATION and DESCRIPTION of AGGREGATE MATERIAL

LOCATION AND SAMPLE DETAILS	Quarry/Pit address: Wragg Quarry, Wraggsey, Daleshire Operator: Greenwood Stone Co.Rondal, Sample: Type Site Stockpile /Daleshire Size 20mm Nominal Preparation Crushed,screened and Supplied by Alhar Ltd/graded	Grid Ref. SD 308 965 Date of sampling 23/12/1982	Date Rec'd 28/12/82 Sampling Cert. No. Qs/1694

TABLE 6.9.

AGGREGATE FORM	Crushed Rock	~~Gravel~~	~~Natural~~ ~~Crushed~~ ~~Mixed~~	~~Sand~~	~~Natural~~ ~~Crushed~~ ~~Mixed~~	~~Land-won~~ ~~Marine-Dredged~~

CLASS (or MISCELLANEOUS)	Carbonate Class	~~Quartz Class~~	Silicate Class: Igneous	Sedimentary	~~Metamorphic~~	~~Miscellaneous Material.~~ (Correct name to be given below).
Petrological name (if known)	Granite					
GEOLOGICAL AGE/ COLOUR/ GRAIN SIZE/ FISSILITY	Pinkish-Grey, Coarse					
Comment (if any)	Some large Pale Red Crystals Present (Phenocrysts)					

Compiled by: L.A.Boratory
Wellmeant Laboratores
Date: 3/1/1983

CADAM — CLASSIFICATION and DESCRIPTION of AGGREGATE MATERIAL

LOCATION AND SAMPLE DETAILS	Quarry/Pit address: Shapeley Quarry Pentown, Culumshire Operator: Roadstone Supplies Inc.Kenrick /Culumshire Sample: Type Quarry Face Size Lumprock & 25mm Nominal Preparation Crushed,screened & graded Supplied by Operator	Grid Ref. NY 755 804	Date Rec'd 28/12/1982
		Date of sampling /12/1982	Sampling Cert.No. Qu/1695

TABLE 6.10.

AGGREGATE FORM	Crushed Rock	~~Gravel~~	~~Natural~~ ~~Crushed~~ ~~Mixed~~	~~Sand~~	~~Natural~~ ~~Crushed~~ ~~Mixed~~	~~Land-won~~ ~~Marine-Dredged~~

CLASS (or MISCELLANEOUS)	~~Carbonate Class~~	~~Quartz Class~~	Silicate Class: ~~Igneous~~	Sedimentary	Metamorphic	~~Miscellaneous Material.~~ (Correct name to be given below).
Petrological name (if known)	Micaceous Schist					
GEOLOGICAL AGE/ COLOUR/ GRAIN SIZE/ FISSILITY	Greenish-Grey, Coarse to Medium Evidence of foliation NOT apparent in sample					
Comment (if any)						

Compiled by: Hugh Stone
Wellmeant Laboratores
Date: 3/1/1983

CADAM — CLASSIFICATION and DESCRIPTION of AGGREGATE MATERIAL

LOCATION AND SAMPLE DETAILS	Quarry/Pit address: Downside Quarry Flatland, Upshire Operator: C.R.M.Quarries Ltd.Upshire Sample: Type Delivery Vehicle Size 20mm Nominal Preparation Crushed,screened and /graded Supplied by McTar Con- tract Ltd	Grid Ref. SX904.270	Date Rec'd 28.12.1982
		Date of sampling 23/12/1982	Sampling Cert.No. Qs/1966

TABLE 6.11.

AGGREGATE FORM	Crushed Rock	~~Gravel~~	~~Natural~~ / ~~Crushed~~ / ~~Mixed~~	~~Sand~~	Natural / ~~Crushed~~ / ~~Mixed~~	Land-won / ~~Marine Dredged~~		
CLASS (or MISCELLANEOUS)	~~Carbonate Class~~	~~Quartz Class~~	Silicate Class: ~~Igneous~~	Sedimentary	~~Metamorphic~~	~~Miscellaneous Material.~~ (Correct name to be given below).		
Petrological name (if known)	Greywacke							
GEOLOGICAL AGE/ COLOUR/ GRAIN SIZE/ FISSILITY	Silurian Mid-Grey, Medium to Fine Some fissility evident from sample provided.							
Comment (if any)								

Compiled by: N. Hance
Wellmeant Laboratories Limited
Date: 3/1/1983

CADAM – CLASSIFICATION and DESCRIPTION of AGGREGATE MATERIAL

LOCATION AND SAMPLE DETAILS	Quarry/Pit address: Operator: Sample: Type Size Preparation Supplied by	Grid Ref. Date of sampling	Date Rec'd Sampling Cert.No.

TABLE 6.12.

AGGREGATE FORM	~~Crushed Rock~~	Gravel	~~Natural~~ / ~~Crushed~~ / Mixed	Sand	~~Natural~~ / ~~Crushed~~ / ~~Mixed~~	Land-won / ~~Marine Dredged~~		
CLASS (or MISCELLANEOUS)	Carbonate Class	Quartz Class	~~Silicate Class~~: ~~Igneous~~	~~Sedimentary~~	~~Metamorphic~~	~~Miscellaneous Material.~~ (Correct name to be given below).		
Petrological name (if known)	Limestone -- 30% Flint -- 70%							
GEOLOGICAL AGE/ COLOUR/ GRAIN SIZE/ FISSILITY	Jurassic Light-Grey Dark-Grey							
Comment (if any)								

Compiled by: C.A.P. Rock
Wellmeant Laboratories
Date: 3/1/1983

CADAM – CLASSIFICATION and DESCRIPTION of AGGREGATE MATERIAL

LOCATION AND SAMPLE DETAILS	Quarry/Pit address: Operator: Sample: Type Size Preparation Supplied by	Grid Ref. Date of sampling	Date Rec'd Sampling Cert.No.

TABLE 6.13.

AGGREGATE FORM	~~Crushed-Rock~~	Gravel	Natural ~~Crushed~~ ~~Mixed~~	~~Sand~~	~~Natural~~ ~~Crushed~~ ~~Mixed~~	~~Land-won~~ Marine Dredged	
CLASS (or MISCELLANEOUS)	Carbonate Class	Quartz Class	Silicate Class: Igneous	~~Sedimentary~~	Metamorphic	~~Miscellaneous~~ Material. (Correct name to be given below).	
Petrological name (if known)	Limestone -- 80%	Vein Quartz	Granite		Gneiss		
GEOLOGICAL AGE/ COLOUR/ GRAIN SIZE/ FISSILITY	Carboniferous Light-Grey	White, Fine	Light-Grey, Coarse		Mid-Grey, Coarse Banded, but NO fissility apparent		
Comment (if any)							

Compiled by: L.E.Vaite, Wellmeant Laboratories Ltd Date: 3/1/1983

CADAM – CLASSIFICATION and DESCRIPTION of AGGREGATE MATERIAL

LOCATION AND SAMPLE DETAILS	Quarry/Pit address: Operator: Sample: Type Size Preparation Supplied by	Grid Ref.	Date Rec'd
		Date of sampling	Sampling Cert.No.

TABLE 6.14.

AGGREGATE FORM	~~Crushed-Rock~~	~~Gravel~~	~~Natural~~ ~~Crushed~~ ~~Mixed~~	Sand	Natural ~~Crushed~~ ~~Mixed~~	Land-won ~~Marine Dredged~~	
CLASS (or MISCELLANEOUS)	~~Carbonate Class~~	Quartz Class	Silicate Class: ~~Igneous~~	Sedimentary	~~Metamorphic~~	~~Miscellaneous Material~~. (Correct name to be given below).	
Petrological name (if known)	Quartz, with occasional shell fragments						
GEOLOGICAL AGE/ COLOUR/ GRAIN SIZE/ FISSILITY	Light Greyish-Yellow						
Comment (if any)	Derived from Marine sand environment						

Compiled by: C.Shaw, Wellmeant Laboratories Ltd Date: 3/1/1983

CADAM – CLASSIFICATION and DESCRIPTION of AGGREGATE MATERIAL

LOCATION AND SAMPLE DETAILS	Quarry/Pit address: Operator: Sample: Type Size Preparation Supplied by	Grid Ref.	Date Rec'd
		Date of sampling	Sampling Cert.No.

replace the existing grade terms for soils and uncemented sediments etc., but would be complimentary to them (Table 6.4).

In order to maintain conformity with the existing code of practice for soils, the mesograde and macrograde intervals could be subdivided into: fine, medium and coarse as in the sand and gravel grades. Thus, a limestone in which the components are of medium sand grain size would be described as a "medium mesograde limestone", and a coarsely crystalline pegmatite in which the crystals are the size of coarse gravel would be designated "coarse macrograde".

In many rocks the shapes of individual components are far from being equidimensional and in consequence visual assessment of size is at best only a crude approximation. This does not render the use of grade size terms invalid if it is accepted that their prime purpose is for first order description. The terms thus have their use in the recording of geological sections in the field and in the logging of borehole cores etc., and it is with the object of ease and consistency of description that the proposals are submitted.

Most rocks comprise aggregates of crystals but, in terms of texture, they can be thought of as being either granular or crystalline in the sense that in granular rocks the components were discrete physical entities brought together by mechanical processes, whereas in crystalline rocks the crystals grew more or less *in situ* by direct crystallisation, recrystallisation or by replacement. Thus a basalt is crystalline but a basaltic conglomerate is granular and similarly a marble is crystalline but most limestones are granular. It could be useful therefore to qualify the grade term with a statement as to whether the rock is granular or crystalline. This could be done by appending, as appropriate, the abbreviation "/gr" or "/xl" to the root of the grade size term, i.e. c-meso/gr sandstone and m-macro/xl gneiss. The suffix "grade" would be used where there is uncertainty as to whether the rock should be designated granular or crystalline: e.g. "a micrograde limestone".**

* Note 1. As noted above Sherman uses the suffix "grade". This has been replaced by the suffix "size" in Table 6.4.

** Note 2. Hartley (1981) has suggested the use of the terms "discrete", "granular" and "crystalline", in which discrete is applied to soils, uncemented sediments and terriginous sedimentary materials, as a result of which the three root additions "ds", "gr" and "xl" then indicate the nature of the soil rock fabric.

References

HARTLEY, A. 1981. Private communication with Prof. D. J. Shearman.

SHERMAN, D. J. 1980. Grade size terms for the description of igneous sedimentary and metamorphic rocks. Pers. Comm.

Chapter 7. Sampling and testing

7.1 Introduction

This Chapter considers the tests and procedures used to describe or evaluate the physical, mechanical and chemical characteristics of aggregates, for the purposes of, a) prediction of the likely 'in service' behaviour of the material, b) comparison between competing materials, c) specification compliance, or d) quality control. Individual limits are not discussed here but are considered in the appropriate Chapter. Sub-base materials (Chapter 10) sometimes have to use tests in BS 1377, but this section confines itself to BS 812 and the relevant standards from other countries.

7.2 Sampling

7.2.1 General considerations. The first step is the collection of samples. Statistically, a sample can be defined as an individual or group of individuals drawn from a large or infinite population. The infinite population represents the sum and interaction of all sources of variability in the materials sampled, together with the variabilities inherent in the method of obtaining sample measurements. *Information obtained from samples is only as representative of the material as the samples on which they are performed.* If observations reveal little variation and there has been no bias in collecting, then a small sample or small number of samples may be highly representative of a population. If the variation is large then more and/or larger samples will be required. Accurate sampling however, is perhaps the most difficult of the control operations to perform satisfactorily. Sampling together with all types of test contains sources of variation and error, so that judgements of materials based on infrequent random tests are fraught with difficulties.

A random sample is one in which each potential observation has an equal chance of being selected. In a geological context there is no particular difficulty if the material is uniformly available, but often this is not the case. For example, rock exposed in natural outcrops is often sampled but frequently it is only the harder parts of a formation which form conspicuous outcrops, while the softer material remains unexposed. In consequence it is often impossible to obtain a random sample.

In reconnaissance investigations single spot samples provide estimates of the characters of the rock. These are only representative, however, if the rock is very uniform. To offset this a number of spot samples can be combined and split to a convenient size, in order to obtain greater representation. *A single sample will yield only a single value for any characteristic, which need not be representative of the whole deposit or rock suite.* The only estimate of variability it provides is the variability which occurs within the sample. If this is less than the variation between samples then it will be a poor approximation to the whole deposit. Variability is inherent in all types of measurement, arising from the raw materials, nature of the plant and the processing, handling practices and methods of sampling and testing. Of these sampling and testing are commonly a major source of variability. Goodsall and Matthews (1970) demonstrated that splitting and the method of splitting lead to variability in respect of the aggregate content, as a large sample is further and further reduced, to the amounts used in many laboratory tests (Table 7.1). From this the suggestion is that quartering introduces higher variability in sub-samples than riffling.

TABLE 7.1. *Effect of sample reduction method on variability*

Passing sieve (mm)	Standard deviation of aggregate content			
	6-stage reduction		5-stage reduction	
	Quartered	riffled	Quartered	riffled
25.4	5.21	4.45	4.02	2.73
12.7	5.48	3.61	3.88	2.70
3.2	3.1	1.95	2.24	1.55
0.075	0.2	0.15	0.2	0.26

After initial exploration and preliminary sampling has indicated the existence of a body of material suitable for development as an aggregate source, a programme of unbiased primary sampling attempts to prove its dimensions and quality. Primary sampling, a) delimits and establishes the dimensions of the deposit, and b) explores geological variability. Sub-sampling reveals the physical, mechanical and chemical features of the material. Physical properties are intrinsic physical characters of a material, e.g. density, porosity. Mechanical properties are a measure of the physical response of the material to external stimuli, e.g. compressive, impact or shear stresses. Further sub-sampling permits investigation of the causes of variation in these features.

Primary sampling in a deposit of large dimensions

involves integration of samples from individual points or localities, to provide a composite picture of the whole deposit. In quarries or potential quarries, on the other hand, a limited area is sampled to estimate the nature of the final product. In a producing quarry it is a bulk aggregate product which is usually sampled, e.g. stockpiles, etc. The properties or test-values obtained from the sub-samples are then attributed to the bulk.

The categories of sample which can be defined are as follows (Fookes *et al.* 1971):

Type sample—composite sample from a groove cut across the complete face, zone or bed.

Typical sample—similar to type samples but not composite.

Single samples—those selected as representative of a zone or bed of homogeneous material of uniform physical character.

7.2.2 Sampling procedures. Samples from existing quarries or pits may be obtained as rock lumps or gravel from the working face, or as finished aggregate products.

7.2.2.1 Quarry face rock. In sampling directly from a quarry face as with a natural outcrop, care must be taken to ensure that the samples reflect the range of variation present. Where there are obvious differences in the nature of the materials each should be sampled separately. BS 812:1975 specifies the minimum number and dimensions of samples which should be collected.

Another method is to take samples composed of pieces of rock cut out of a small trench or channel, e.g. channel samples, 100×25 mm or 150×12 mm deep and approximately 1 m long. By sampling at regular and close intervals an accurate estimate of quality can be obtained (Northern Mineral Press, 1955).

Variation in terms of composition, quality or state is not confined to originally layered or banded rocks, i.e. sedimentary and banded metamorphic rocks. For example a network of veins, dykes or sheets of totally different composition and degree of alteration may intrude a rock body and demonstrate quite different physical properties from those of the host rock.

Uniform weathering (Chapters 2, 3 and 4) may produce a progressive change in rock state inwards from the surface, giving rise to a surficial anisotropy. Non-uniform weathering preferentially controlled by joints, e.g. the common flat-lying joints in granite, can impart a rude form of layering based on variable states of freshness. Hydrothermal alteration may produce similar patterns.

In addition to inhomogeneities resulting from primary sedimentary banding, variability in anisotropic rocks includes:

a) Igneous rocks—zones of highly vesicular or amygdaloidal texture; flow banding.
b) Sedimentary rocks—clay layers or films in limestone; micaceous films or aggregates on bedding-planes of sandstones; layer parallel jointing.
c) Metamorphic rocks—seams of chlorite, mica etc; mica schist bands in gneiss, marble and psammite; schistose horizons in massive amphibolite or granulite.

To obtain material for testing, rock samples may have to be crushed to aggregate in the laboratory. In the controlled and somewhat atypical conditions of the laboratory however, the available crushing equipment generally produces an aggregate whose character may differ markedly from that in commercial production. Shergold (1963) reports that laboratory-produced aggregates display rougher surfaces, sharper edges and a more cuboidal shape. These differences in turn affected mechanical properties like Polished Stone Value (PSV) (7.5.4.1) and Ten Per Cent Fines Value (7.5.3.4), where higher (superior) values were obtained from the laboratory-crushed material. He cautioned therefore, that such material should only be used in mechanical testing when commercially-produced finished material is not available.

7.2.2.2 Sand and gravel deposits. In sand and gravel pits sampling distinguishes:

a) Composition, e.g. lithological nature and state of the constituent pebbles.
b) Grading and shape distribution in the bulk deposit and the size range of the individual components.
c) Uniformity, e.g. presence of unsuitable material and the possibilities for isolation.
d) Other physical, mechanical and chemical properties.

Sufficient material is taken to allow mixing and quartering down to a suitable size. In the course of sieving the bulk sample the percentage of material retained on 75 mm and 150 mm sieves is an indicator of the crushing which will be required.

7.2.2.3 Aggregate products, sand and gravel. In a working sand and gravel pit the sites for aggregate sample collection are conveyor belts, trucks, hoppers, bins and stockpiles. If stockpiles are sampled the samples should not be taken from the edges or the top, although these are the easiest places from which to do so. In top-loading stockpiles coarser fragments concentrate towards the base of the slope. Samples should be collected from the middle of stockpile faces and points around the pile, and at least 200 mm below the surface. The sampling procedure may also need to recognise other forms of segregation and any variations due to drainage.

The principal causes of variability in natural gravel aggregates include pebble lithology, grading, shape and distribution, and weathering. In evaluating the quality of an aggregate for concrete, and to some extent for asphaltic concrete, important considerations include the quality of pebbles, shape and cleanness, i.e. presence of coatings (Chapter 2). Grading can be corrected to comply with specifications.

7.2.2.4 Aggregate products, crushed-rock quarries. Crushed rock aggregate should be sampled while in motion, e.g. from conveyor belts or at discharge from bins etc. (Shergold, 1963). A minimum of 8 increments is collected over a period of a day, the weight depending on the size of the material (Table 7.2).

TABLE 7.2. *Minimum weights for sampling*

Maximum size present in substantial proportion (85% passing) mm	Minimum weight of each increment	Minimum number of increments	Minimum weight despatched
	kg	kg	kg
64	50	16	100
50	50	16	100
38	50	8	50
25	50	8	50
19	25	8	25
13	25	8	25
10	13	8	13
6.5 and less	13	8	13

If collected at a delivery point the samples should be taken from at least 8 vehicles, arriving at intervals. Collection is made at the time of discharge. Increments are mixed to form a composite sample and then reduced by quartering and/or riffling to the amounts required. In this it should be remembered that such sample reduction is attended by an accumulating variability in aggregate grading analyses (Cox, 1973).

If the main sample is taken from a stockpile or other stationary source it should be an aggregate of at least 10 increments drawn from different parts of the bulk. Shergold (1963) stressed the difficulties of sampling stockpiles, stationary vehicles or top surfaces of bins. Stockpiles should be studied at intervals to establish any fluctuation in the product, resulting from plant operation or intrinsic variability. These are major factors which make it difficult to secure representative samples and affect compliance and 'in service' performance.

TABLE 7.3. *Sample sizes complied from BS812: 1975 and BGWF INF/9,1966*

Test	Minimum amount to be available at the laboratory
1. Durability	
i. Nominal maximum size > 12.5 mm	80 pieces
ii. Nominal maximum size < 12.5 mm	3.6–5.4 kg
2. Sieve analysis	
Maximum size up to 63 mm	50 kg
50 mm	35 kg
40 mm	15 kg
20 mm	2 kg
14 mm	1 kg
10 mm	0.5 kg
3. Shape Indices of flakiness and elongation	800 pieces
4. Proportion of impurities	As for sieve analysis

7.2.3 Sample size. The size of samples which must be collected for physical testing is given in Table 7.3.

7.3 Statistical considerations

This section presents a brief review of the significant expressions used in analysis of sampling and test data. Further information may be obtained by reference to Grant (1952), Griffiths (1967), Harr (1977).

The commonest value obtained from a set of observations is the average or Arithmetic Mean,

$$\bar{X} = \sum_{1}^{\eta} \frac{x_i}{\eta} \qquad x_1\text{—observations} \quad \eta\text{—number of observations}$$

The recording of highest, lowest and average readings fails to give an adequate picture of the property under study. It is important to know the amount by which each value deviates from the average and the measurement of this is given by the Variance (σ^2) of the values, or their spread around the mean,

$$\sigma^2 = \sum_{1}^{\eta} \frac{(x_i - \bar{X})^2}{\eta - 1}$$

It is necessary to square the deviations before summing, as the sum of deviations around the mean is always zero, a property of the mean. To reduce the Variance figure to linear units the square root or Standard Deviation (σ) is used,

$$\sigma = \sum_{1}^{\eta} \left(\frac{(x_i - \bar{X})^2}{\eta - 1}\right)^{1/2}$$

This measure of dispersion is in absolute values and so is independent of relative assessments.

A relative measure of variation is given by the Coefficient of Variation (CV), i.e. percentage Standard Deviation,

$$\mathrm{CV} = \frac{\sigma}{\bar{X}} \times 100$$

In general the average gives the best value of the characteristic being measured, the Standard Deviation shows the spread in data caused by random variations and by variations in the materials being tested. The number of observations indicates the completeness and accuracy of the test, when considered in conjunction with the Standard Deviation.

When a series of readings is taken occasional values may have an abnormally large deviation from the mean. These may be suspected of containing an overall error representing a condition very different from that of the remainder of the test. A criterion which determines whether or not this suspect reading may be rejected is based on showing that the reading is not likely to be a random variation. The degree of likelihood of this however, depends on the distribution of the data. Readings that deviate by more than three times the Standard Deviation may ordinarily be rejected, because such data will not exceed 11% of the total data regardless of the distribution, and in the case of a normal (Gaussian) frequency distribution there is less than one chance in a hundred that a reading would fall outside the same range due to random variation.

The averages of samples of size tend to a normal (Gaussian) distribution as grows large. In such a distribution 99.7% of the population will fall within a distance of 3 standard deviations on either side of the average for the whole population. If random samples of size are taken it may be expected then, that 3 means in every 1000 will fall outside these limits, i.e. 99.7% of the means will fall in the range

$$X+\frac{3\sigma}{\eta^{1/2}} \quad \text{to} \quad \bar{X}-\frac{3\sigma}{\eta^{1/2}} \qquad \text{(Confidence Limits)}$$

If the shape of a frequency distribution is not known, an indication of the minimum proportion of a distribution included in a specified range of Standard Deviations about the mean is given by Tchebycheff's inequality (Woodruff, 1966). This theorem states that more than $1-(1/t^2)$ of any set of observations must fall within the range of $\bar{X} \pm t\sigma$, where $t > 1$. For example, if $t = 3$ this means that more than $1-(1/3^2)$ or $\frac{8}{9}$ of a set of numbers must fall within the limits $\bar{X} \pm 3\sigma$. The fraction falling outside the limits may well be less than $\frac{1}{9}$, the theorem simply states that it cannot be as much as $\frac{1}{9}$ (Grant, 1952). In addition if all of the observations are to fall in the range $\bar{X} \pm t\sigma$, then t can be determined from the expression $t = \eta^{1/2}$ (Woodruff, 1966), e.g. if $\eta = 2$ all observations must lie between the limits $\bar{X} \pm \sqrt{2}\sigma$. The normal distribution with >99% of its area in the range of 3 Standard Deviations is well within the limit of inequality.

In the analysis of test results Confidence Limits for a probability of 0.1 only become acceptably close with a large number of tests, too high for common practice. In single samples of production aggregate the limits within which test values for strength and durability fall are too high to be representative of the quarry (Shergold, 1963). This indicates that a single sample is not likely to be representative of even a single day's output. He suggested that at least five samples should be tested for the mean results to be representative of the average output at any given time. To offset the large number of samples which would be involved he proposed the combining of samples and mixing to form a composite sample. This could then be reduced to appropriate quantities for testing.

Some tests, like those for aggregate grading and shape, normally apply only to limited consignments and are required for compliance or quality control. In other tests there is a tendency to regard results obtained from one sample as being valid indefinitely. Because of the variability of most aggregates only recent test-values are really valid. In addition, results from one part of a quarry may differ from one another, even when sampled at the same time.

7.4 Repeatability and Reproducibility

Important considerations in any tests programme, whatever its aims, are the notions of repeatability and reproducibility (BS 812:1975). A number of quite different factors can contribute to a spread in test values and to allow for this in comparing results obtained by the same operator or by different operators in different laboratories, some estimate of the distribution of values is necessary.

7.4.1 Repeatability (r). This is the quantitative expression of the random error associated with a single test operator obtaining successive results on identical material, with the same equipment and constant operating conditions. It is the difference between two single results which would be expected in only one case in twenty, assuming normal and correct operation of the test (BS 812:1975). This is expressed by,

$$r = 1.96\sqrt{2}\,\sigma_1$$

is the single operator Standard Deviation within a laboratory.

7.4.2 Reproducibility (R). This is an expression of the random error associated with test operators working in different laboratories, each obtaining single results on identical test materials when applying the same method. As with r, it is the difference between two single and independent results which would be expected in only one case in twenty. This is expressed by,

$$R = 1.96\sqrt{2}\,\sigma_1^2 + \sigma_2^2$$

where σ_2 is the Standard Deviation applicable to all causes of variability other than repeatability, when the results from different laboratories are compared.

In Table 7.4 (BS 812:1975) values of r & R are given for the certain standard physical and mechanical tests. This indicates the greatest differences expected between tests on separate samples, and more importantly the minimum number of tests to provide a representative value for the material.

7.5 Testing

As the several demands on aggregate call for different characteristics, a wide range of tests has been devised to describe the material and assess its potential value. These tests have the principal functions of assisting predictions of 'in service' performance and enabling

TABLE 7.4. *Estimates of repeatability for some standard aggregate tests*

Test	Repeatability (r)	Reproducibility (R)
Relative density:		
i. most aggregates	0.02	0.04
ii. low density porous aggregates	up to 0.4	up to 0.08
Water absorption	5% of value recorded	10% of value recorded
Bulk density	$10\ kg\ m^{-5}$	$20\ kg\ m^{-3}$
Aggregate Impact Value	1.0	2.0
Aggregate Crushing Value	0.8	1.5
Ten Percent Fines Value	7 kN	14 kN
Aggregate Abrasion Value	1.5	3.0
Polished-Stone Value	4.9	6.0

materials from different sources to be compared. Most assess some physical or mechanical attribute of the material, while a few investigate particular chemical characteristics. The majority are performed on the finished aggregate product, but some can profitably be performed on the ledge rock and are therefore of value in the exploration stage.

The literature is extensive so it has been necessary to abstract considerably and the various standard sources should be consulted for more comprehensive discussion. The tests described are those which are standard in Britain and USA, but some discussion of non-standard variations has also been included.

Physical tests:

a) Aggregate grading — BS 882 & 1201:1973; BS 812:1975; ASTM Designations C33 and 136

b) Aggregate shape, angularity, sphericity, roundness, surface texture — BS 812:1975:Part 1

c) Relative density, bulk density, unit weight — BS 812:1975; ASTM Designations C29 and 127

d) Water absorption — BS 812:1975; ASTM Designations C127 and 128

e) Aggregate shrinkage — BRS Digest 35

Petrographic examination — ASTM Designation C295

Mechanical tests:

Strength

a) Aggregate Impact Value — BS 812:1975

b) Aggregate Crushing Value — BS 812:1975

c) Ten Percent Fines Value — BS 812:1975

d) Franklin Point Load Test — Franklin 1970

e) Schmidt Rebound Number — Duncan 1969

Durability

a) Aggregate Abrasion Value — BS 812:1975

b) Aggregate Attrition Value — BS 812:1943

c) Los Angeles Abrasion Value — ASTM Designation C131

d) Polished Stone Value — BS 812:1975

e) Slake Durability Value — Franklin 1970

f) Sulphate Soundness — ASTM Designation C88

Chemical tests:

a) Chloride content — BS 812:Part 4:1976

b) Sulphate content — BS 1377

c) Organic content — BS 1377

d) Adhesion Tests — HMSO 1962

7.5.1 *Physical tests*

7.5.1.1 a) Aggregate Grading. Aggregate, both crushed rock and natural, is differentiated at the 5 mm size range into coarse and fine aggregate (BS 882: 1973). In the laboratory a suitably reduced sample is sieved on a standard group of BS sieves and the grading of the sample is obtained (BS 812:1975). Fine aggregates for concrete are presently classified into four grading Zones (Table 7.5) becoming progressively finer from Zones 1 to 4 (BS 882:1973)

TABLE 7.5. *Grading of fine aggregate by Zones, BS 882:1973*

Test sieve mm.	Percentage by weight passing BS sieves: Grading Zone 1	Grading Zone 2	Grading Zone 3	Grading Zone 4
10.0	100	100	100	100
5.0	90–100	90–100	90–100	95–100
2.36	60–95	75–100	85–100	95–100
1.18	30–70	55–90	75–100	90–100
μm				
600	15–34	35–59	60–79	80–100
300	5–20	8–30	12–40	15–50
150	0–10	0–10	0–10	0–15

In engineering usage the terms describing sediment grading differ from those employed in sedimentology, e.g. a sediment containing a range of sizes from fine to coarse (Table 7.5) would be classified as well-graded in engineering terms but the same term would mean poorly sorted to a sedimentologist.

In sieving an aggregate sample the product is influenced by the shape of the constituents. The intermediate dimension in flaky particles get larger as they become more flaky, and this can affect the shape proportions of a fraction when using sieves with circular as against square holes (Lees, 1964).

The volume and size of the particles retained on a particular sieve are conditioned by shape, e.g. elongate fragments in any size approximate in size to the flaky fragments of the next coarser size (Lees, 1964). Sieves therefore, do not rigorously size the fragments where the shapes are very different. This implies that in an aggregate with a high proportion of elongate grains the particle size would be coarser than one which is rich in flakes. One consequence of this size variability is that in flaky aggregates there is a higher surface area for a given weight than in elongate or equidimensional samples. These features affect the void-filling characteristics, permeability and strength of concrete and road base aggregates. They are also important considerations in filter design.

Grading data are commonly plotted on semi-logarithmic charts (BS 812:1975). An alternative method is the employment of parameters which describe the coarseness of each grading and determine whether samples are well-graded, single-sized or gap-graded, e.g. D_{60} the sieve aperture size through which 60% passes. This is a measure of modal size, distinguishing between coarse, medium and fine materials, i.e. $D_{60} > 10$ mm implies coarse grain, $D_{60} > 5$ mm implies medium grain.

An index of grain-size dispersion is the uniformity coefficient, $CU = D_{60}/D_{10}$. When CU is low particles are more or less uniform in size. A well-graded mixture of gravel-sand-silt-clay may exhibit values as high as 1000. Another value, the coefficient of curvature, $CZ = D_{30}^2/(D_{60} \times D_{10})$ together with CU can be used to distinguish well-graded and gap-graded material. Where $CU > 35$ and $CZ < 6$ the material is well-graded: $CU > 15$, $CZ < 6$ is medium-graded; $CU < 15$, $CZ < 6$ is poorly-graded or uniformly graded in engineering terms and $CU > 35$, $CZ > 6$ is gap-graded.

In natural gravels the grading of the material as-dug is an intrinsic characteristic of the deposit, reflecting the environment of formation. In consequence the material may not form a regular graded assemblage conforming to a particular grading zone, but rather have deficiencies or excesses in particular size ranges. As the grading required varies with the project there is no unique system of crushing which will satisfy all requirements. Crushed rock or gravel therefore may be screened into a series of 'single size' ranges and subsequently re-mixed in required proportions to satisfy specifications.

The mean grading of a deposit is the average of individual samples. Cox (1973) demonstrated a high repeatability for sieve analyses performed by repeated sieving of a single sample. On the other hand, careful subsampling of a larger bulk sample, even by the same operator, introduces considerable variability into the grading.

In evaluating the variability induced by sample splitting, Pike (1978) concentrated on particular but significant size levels, e.g. the proportion passing the 5 mm sieve. Successive sample reduction showed a progressive increase in the proportion of this size range present. From a bulk sample with 40% passing 5 mm sieve successive reduction resulted in a range of values between 34% and 40% by the fifth riffling stage.

The properties of the aggregate which affect the degree of variability induced by the sample reduction include moisture content, absolute grading, particle shape and texture. The variability is lowest in fine-grained aggregates and highest in gap-graded, coarse sub-grade material (Chapter 10).

b) Sand Equivalent Value (ASTM 2419). A rapid field test to ascertain the proportions of clay-like or plastic fines in granular soils and fine aggregate passing ASTM Sieves No. 4 (4.75 mm) is provided by the Sand Equivalent Value. A small quantity of flocculating solution at 22 ± 3°C (solution of calcium chloride, glycerine and formaldehyde) and a measured volume of oven-dried fine aggregate are poured into a graduated plastic cylinder. Agitation loosens the clay-like coatings from the coarser grains. After irrigation with additional solution the clay-like material is forced into suspension above the sand. After a prescribed sedimentation period the heights of accumulated clay and sand are measured and the sand equivalent is expressed as the ratio of the height of sand to clay times 100.

7.5.1.2 a) Aggregate shape. In both natural and crushed rock aggregates the constituent particles within a particular size fraction may display a range of shapes. The distribution of shapes reflects intrinsic petrological-petrographic characteristics, together with environmental factors in formation (natural gravels) and production factors. (crushed rock aggregate). British Standard BS 812 (1975) groups these shapes into six categories, rounded, irregular, angular, flaky, elongated, flaky and elongated (Fig. 7.1a). In the rounded, irregular and angular categories the particles approximate to equidimensional and for many assessments are grouped together as one, equidimensional (Lees, 1964) or cuboidal (Ramsay, 1965).

Several schemes exist for establishing classificatory limits to the proportions of the four shape categories (BS 812:1975; Zingg, 1935; Rosslein, 1941; Lees, 1964) and for the actual measurements of the shape of individual fragments, as illustrated in Fig. 7.1b. In the description of shape BS emphasises departure from

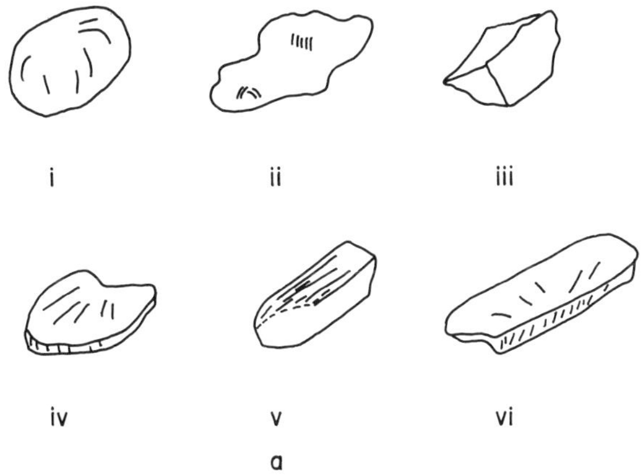

FIG. 7.1a. Particle shape. i, rounded; ii, irregular; iii, angular; iv, flaky; v, elongate; vi, elongate and flaky.

equidimensional in Indices of Flakiness and Elongation.

b) Flakiness Index, I_F (BS 812:1975). This is restricted to aggregate coarser than 6.5 mm and is an expression of the weight-percentage of particles, in a minimum sample of 200 pieces, whose least dimension is less than 0.6 times the mean dimension. By mean dimension is meant the arithmetic mean of the side dimensions of the delimiting square-holed sieves. For example, in 14–10 mm aggregate this would be 11.125 mm. Measurement has been facilitated by the design of the standard shape gauge (Fig. 7.2a).

c) Elongation Index, I_E (BS 812:1975). This is the weight-percentage of particles whose long dimension is greater than 1.8 times the mean dimension. Again measurement can be made with a standard gauge (Fig. 7.2b).

d) Discussion. The most generally desirable aggregates are those with a high proportion of roughly equidimensional (cuboidal) particles. Intrinsic factors such as rock composition and fabric affect the shape of crushed rock aggregate. In the main, strong and hard or brittle rocks produce a higher proportion of flakes than weak types, although the latter generate more fines in crushing. It is chiefly at the extremes of strength and fabric anisotropy that grading and shape of the finished product is significantly affected.

In a critical review of aggregate shape and classification Lees (1964) followed the sedimentological practice of describing particle shape in terms of axial ratios obtained from three orthogonal dimensional axes (longest, intermediate and shortest). Elongation and flatness ratios are

$$\frac{\text{intermediate}}{\text{longest}} \quad \text{and} \quad \frac{\text{shortest}}{\text{intermediate}}$$

respectively (Fig. 7.1b). In this scheme the BS limits for elongation would be 0.55 and flakiness 0.6. These limits divide up a plot of elongation vs flatness ratios into four fields, equidimensional, elongate, flakes, flaky-elongate, within which all particles may be classed. Lees indicates a weakness in the BS

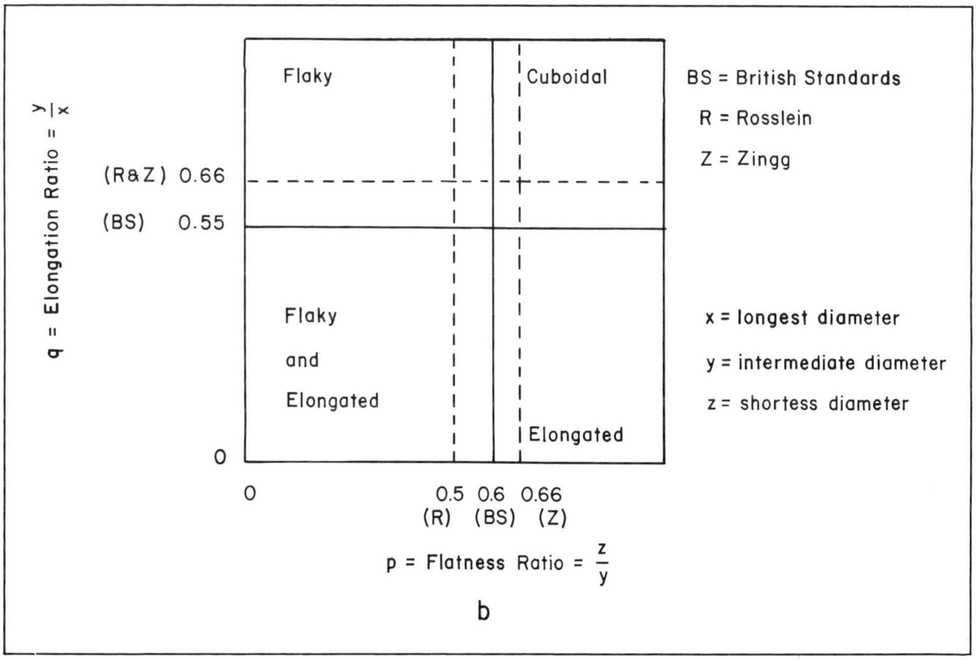

FIG. 7.1b. Three dimension shape categories bounded by arbitrary limits of p and q (Lees 1964).

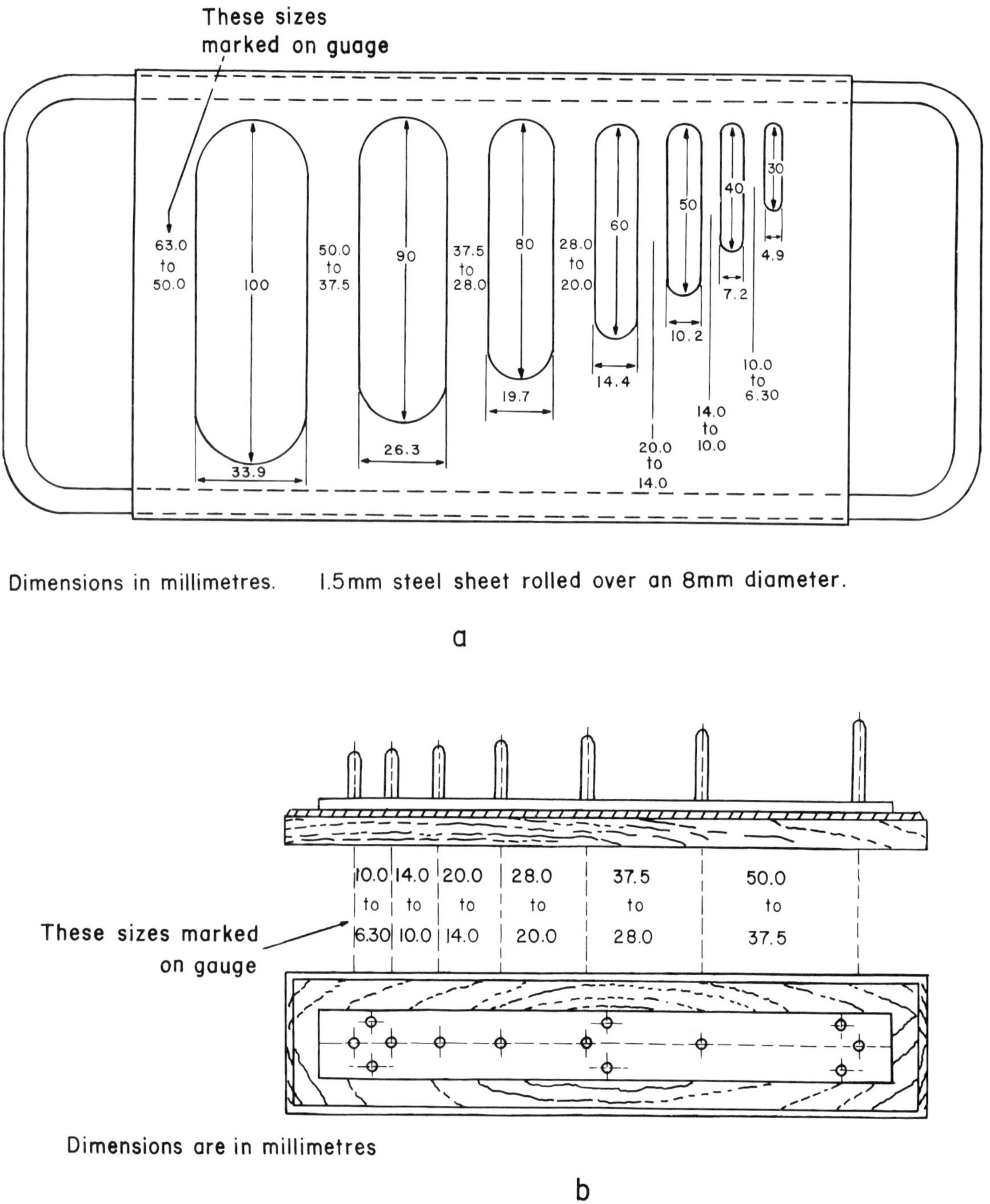

FIG. 7.2. Diagrams of B.S. thickness (a) and length (b) gauges. (BS 812:1975).

classification scheme, where simplification of the measurement procedure has led to very broad shape categories. This allows a wide range of shape characters for different populations which have the same values of I_F and I_E. This could affect the behaviour of aggregates in a number of situations and distort comparisons between products of the same rock type. To overcome this deficiency, especially in the more rigorous conditions of a research programme, more precise means of particle measurement have been devised (Rosslein, 1941; Lees, 1964).

In commercial use flakes lead to an increase in the

specific surface of the aggregate and control the packing pattern and therefore porosity and permeability in the unbound condition. These features, in varying degree, can also affect shear and crushing resistance, tensile and flexural strength in concrete, asphalt and macadam and workability in concrete.

In road surface-dressing good cuboidal chippings are essential but, however, many desirable rocks are difficult to process for good shape. All rocks do not behave alike under size reduction. Flakiness can be induced in the crushing process and to minimise this, in the more difficult instances it is important to keep the Reduction Ratio

$$\frac{\text{mean size of feed}}{\text{mean size of product}}$$

as low as possible, i.e. 3:1 or 4:1, Shergold (1959) reviewed the performance of crusher types and listed them in descending order of cuboidal production, i.e. impact, single toggle jaw, gyratory, roll, disc and cone crushers (Shergold & Greysmith 1947).

While the strength of a rock is an intrinsic characteristic, the effective strength of an aggregate particle is modified by its shape and size (7.5.3). The greater degradation and preferential elimination of flaky particles (Ramsay *et al.* 1974) in a particular size range is related to higher bending moments which can be applied (Lees, 1964).

e) Angularity. This is a measure of the lack of rounding of aggregate particles. This characteristic adversely affects the workability of concrete but can improve the stability of interlocking fragments. As seen in previous sections, determination of shape characteristics rests heavily on the measurements on individual particles. As this is time-consuming several attempts have been made to devise more rapid methods for determining the degree of rounding of gravels through its effect on bulk properties.

One such test, adopted in BS 812:1975, is based on computation of the percentage of voids in a compacted aggregate. As pebbles in a gravel become less angular there is a progressive reduction in the proportion of voids (V) as expressed by,

$$V = 100\left(\frac{1-W}{cG_a}\right)$$

where V is the percentage of voids, W is the weight of a standard volume of aggregate, c is the volume of the cylinder and G_a is the apparent dry relative density of the material.

Since the Angularity Number (AN) = V-33, where 33 is the percentage of voids in a very well-rounded river gravel, the expression for V can be substituted to express AN,

$$\text{AN} = 67 - \frac{100W}{cG_a}$$

AN varies from 0–9, the higher number being typical of freshly crushed rock.

Lees (1963) devised a scheme for the determination of the degree of angularity in grains and crushed-rock fragments. This involves either measurement and calculation in accordance with prescribed methods and formula or, more simply, visual comparison with a descriptive chart similar to that proposed by Krumbein (1941) for roundness.

f) Sphericity and Roundness. Wadell (1933) proposed these terms for describing particle shape which varied from spherical to less regular. Sphericity (Ψ) is the deviation of a particle from a sphere, i.e. Ψ is the ratio of the surface area of a sphere with the same volume as the particle to the surface area of the particle. Because surface area and volume are difficult to measure in small grains he suggested $\psi = d_p/D$ where d_p is the projected diameter and D the diameter of the smallest circle circumscribing that projected area. Alternatively $\psi_2 = d_e/L$ where d_e is the diameter of particles obtained as the diameter of a sphere having the same volume as the particle and L is the length of the particle. Sphericity Number ranges between 0–1.

Roundness is a measure of the sharpness of corners, i.e. smoothness,

$$p = \frac{\sum \frac{r}{R}}{N}$$

Where r is the radius of curvature of a corner of the particle surface, R is the radius of the maximum inscribed circle in the projected plane and N is the number of corners. With wear r approaches R and p approaches 1 (Harr, 1977). In genetic terms roundness is a function of wear while surface texture reflects environmental differences.

The simplest routine estimation of these parameters is facilitated by visual reference to a calibrated set of standards (Krumbein, 1941; Lees, 1963). Griffiths (1967) however, cautioned that there are psychological deficiencies in the method, owing to the subjective nature of the method and the likelihood of operator error.

g) Surface texture. This expression of the microtopography of aggregate particles (Krumbein and Pettijohn, 1938) reflects petrographic and environmental factors of the original material. This property affects frictional properties and intergranular slip in unbound aggregate and the adhesion of binders or cement. BS 812:1975 outlines six broad and qualitative categories e.g., glassy, smooth, granular, rough, crystalline, honeycombed and porous.

A more objective and quantitative scheme was proposed (Wright, 1955), but this has not been adopted in the aggregate industry. This method entails analysis of enlargements (×125) of particle profiles in thin-section. This provides,

$$\text{Roughness Factor} = \frac{100Y}{\text{sample length}}$$

where Y is the excess in length of the actual profile (roughness line) over the length of the segmented curve (unevenness line) which represents the general undulations of the roughness line.

7.5.1.3 Relative density, bulk density and water absorption. The relative density of samples of the raw material can be established by one of several conventional techniques using balance or steel-yard and the reader is referred to any text-book of physics or physical mineralogy (e.g. Hurlbut and Klein, 1977).

The relative density of aggregate is obtained in broadly similar fashion (BS 812:1975) but at the same time a measure of the water absorption can be obtained. For aggregate coarser than 10 mm a thoroughly saturated sample (e.g. 2 kg) is weighed in water (Mass B). It is then surface dried and weighed (Weight A). Finally it is oven-dried for 24+ hours at 110–110°C, cooled and weighed (Weight C).

Relative density (oven-dried) $\dfrac{C}{A-B}$

Relative density (Saturated and surface dry) $\dfrac{A}{A-B}$

Apparent relative density $\dfrac{C}{C-B}$

Water absorption (% dry mass) $100\,\dfrac{(A-C)}{C}$

There are variations in the method for finer aggregates using pycnometer, gas jar or specific gravity bottle (BS 812:1975).

7.5.1.4 Bulk density (unit weight). This is a measure of the weight of material in a unit volume, e.g. lb ft^{-3}, kg m^{-3} etc. This test is described by BS as a rapid means of comparison between different aggregates. In the same test one can obtain the percentage of voids:

$$\%\ \text{voids} = 100\left(\frac{a-\dfrac{b}{1000}}{a}\right)$$

where a is the relative density of the aggregate on an oven-dry basis and b is the bulk density of an oven-dry aggregate.

7.5.1.5 Aggregate shrinkage. Some rocks exhibit the property of absorbing water with attendant change in dimension. This in turn can have a very significant influence on the shrinkage characteristics of concrete made from this aggregate (Fig. 7.3a) (Edwards, 1967). This relationship has been extended to the drying shrinkage of such aggregates and the shrinkage of concrete (Fig. 7.3b) and possibly to bituminous mixes made from it (Edwards, 1970b).

In addition to bulk dimensional change it has been suggested that there may be a connection between the property of shrinkage and unsoundness in concrete (Edwards, 1967). This unsoundness is attributed to the internal strains attending cycles of slaking, i.e. wetting and drying.

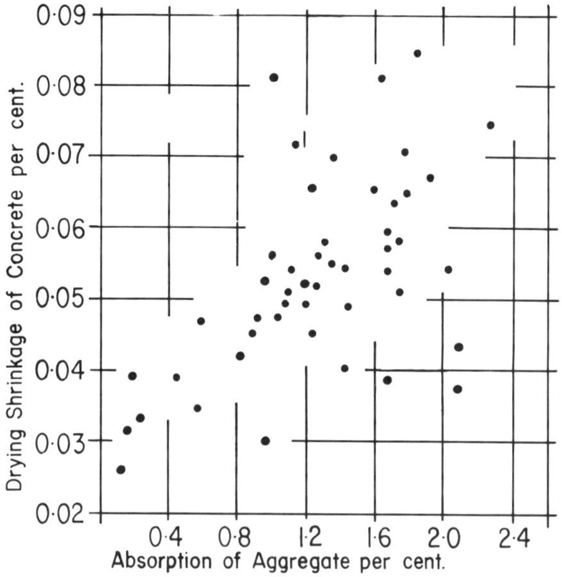

FIG. 7.3a. Relationship between absorption of aggregate and drying shrinkage of concrete.

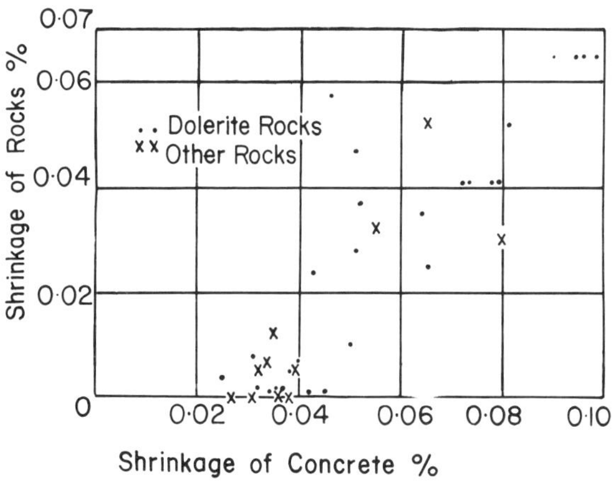

FIG. 7.3b. Relationship between shrinkage of rocks and concrete.

Higher values of shrinkage, e.g. >0.065%, are frequently associated with dolerite, some basalts, felsites and shales, while lower values are obtained for highly siliceous rocks such as granite, flint, and the calcareous rocks, e.g. limestone. When natural gravels contain significant proportions of high shrinkage rocks then they too may exhibit high shrinkage (Edwards, 1970a).

The greatest internal stresses in concrete, with resultant cracking, occur when the differential shrinkage between aggregates and the mortar reaches a maximum, e.g. highly shrinkable sand and non-shrinkable aggregate.

A test for concrete shrinkage has been described by Edwards (1963) in which 28 day old prisms of concrete are immersed in water for 4 days, then surface dried and measured (l_c). The prisms are then oven-dried at a specified temperature for 3 weeks, cooled and measured. Drying is continued for a further week followed by measurement of length. This procedure is repeated until a constant length is obtained (l_d)

$$\text{Drying shrinkage} = \frac{l_c - l_d}{l_d} \times 100\%$$

7.5.2 Petrographic examination of aggregates (ASTM Designation C295). An important preliminary aspect of any aggregate investigation is a petrographic examination of representative samples by an experienced geologist, (Chapter 6). The extent of the study and the procedures employed depend on the purpose of the examination and the nature of the material which may only become apparent in the course of the examination. In general terms a petrographic examination is made for the following purposes:

a) To describe and classify the constituents of the sample.
b) To assess the physical and chemical character of the constituent material.
c) To determine the relative amounts of the constituents of the sample. This is necessary where the constituents differ significantly in character or properties and which may have a bearing on the quality of the material, e.g. the amount of secondary clay minerals which might cause swelling on wetting.

7.5.2.1 Natural sand and gravel. Representative samples of gravel and natural sand, reduced from larger samples are sieved to provide populations of each sieve size with at least 300 particles in each. The size fractions are examined separately, commencing with the largest. The first stage of the study begins with a megascopic examination. More difficult material, such as dense, fine-grained material, is examined with a stereoscopic microscope, and if this is inadequate thin-sections are prepared for study using a petrological microscope. The aspects studies in gravel include the following:-

a) Coatings —Their presence, whether potentially deleterious (e.g. opal, gypsum, very soluble salts, organic material) and persistence.
b) Rock types —The fraction is separated into rock types by visual examination and with the stereoscopic or petrological microscope. For particular purposes additional and more definitive techniques such as X-ray diffraction (XRD), X-ray fluorescence (XRF), differential thermal analysis (DTA) and infra-red spectroscopy may be employed.
c) Rock state —Each group is further subdivided according to physical conditions, e.g. fresh or grade of weathering (Fookes *et al.* 1971), dense, porous or friable. The number of these subdivisions should be held to a minimum, e.g. two or three.
d) Particle shape —Notes should be taken on particle shape, surface, grain-size, texture and structure, colour, heterogeneities, presence of deleterious constituents.

7.5.2.2 Sand. A sample is split on the ASTM sieve No. 30 (0.595 mm) and riffled down to approximately 300 particles in each fraction. These are examined by stereoscopic microscope, either dry or immersed in water to lessen reflections from outer surfaces. Where necessary a polarising microscope may be used.

7.5.2.3 Rock cores. Each core is examined and a log prepared showing, length recovered and core loss, location and spacing of fractures, parting planes, lithological types and alternations, physical state, toughness, coherence, obviously porosity, grain-size and texture, types of breakage and presence of deleterious constituents (e.g. opal, chalcedony, natural glass) (Geol. Soc. Rep. 1970). Some parts may require sectioning for study with a petrological microscope. If clays are present, with the possibility of montmorillonite (swelling clay), DTA and other techniques may be necessary.

The procedures for the study of ledge rock and crushed stone are similar to those for cores.

7.5.2.4 Calculations. The calculations of each sieve fraction of a heterogeneous sample and the weighted average composition of the whole sample are calculated as follows:-

a) For all fractions the number of pebbles of each constituent type, in each condition (e.g. fresh, weathered, etc.) is expressed as a percentage of the total in the size population (a)
b) The percentage by weight of the sieve fraction of the whole sample (b)
c) The weighted percentage of constituents in sieve fractions, i.e. (a) multiplied by (b)
d) By adding the weighted percentage of each constituent in each sieve fraction, the weighted percentage of each constituent in the whole sample is obtained.

7.5.2.5 Report. The report follows that described in Appendix A6.2 of Chapter 6.

7.5.3 Mechanical tests

7.5.3.1 Strength. At one time the strength tests for potential aggregates were performed on one-inch cores of intact rock, e.g. Uniaxial Compressive Strength. The correlation between the performance of rock in the intact state and the aggregate form is not consistent however, in other than general terms (Dhir *et al*. 1971). To simulate the natural conditions more closely, a suite of tests was designed which express the various aspects of aggregate strength namely, Aggregate Impact Value, Aggregate Crushing Value and Ten Percent Fines Value. These are embodied in British Standard 812 (1975).

7.5.3.2 Aggregate Impact Value (AIV). In this test a standard sample in the size range 12.25 mm to 9.5 mm is subjected to a discontinuous loading in the form of 15 blows from a hammer or piston (13.5 kg to 14.1 kg) falling through 381 ± 6.5 mm (BS 812: 1975) (Fig. 7.4). The sample suffers degradation to a graded assemblage of fines. An arbitrary sieve size, 2.40 mm, is chosen as the diagnostic cut-off level and the percentage of material passing, relative to initial weight gives the Aggregate Impact Value, and is used as the measure of resistance to granulation. Two tests are performed and the results should be within a numerical value of 1. In this test a lower numerical value indicates a more resistant rock. The reproducibility of results in the test is high (Table 7.2) so that 2 values per sample are sufficient. The apparatus used in this test is relatively portable and cheap to operate, allowing both laboratory and field testing of aggregate.

7.5.3.3 Aggregate Crushing Value (ACV). In this test a sample of approximately 2 kg is subjected to a continuous load transmitted through a piston, in a compression test machine (Fig. 7.5). A total load of 400 kN is achieved in 10 minutes. As in the AIV the fines passing sieve BS 2.40 mm are calculated as a percentage of the initial sample weight. This is the Aggregate Crushing Value. Two values are produced for each test material and values should be within a value of 1. Once again a lower value indicates a more resistant rock.

FIG. 7.4. The Aggregate Impact Test Apparatus.

FIG. 7.5. The Aggregate Crushing Test Apparatus.

7.5.3.4 Discussion. The approximate similarity in numerical values for the ACV and AIV tests was built into the design of the AIV, which was intended to be auxiliary to ACV. Some values of AIV and ACV from a range of rock types currently exploited in Scotland (Ramsay *et al*. 1974, Edwards, 1967) are given in Table 7.6.

TABLE 7.6. *Some values of AIV and ACV for rock types exploited in Scotland*

Rock type	AIV	range	ACV	range
Basalt	11	10–13	16	16–17
Andesite	13	11–16	16	15–17
Dacite	12		13	
Porphyry	13	12–14	12	
Felsite	13	12–15		
Dolerite	13	10–17	19	
Teschenite	22			
Granite	19	17–21	26	23–30
Limestone	17	15–20		
Greywacke	9			
Marble	19	16–21	26	25–28
Psammite	14	14–15		

These tests do not measure strength in the sense of compressive strength in intact rocks, rather they are indices of the resistance to pulverisation over a prescribed interval or sequence of loading.

The range of values in Table 7.6 is much smaller than other published accounts, e.g. Shergold, 1950; Road Research, 1959. The difference stems from the use of more rigorously defined rock types rather than the much broader rock associations or Trade Groups (Chapter 6).

The variation in Aggregate Impact and Crushing Values within specific rock type categories, or between broader rock groups, can be attributed to the influence of aggregate particle shape (Ramsay 1965) and geological features such as bulk composition, e.g. silicate and carbonate rocks etc.; grain size; texture and structure; alteration (Dhir *et al.* 1971, Ramsay *et al.* 1974). In addition methodological factors like apparatus rigidity and the nature of the test substrate can introduce considerable variation into the values obtained for AIV (Ramsay *et al.* 1973).

The shape of the aggregate particles, in particular the index of flakiness affects both AIV and ACV in a simple rational fashion (Figs 7.6 and 7.7). This is

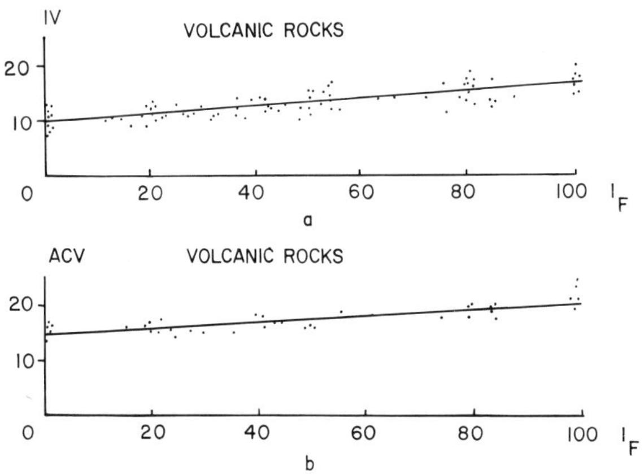

FIG. 7.6. (a) Aggregate Impact Value vs. Flakiness index (I_F) for a suite of volcanic rocks. (b) Aggregate Crushing Value (ACV) vs. I_F for volcanic rocks.

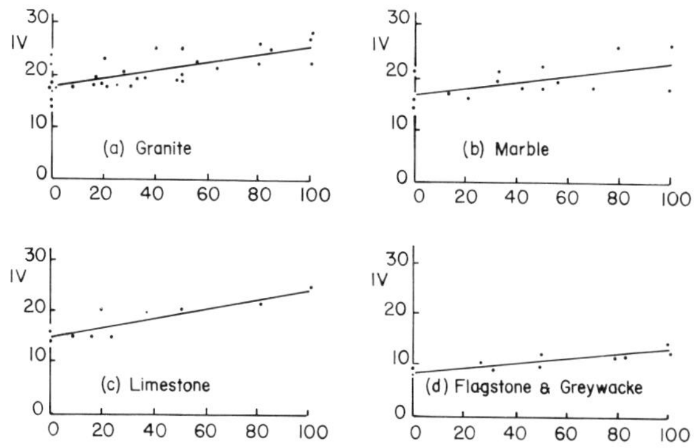

FIG. 7.7. Influence of Flakiness Index on impact value for (a) granites, (b) marble, (c) limestones, (d) greywacke-flagstone. The upper points in 4b belong to a Schistose marble, the remainder are from granoblastic types.

expressed by:

$$AIV = C + nI_F$$

$$ACV = C + nI_F$$

where n is the coefficient of flakiness, ranging between 0.04 and 0.1, I_F is the flakiness index, C is a constant reflecting the other geological parameters. Flakiness can introduce differences in AIV and ACV of 30% to 60%, so that comparison of particular values without details of I_F can be misleading. The preferential elimination of flaky particles in the course of testing can be appreciated from Fig. 7.8. This destruction of flakes characterises both of the standard aggregate strength tests and the 'in service' behaviour as surface dressing and open-textured macadam aggregates. In dense bituminous mixes and densely unbound aggregate the mortar or matrix affords particle support, thereby reducing flake failure (Chapters 10 and 11).

Figure 7.6a is a synoptic plot of AIV vs I_F from volcanic rocks in the compositional range basalt to

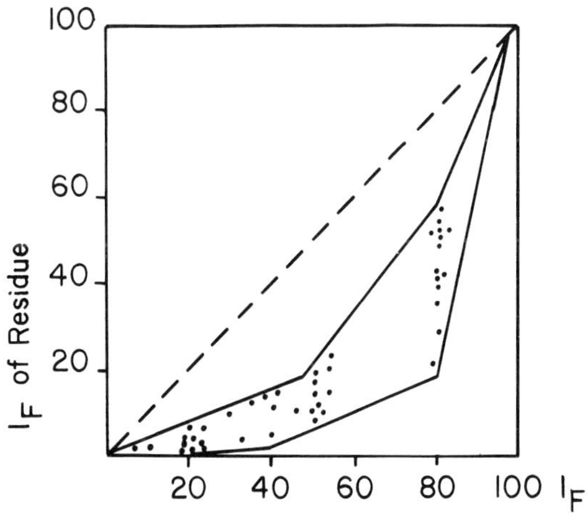

FIG. 7.8. Relationship between pre-test Flakiness Index (I_F) and post-test Flakiness Index (RI_F).

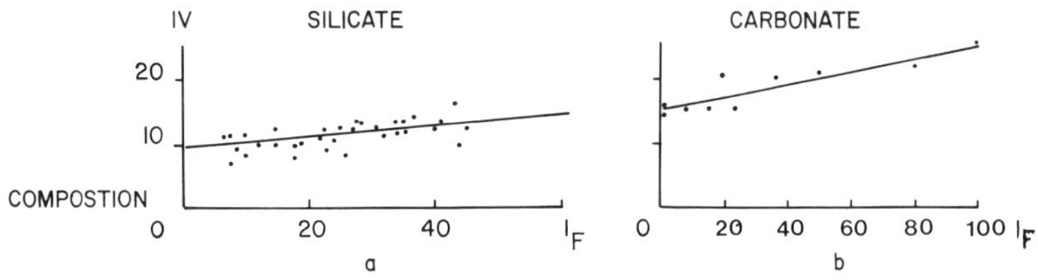

FIG. 7.9. AIV vs. I_F for silicate rocks (volcanic) (a) and carbonate rocks (b).

felsite. Values lie within a narrow band indicating similar deformational performance despite the considerable variation in chemistry-mineralogy. This implies that the principal rock-forming silicate minerals of which these rocks are composed have broadly comparable physical properties, within the sensitivity of these tests.

The geological parameter C in the above equations is the sum of several characteristics, each of which may have a major or minor influence on the deformational behaviour. The role of each of these features can be deduced from the intercept with the Y-axis (e.g. at $I_F = 0$) in Figs 7.9 to 7.11. For example if the AIV for coarse and fine-grained igneous rocks are compared there is a marked difference. In normal fresh material the variables like structure, bulk petrology and alteration etc. are absent or insignificant, so the principal cause of variation in AIV is grain size (Fig. 7.10). The effect of other features can be appreciated from Figs 7.9 and 7.11.

When values of AIV exceed 26 the sensitivity of the test may be impaired, due to the filling of pore spaces between particles with crushed fine material, produced in the course of the test. This fine material effectively buffers the remaining fragments from the impact stresses. In the deformation of such material it has been shown (Ramsay *et al.* 1977) that during the earlier stages of the test increase in the AIV results both from the granulation of the aggregate and pulverisation of already-formed coarser debris (i.e. > 2.40 mm) At a certain stage in the test the pattern of behaviour changes and AIV increased dramatically, but chiefly at the expense of the broken material in the size range 9.5 mm to 2.40 mm. The insensitivity of AIV in such cases has been countered by modifications to the tests, e.g. Ten Percent Fines Value and Modified Impact Value (7.11.4 and 7.11.5).

7.5.3.5 Ten Per Cent Fines Value. This variation of the Aggregate Crushing Value presents the load required to produce 10% fines rather than the amount of crush for a specific load. A uniform loading rate is applied to cause a total penetration of the plunger of approximately 15 mm (gravels), 20 mm (normal crushed rock), 24 mm (honeycombed aggregates) in 10 minutes. The fines less than 2.40 mm should fall within 7.5% to 12.5% of the initial weight, the force required to produce ten per cent fines $= 14x/y + 4$ where x is the maximum force (kN); y is the mean percentage fines from two tests at xkN force.

7.5.3.6 Modified Aggregate Impact test (Shergold & Hosking, 1963; Hosking & Tubey, 1969). For this test the aggregate is saturated, although surface dry. After testing the sample is oven-dried (12 hours at 100–110°C) before sieving. The test procedure is similar to that of the standard Aggregate Impact test (BS 812) except that the number of hammer blows is

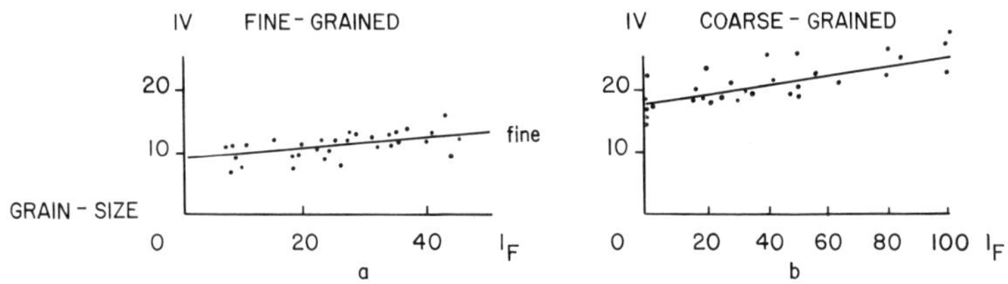

FIG. 7.10. AIV vs. I_F for fine-grained (a) and coarse-grained (b) igneous rocks.

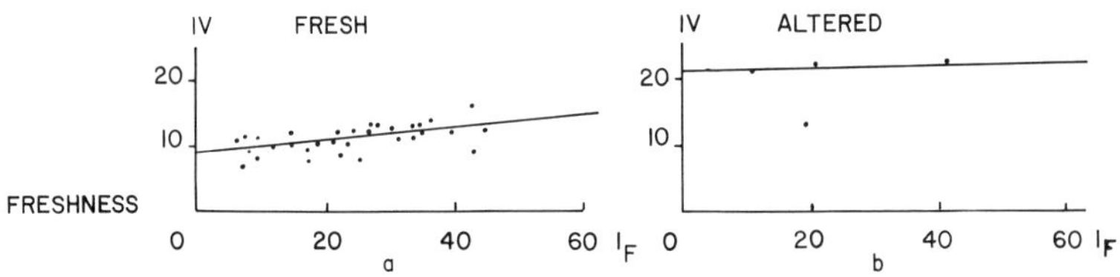

FIG. 7.11. AIV vs. I_F for fresh (a) and altered (b) volcanic rocks.

limited to that which will yield between 5% and 20% fines. The "modified AIV" is obtained by multiplying the percentage finer than 2.40 mm by 15/*X* where *X* is the actual number of blows, e.g. if 5 blows yield 10% fines:

$$\text{modified AIV} = 10 \times \frac{15}{5} = 30.$$

An alternative test for such weak rocks is the Ten Percent Fines Test on water-saturated samples.

7.5.3.7 Other non-standard values. In both AIV and ACV tests the arbitrary cut-off level of 2.40 mm as the sole index of pulverisation gives an imperfect assessment of aggregate response to the rigors of the test (Ramsay *et al.* 1974). Thus an aggregate which gets heavily broken, but not finely pulverised, can apparently compare favourably in AIV and ACV with material less heavily crushed. For example, a basalt with $I_F = 53$ gave an AIV of 16, apparently not much inferior to the average basalt with $I_F = 30$ and AIV of 13 (Ramsay, 1965). Sieving the crushed sample to obtain the plus 9.5 mm material established the proportion still remaining within the original size range. In the first basalt this residue was 13.9% as against 35% for the second. These values, Aggregate Impact Value Residue (AIVR; Ramsay, 1965) and Aggregate Crushing Value Residue (ACVR, Dhir *et al.* 1971) have been included as routine values in all subsequent testing by these authors. Both AIVR and ACVR are more sensitive to I_F than the standard values (Fig. 7.12), although the relationship is a simple linear one:

$$\text{AIVR} \quad \text{or} \quad \text{ACVR} = C - nI_F$$

where *n* ranges from 0.11 to 0.33.

When considering the significance of AIV and ACV it should be remembered that an aggregate should exhibit minimal break-up, to particles of any size. The standard tests take no account of the coarser break-down products (i.e. 9.5 mm to 2.40 mm).

7.5.3.8 Tests on ledge or lump rock

a) Franklin point load strength. The Franklin Point Load apparatus is a simple, portable device (Fig. 7.13) for obtaining an indirect measure of the compressive strength of rock cores in the field (Brock and Franklin, 1972). In can be used, however, with small irregular lumps as might be obtained from the ledge rock.

A load is applied through round-end conical plattens, the ram being driven by a small hand-operated hydraulic pump. In loading through point contacts the specimens fail in tension at a fraction of the load required in the standard laboratory compression tests. The values obtained in this test correlate well with the laboratory uniaxial compression test so that an estimated value for uniaxial strength can be easily and rapidly obtained. Used in conjunction with an index test like fracture spacing this value forms the basis of a rock quality classification (Fookes *et al.* 1971).

b) Schmidt Rebound Number (BS 1881). An experienced geologist can obtain a qualitative impression of the toughness, elasticity and state of freshness of a rock from the impact and sound of his hammer striking the rock. The Schmidt hammer is a simple quantitative extension of this test (Fig. 7.14). Although originally designed as a non-destructive test for concrete it has been applied unmodified to natural rock materials. The results obtained however, can be strongly influenced by methodological factors such as surface preparation for the plunger and the unsuspected presence of open, face-parallel cracks close behind the rock surface.

The instrument is held tightly against the rock and a spring-loaded hammer or piston, travelling through a fixed distance, applies a known energy input to the rock. The hammer rebounds from the rock, the actual amount being influenced by the elasticity of the rock. This rebound is recorded on a scale as a percentage of the forward travel. Fresh igneous rocks give values of 50 and greater. Where weathering is present and the rock is less elastic and more porous, the Schmidt hammer is sensitive to this change. A number of published studies provide examples of the use of this technique on rocks (Duncan, 1969; Geol. Soc. Rep. 1977).

An advantage of this equipment is its portability and ease of operation in the field situation, many readings being obtained in a short time. It is perhaps used to best advantage however, in combination with other tests or observations such as specific gravity, point load strength, joint spacing etc.

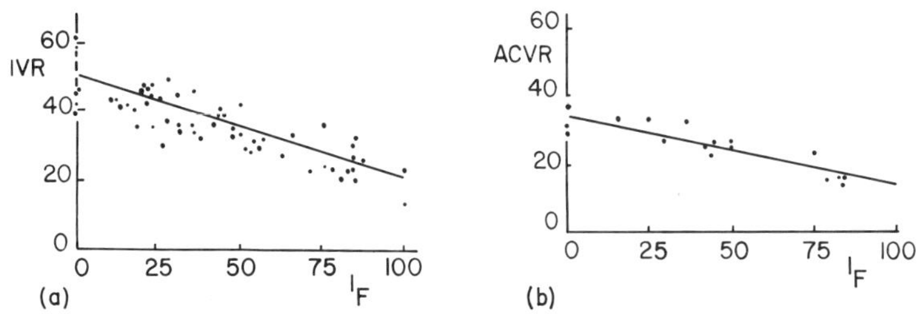

FIG. 7.12. Influence of Flakiness Index on impact value residue (a) and crushing value residue (b) for basic volcanics.

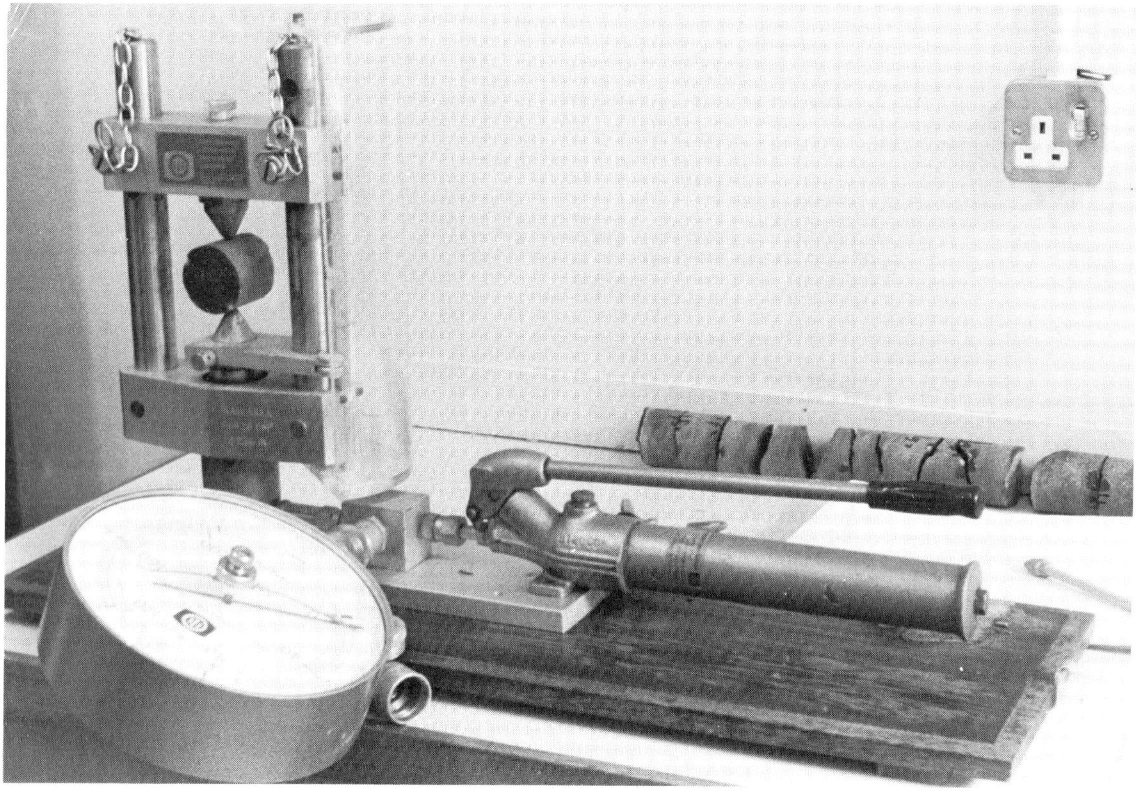

FIG. 7.13. Portable point-load testing machine.

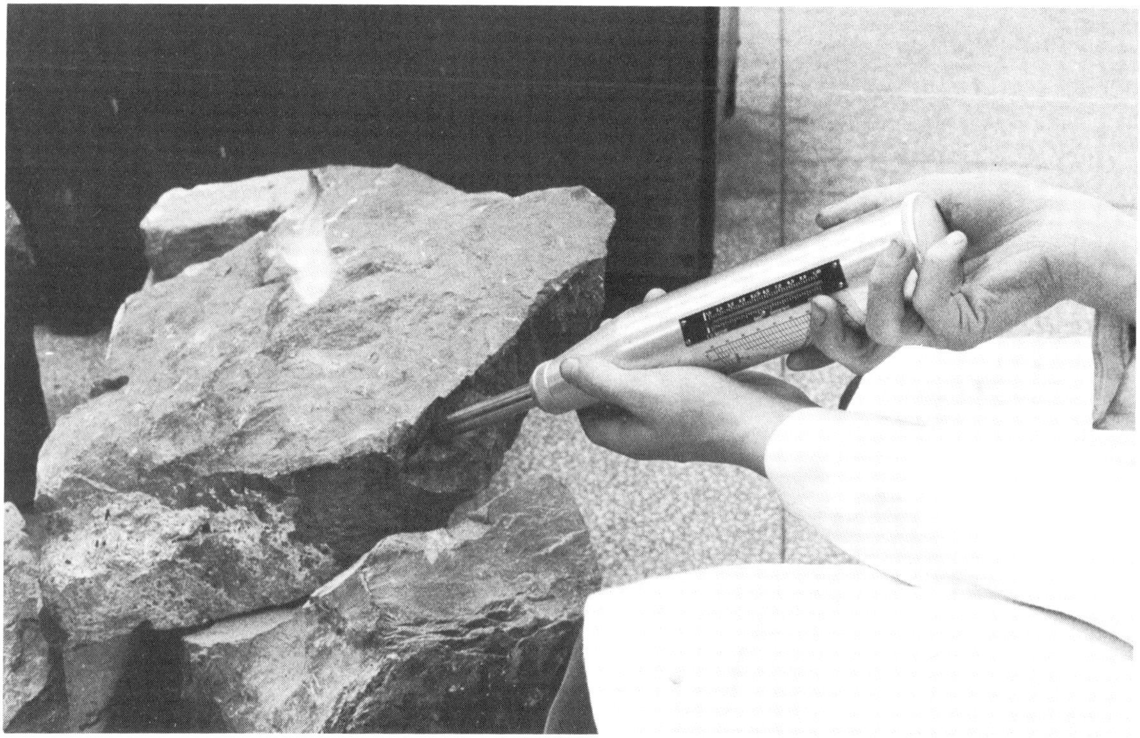

FIG. 7.14. Schmidt concrete test hammer.

7.5.4 Durability and frost susceptibility. Durability or resistance to wear or decay is a significant requirement of many aggregates. Some aggregates with satisfactory values from the standard strength tests deteriorate with time in the stockpile or 'in service' on roads. The bond between the constituent grains or particles may be intrinsically strong or weak, but frequently is weakened as a result of subaerial processes, past or present, e.g. weathering, frost etc. There are therefore two aspects of durability:

a) Mechanical deterioration or wear.
b) Physico-chemically activated deterioration, e.g. soundness.

7.5.4.1 Mechanical tests

a) Aggregate Abrasion Value (AAV). This test provides an estimation of surface wear (BS 812:1975). 33 cm^3 of clean 14 mm to 10 mm non-flaky aggregate are set into a polyester resin backing, allowing 6 mm protrusion of the aggregate (Tubey and Szafran, 1970). This sample is held against a rotating lap (28 to 30 revs. $min.^{-1}$) by a total load of 2 kg, for some 500 revolutions. Leighton Buzzard sand is fed in front of the sample at 0.7 kg to 0.9 kg min^{-1}. The loss in weight during the test is expressed as a percentage of that of the original sample, to give the Aggregate Abrasion Value. The mean of 2 tests is presented to the first decimal place. In this test a lower numerical value indicates a more resistant rock. Values range from 1 for hard flints to over 15 for aggregates which would normally be considered too weak for use in road surfacings (Hawkes & Hosking, 1972).

The AAV reflects hardness and brittleness of the constituent minerals, the influence of mineral cleavage and the strength of intergranular bond. Igneous and non-foliate metamorphic rocks yield the highest values within comparable composition ranges. Aggregate abrasion value is related in a simple and rational fashion to AIV and PSV for arenaceous rocks (Fig. 7.15) (Hawkes & Hosking, 1972).

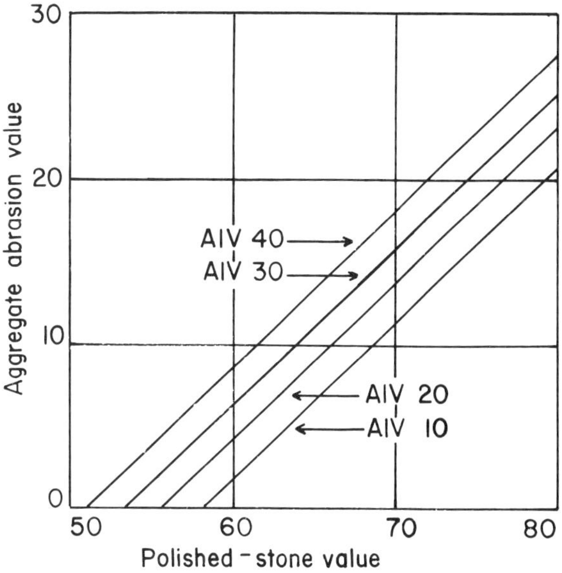

FIG. 7.15. Graph of relationship between PSV, AAV, and AIV for arenaceous roadstone. (Hawkes & Hosking, 1972).

Inadequate resistance to abrasion of road surfacing aggregates causes an early loss of the texture depth that is necessary to maintain high-speed skid resistance (Hawkes & Hosking, 1972).

b) Aggregate Attrition Value (BS 812:1943). This test, sometimes referred to as the 'Deval' test, is little used nowadays and is no longer in BS 812. Reproducibility is poorer than other aggregate tests including the Los Angeles test, which is more widely favoured (ASTM C131).

Samples are rotated in a closed, inclined cylinder resulting in mutual attrition of the aggregate. The percentage of material removed from the sample during the test is the Aggregate Attrition Value. The test is performed both dry and wet.

c) Los Angeles Abrasion Value (ASTM Designation C131 and C535). A sample charge with 6 to 12 steel balls is rotated in a steel cylinder for 500 or 1000 revolutions at 33 rpm for material less and greater than 19 mm respectively. This causes attrition through mutual tumbling and impact of the particles and the steel balls. After the test the sample is screened on 1.68 mm sieve. The coarser fraction, washed and oven dried, is weighed and the loss in weight, as a percentage of the original weight is the Los Angeles Abrasion Value. Values for this test are numerically similar to the ACV of similar sized charges for values up to about 30, i.e. for all aggregates that would normally be used in bituminous surfacing (Shergold, 1960).

d) Polished Stone Value (PSV). The Polished Stone Value was designed as a predictive measure of the susceptibility of a stone to polishing, when used in the wearing surface of a road (BS 812:1975). An accelerated polishing apparatus simulates the action of dust-laden pneumatic tyres on samples of aggregate set in a cement-mortar or polyester resin backing mounted in standard moulds on a rotating 'road wheel'. A 200 mm diameter × 50 mm broad tyre bears on the aggregate with a total load of 40 kg. Corn emery and water are continuously fed to the surface of the tyre. This is repeated using emery flour as abrasive.

The polish of the specimens is then measured using a standard pendulum arc friction tester. The deflection of a pointer on a calibrated scale yields the coefficient of friction expressed as a percentage, and this is the Polished Stone Value. A higher value signifies greater resistance to polishing.

Cox (1973) provided a perceptive review of this test, highlighting the procedural problems, and providing a perspective on numerical reproducibility in what is a relatively crude procedure.

e) Discussion. Skid resistance is provided by the roughness of a road surface and a suitable aggregate should retain a high degree of roughness in service. In the PSV test hard and tough rocks do not necessarily yield the best results as such rocks, because of their dense texture and hardness, can take on a good polish. Igneous rocks with a small proportion of soft minerals have a high polish resistance. Foliation in basalt also increases resistance to polish while fresh, fine-grained igneous rocks accept a high polish. Secondary minerals, if their hardness differs from the bulk mineral content, increase resistance. Gritstones tend to be resistant as a result of grain plucking continually renewing the rough surface. Limestone, especially with a small volume of insoluble residue, has a low resistance to polishing although this increases with the volume of the residue. Monomineralic, well-bonded quartzite, hornfels and flint take a high polish (Fig. 7.16) (Hartley, 1970, 1971).

Values over 65 indicate a highly polish-resistant rock especially suitable for road wearing courses in high-risk areas. The number of sources of material of this quality in the UK is limited (Chapter 11).

For a suite of greywackes, meta-greywackes and arkoses of precambrian to Upper Carboniferous age, Hawkes and Hosking (1972) demonstrated a broad correlation between the values obtained from the mechanical tests concerned with aggregate strength properties, e.g. crushing and impact strength and abrasion resistance, but an inverse relationship between abrasion resistance and polishing resistance (Fig. 7.15).

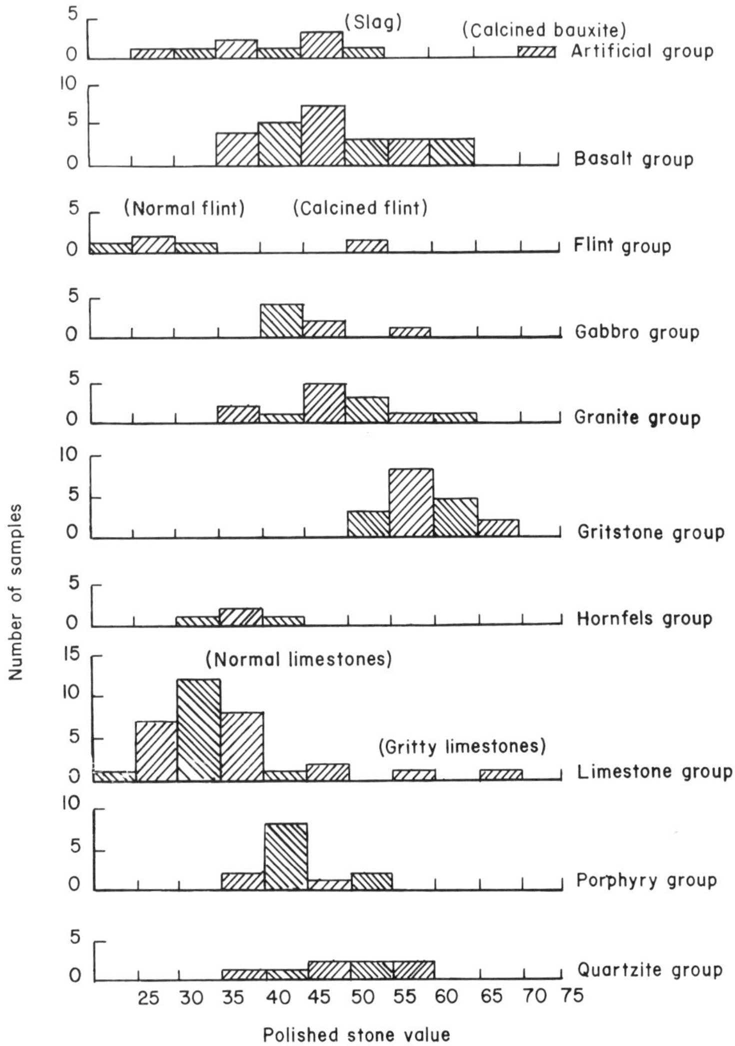

FIG. 7.16. Distribution of polished stone values in different groups of rock. (Hartley, 1970, 1971).

7.5.4.2 Physico-chemical tests

a) Soundness. This is the term used for the ability of an aggregate to resist excessive volume changes, as a result of changes in the physical environment, e.g. freeze-thaw, thermal changes at temperatures above freezing, slaking, i.e. alternate wetting and drying. The rocks most susceptible to such volume changes are dolerite, porous chert, shales and any rocks with a significant content of clay minerals, e.g. montmorillonite or illite. Used in concrete such material could lead to deterioration through scaling or more extensive cracking.

The soundness of a rock may be decreased through weathering or alteration. Weathering is a phenomenon most commonly developed on a geological time scale, but in some situations can occur in service over a period of months or years in some freshly exposed rock surfaces. The weathering is manifested by an increase in porosity and water absorption and this in turn can affect the response of the rock to the physical environment.

b) Slake Durability Index (Franklin, 1970). This is a measure of the resistance of a rock to slaking and was developed as an index test for mud rocks. A number of small samples of known weight are placed in a wire-mesh drum and immersed in water. The drum is rotated at a fixed speed for 10 minutes, causing the specimens to rub and abrade. After 10 minutes, the specimens are dried and weighed and any loss in weight is expressed as a percentage of initial weight, the Slake Durability Index.

For some rocks, e.g. compacted shales, weathered rocks etc., slaking causes internal swelling pressures or softening. The resultant weakening is accentuated by the rotation and mutual collisions. In aggregates this test is only relevant to the lowest grade materials whose weakness is self-evident and which might be employed in road sub-base material.

An alternative to the Slake Durability Value is the Washington Degradation Test. The Degradation Factor is a measure of the quantity of fines produced by rock particles when they are abraded in the wet condition. A sample is sieved into size fractions 13.20 to 9.50 mm; 9.50 to 6.70 mm; 6.70 to 4.75 mm and 4.75 to 2.00 mm, which are washed and oven-dried. Equal proportions of each fraction are combined into a representative test sample which is placed in the test canister with water. This is agitated for 20 min in a sieve shaker. After shaking the sample is washed over 2.00 mm and 75 μm sieves into a measuring cylinder. The solids in the wash water are brought into suspension by 10 rapid inversions of the cylinder and transferred to a 380.0 mm graduated tube. After further agitation the fines are allowed to settle for 20 min. The height of the sediment column is measured (H) and the Degradation Factor given by

$$D = \frac{(380 - H)}{(380 + 1.75H)} \times 100$$

c) Modal analysis. Determination of the secondary minerals created by weathering or alteration is achieved by means of point-count or modal analysis of thin sections (Chayes, 1949). Weinert (1964) established a link between the soundness of dolerite and the proportion of these secondary minerals. He found that in satisfactory dolerites less than 30% of the ferro-magnesian minerals were identifiably altered. This proved reliable for prediction of performance, save in the borderline range of 30% to 33%. This was considered adequate as a first indication of the degree of weathering.

Irfan and Dearman (1978) characterised weathering of a granite by a micropetrographic index, similar to that of Mendes (1966), calculated from the ratio of sound to unsound constituents determined from modal analysis. This index is correlatable with the widely-used Weathering Grade scheme (Geol. Soc. Rep. 1977).

d) Sulphate Soundness test (ASTM C88). This test determines the resistance of an aggregate to disintegration by saturated solutions of magnesium or sodium sulphate. The information obtained is thought to be a qualitative indicator of the soundness of an aggregate subjected to weathering action, freeze and thaw conditions, or deterioration attending salt crystallisation (Fookes & Higginbottom, 1980).

In this test samples of graded aggregate are subjected to alternate cycles of immersion in a saturated solution of sodium or magnesium sulphate and oven drying. Crystallisation in the pores may set up bursting pressures within the rock. The loss of material after 5 cycles of immersion and drying denotes the soundness. The value of this test as an index of freeze-thaw action etc. is still somewhat uncertain. British tests exist which are basically similar (Property Services Agency, 1979a and b), but the ASTM C88 Soundness Test (ASTM C88) is the one most widely chosen internationally.

In reporting the performance of a series of low grade roadstone aggregates from UK, South Africa and Mauritius, Hosking and Tubey (1969) employed sulphate soundness in conjunction with other standard and modified mechanical tests. An arbitrary limit of 20% was suggested for the sulphate test for the boundary between the 'doubtful' and 'unacceptable' categories.

For the poorer material the sulphate test may be a valid means of differentiating them into sound and unsound, but how far this is applicable is uncertain. Weinert (1964) cautioned that absolute meanings should not be attached to test values.

e) Frost Susceptibility. The T.R.R.L. test (T.R.R.L. 1981) is undertaken on 102 mm by 152 mm high specimens of aggregates compacted at or near the optimum moisture content and maximum dry density as determined by BS 5835: Part 1 1980 (Temporarily Test 14, BS 1377: 1975 will be permitted as an alternative). Nine such specimens are placed in a self refrigerated unit (SRU) (Lomas & Jones, 1981) so that the air temperature above is maintained at $-17 \pm 1°C$. The specimens rest on porous ceramic discs within copper carriers. The discs are in contact with water which is maintained at constant level and at a temperature of $+4 \pm 0.5°C$. The sides of the specimens are wrapped with waxed paper and the intervening space filled with loose dry sand (5 to 2.36 mm fraction). Push rods bearing on caps placed on top of the specimens enable the heave to be measured (Fig. 7.17). The specimens are allowed to stand in the SRU for 5 days, to allow for pore water equilibration (Sherwood, 1981), before the refrigeration is switched on. Heave measurements are then taken daily for 4 days. Usually, a test run will consist of three specimens of three different materials. Currentiy, materials are judged as frost susceptible if, during 96 hours, the average heave of three specimens is more than 10 mm (in England and Wales) or 12 mm (in Scotland). Only if doubt exists over the classification will nine specimens of the same material be tested.

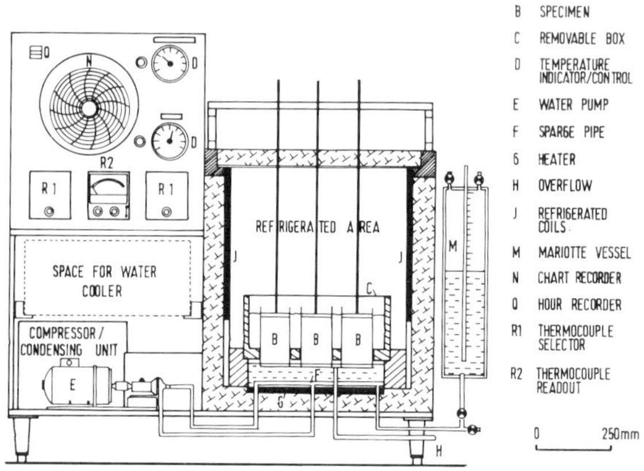

FIG. 7.17. Self refrigerated unit.

7.5.5 Chemical tests

7.5.5.1 Organic content. A variety of tests are employed for determination of the organic content of aggregates, but methodological difficulties led to its omission from BS 812. The presence of such organic material should not be overlooked however, and despite being little used in UK it is more widely employed overseas (BS 1377).

7.5.5.2 Chloride content. A test method for measuring the amount of water soluble chloride present in aggregate is given in BS 812: Part 4: 1975. The method is based on that of Volhard where an excess of standard silver nitrate solution is added to the chloride solution acidified with nitric acid, and the excess silver nitrate is back-titrated with potassium thiocyanate using ferric ammonium sulphate as an indicator.

Other methods are the Mohr direct titration method in neutral solution, using potassium chromate as an indicator; measurement of chloride concentration by specific ion electrode, or, for field work and production control work of saline aggregates, the use of test strips (Figg and Lees, 1975).

The method given in BS 812, and that of Figg and Lees is designed particularly for testing those aggregates, such as marine-dredged flint, where the chloride is a surface contamination only. With porous aggregates, or those of sedimentary or evaporite origin, where the chloride can be disseminated throughout the particles, complete water extraction may not be achieved unless the aggregates are finely ground before test.

In BS 812, the chlorides are expressed in terms of equivalent sodium chloride and reported as the chloride salt content. This mode of expression is not universally adopted however, and when considering the amount of chloride in concrete arising from all sources, there are advantages in reporting the chloride content of the aggregates in terms of the chloride ion (Cl).

The presence of chlorides in concrete can present potential hazards with some cements, e.g. by reducing sulphate resistance, and can increase the risk of corrosion of embedded metal.

7.5.5.3 Sulphate content. The method most widely adopted for the determination of sulphates is the gravimetric procedure (BS 1377). In this method sulphates are extracted with hydrochloric acid, and after removal of interfering substances if necessary, barium chloride is added to the filtered solution to precipitate the sulphate in the form of insoluble barium sulphate. The precipitate is collected on either a filter crucible or on paper, and is either oven-dried or ignited, depending on the method of collection, cooled and weighed.

Turbidimetric and nephelometric methods may be employed for measuring the precipitated barium sulphate, but the accuracy of these methods depends on more rigorous attention to analytical detail than when the gravimetric finish is adopted.

A standard method for the determination of sulphate in artificial aggregates (clinker, foamed slag, pulverised-fuel ash) is given in BS 3681 "Methods for the sampling and testing of lightweight aggregates for concrete". This method can also be used for testing mineral aggregates. Methods for determining sulphate and total sulphur contents are given in BS 1047: Part 2: 1974 "Air-cooled blast furnace slag aggregate for concrete". Methods for the determination of sulphate generally require that the test sample is finely ground. An exception could be made in the case of impervious, insoluble aggregates, such as flint gravel and silica sands. In these cases, the procedure given in BS 812: Part 4: 1975 for extracting chlorides can be used, substituting a measured volume of 10% hydrochloric acid for the water, and determining the sulphate present in an aliquot.

The sulphate compound most likely to be present in mineral aggregates is calcium sulphate (gypsum) but others do occur and must be watched for. Artificial aggregates, in particular blast furnace slag, can also contain sulphides. In slag, these are mainly calcium sulphide, although iron and manganese sulphides may be present in smaller amounts.

The presence of large amounts of soluble sulphate in concrete made with ordinary Portland cement may result in expansion and cracking (see Chapter 8). The sulphates react with the calcium aluminates of the cement to produce calcium sulphoaluminates. The accompanying increase in molecular volume creates forces within the hardened cement paste and leads to its disintegration.

The total amount of sulphate in Portland cement concrete, including that contributed by the cement, should generally not exceed 4% by weight of cement as stated in BS 5328:1976, Appendix A.

7.5.5.4 Potential alkali-reactivity. Disruptive expansion of concrete can result from reactions between certain minerals in aggregates and alkali metals (sodium and potassium) present in cement. The most widely known is the alkali-silica reaction which can occur when the aggregates contain silica as tridymite, cristobalite, opal, chalcedony or in a glassy form. Less common are alkali-carbonate reactions involving certain dolomitic carbonate rocks.

Test methods used to assess potential alkali-silica reactivity are petrographic examination (ASTM Designation C295), measurement of length changes of mortar bars (ASTM Designation C227), chemical tests (ASTM Designation C289) and gel pat tests (Jones and Tarleton, 1958). The Conrow test for potential volume change of mortar bars (ASTM Designation C342) is applicable particularly to certain aggregate-cement combinations widely used in the United States, and is not generally adopted elsewhere.

The mortar bar test gives a direct measurement of expansive reactivity, but is time consuming. Except with a few highly expansive aggregates it is necessary to continue length measurements for at least six months. For siliceous aggregates, an increase of 0.10 per cent at six months, or not less than 0.05 per cent at three months, is said to indicate deleterious expansion. However, minerals other than reactive silica, gypsum for example, can cause expansion of the specimens, and where length increases greater than those given above are recorded, it is recommended that a petrographic examination is made of the aggregate.

In the chemical test the sample is crushed to between 300 and 150 microns, then digested for 24 hours with a known volume and concentration of hot sodium hydroxide solution. After digestion the solution is analysed to determine the amount of alkali which has entered into reaction and the amount of silica dissolved from the aggregate during the reaction. Potential reactivity is assessed from the analytical results by reference to a graph given in the method which correlates test results with the known field performance of a number of aggregates. The chemical test does not give reliable indications when the aggregates contain carbonates or certain silicates and, where marginal reactivity is indicated, confirmatory petrographic examination is recommended. Mortar bar measurements are recommended to confirm chemical test results which indicate 'deleterious' or 'potentially deleterious' degrees of activity.

In the gel pat test pieces of aggregate are cast into cement paste pats, which after curing are ground to expose the aggregate surfaces. The pats are stored in sealed containers in an alkaline solution and arranged so that the aggregate faces are uppermost. The presence of reactive silica is shown by formations of white siliceous liquid or gel on the exposed surfaces. These may be apparent within seven days, but the tests should be extended to four weeks, and can be usefully continued for longer. Some laboratories have claimed success for accelerated indicative testing by storage of gel pats at elevated temperatures (Sims and Poole, 1980). The test gives a fairly rapid positive indication of the presence of reactive silica and does not require the analytical precision of the chemical method. However it does not necessarily follow that an aggregate showing reactivity in the test will be expansively reactive in concrete and it is recommended that positive results are confirmed by petrographic examination or mortar bar tests.

The potential alkali reactivity of carbonate rock aggregates may be assessed by ASTM Designation C586 in which the length change of rock specimens immersed in sodium hydroxide solution is measured. It is emphasised that the test is intended as a fairly rapid screening method, rather than a specification, and that acceptance or rejection of aggregate sources should not be based solely on the results. Aggregates which show expansion in the test should be tested in concrete.

Examination of the mineralogical composition and testing by an experienced petrographer remains the most reliable method for evaluating the potential reactivity of carbonate aggregates.

It should be noted that certain unusual reactive rock types do not always respond to the normal alkali-reactivity tests and other more sophisticated procedures have sometimes been adopted (Gillott, 1975). He emphasises the important role of initial petrographic examination.

Although the mechanisms of the various types of alkali-aggregate reactivity differ they all require a relatively high content of alkalis in the concrete and the presence of moisture. The alkali-silica reaction is attributed to the swelling of a gel, resulting in pressures which can exceed the tensile strength of the concrete and result in expansion and cracking. The expansion of concrete containing alkali-reactive carbonate rock aggregates is believed, at least in some cases, to be due to moisture absorption by clay or clay-like constituents following a dedolomitization reaction. Where susceptible minerals are present in an aggregate, disruptive expansion of concrete generally will not occur if the total alkali content of the cement used does not exceed 0.6 per cent (as equivalent Na_2O); exceptions to this have been reported, especially when alkalis are available from external sources such as sea water.

7.5.6 Adhesion tests. Adhesion tests are not confined to aggregates but are interface tests between aggregates and the bituminous binder and influenced by the properties of both. For discussion of these tests refer to Chapter 11.

7.5.6.1 Static immersion tests. Coated aggregate particles are immersed in jars of distilled water and the degree of stripping is assessed as the visually estimated percentage of uncoated surface area after periods of 1, 3, 24 and 48 hours at 25°C and 40° respectively. This test has only low reproducibility due to the human factor. A variety of static immersion tests appears in Queensland Standards Q16-1966 where the aggregate is not completely coated before immersion. In this test 50 representative pieces of the aggregate sample are pressed into a thin layer of the bituminous binder, left therein in an oven at 60°C for 24 hours and then immersed for 4 days in a water bath at 50°C. The particles are then plucked by hand from the tray and the degree of retention of binder assessed visually. In this test and the immersion tray test described below (7.5.6.6) there is a strong risk of influencing the result by the handling process i.e. dust coatings are removed to varying degrees by handling and varying amounts of oil/perspiration are deposited from the fingers of the handler. These factors contribute to the low reproducibility mentioned above.

7.5.6.2 Chemical immersion tests. As for static immersion tests but a chemical solution is used in place of distilled water, generally Na_2CO_3 in varying concentrations.

7.5.6.3 Dynamic immersion tests. Similar to the static test but agitation or stirring is involved.

7.5.6.4 Immersion mechanical tests. These follow a variety of types of tests on graded and compacted coated material. After soaking for a specified number of hours at a specified temperature the materials are subjected to tests for say compressive strength, flexural strength, tensile strength, cone penetration, abrasion resistance (e.g. modified Deval, modified Los Angeles). It has been commented (H.M.S.O. 1962) that little is known about correlation between the tests and performance in service and that with many comparatively dense and strong mixes which have been assessed by these means it is doubtful whether an important practical problem really exists.

7.5.6.5 Immersion trafficking test. An attempt to introduce the known significant factor of trafficking into adhesion testing came with the development of circular track machines and reciprocating loaded wheels (the Immersion Wheel Tracking Test, H.M.S.O. 1962). The time to failure of the specimens (continuously "trafficked" under water at 40°C) is taken as a measure of resistance to stripping. Fairly good correlation with service behaviour has been established and examples are given in the reference to show the effect of different fillers on laboratory and field resistance to stripping.

7.5.6.6 Coating tests. These differ from those above in that the aggregate is not coated initially and its resistance to stripping subsequently assessed or measured. In coating tests the ability of the aggregate to attain a coating through the medium of water is assessed. The immersion tray test (Lees, 1963) involves the placing of a layer of binder in a tray which is then covered with water. The aggregate particles are then applied by hand to the binder film, left therein for 10 minutes, then removed and the percentage of binder adhering to the face of the stone which has been pressed on the binder surface assessed visually (T.R.R.L. 1981). The test while low in reproducibility has been found useful to assess the degree of effectiveness of adding certain types and proportions of adhesion agent to standard bituminous binders.

Adhesion can also be assessed by more fundamental tests including measurement of contact angle (θ) and interfacial tension between binder and water (γ_{BW}) (Lees, 1963; Hallberg, 1950) and by direct measurement of adhesion tension ($\gamma_{BW} \cos \theta$) (Hallberg, 1950).

7.5.6.7 Discussion. In spite of all the work that has been done on adhesion testing and all the variations of test method that exist, no completely satisfactory method appears to have been developed.

Many of the tests involve handling the aggregate particles and it has been observed that the quality of adhesion varies with both the amount of dust removed and human skin oils deposited in the handling process. As stated in 7.4.4.2 the dustiness of the aggregate has a considerable adverse effect on adhesion. The relative dustiness and angularity of aggregates from different sources may cause them to appear in reverse order of merit in regard to adhesion compared with the order in which they would have been placed if those factors could have been held constant. In spite of these comments some indication can be obtained of the relative order of merit of aggregates and of the effectiveness of adhesion agents by several of the tests described.

References

American Society for Testing Materials 1954. Potential volume change of cement-aggregate combinations. ASTM Designation C342.

American Society for Testing Materials 1956. Water-soluble chlorides present as admixes in graded aggregate road mixes. ASTM Designation C1411.

American Society for Testing Materials 1959a. Specific gravity and absorption of coarse aggregate. ASTM Designation C127.

American Society for Testing Materials 1959b. Specific gravity and absorption of fine aggregate. ASTM Designation C128.

American Society for Testing Materials 1960. Unit weight of aggregate. ASTM Designation C29.

AMERICAN SOCIETY FOR TESTING MATERIALS 1963a. Abrasion of graded coarse aggregate by use of the Deval machine. ASTM Designation C289.

AMERICAN SOCIETY FOR TESTING MATERIALS 1963b. Sieve or screen analysis of fine and coarse aggregates. ASTM Designation C136.

AMERICAN SOCIETY FOR TESTING MATERIALS 1963c. Soundness of aggregates by use of sodium sulphate or magnesium sulphate. ASTM Designation C88.

AMERICAN SOCIETY FOR TESTING MATERIALS 1964a. Resistance to abrasion of small-size coarse aggregate by use of the Los Angeles machine. ASTM Designation C131.

AMERICAN SOCIETY FOR TESTING MATERIALS 1964b. Standard specifications for concrete aggregates. ASTM Designation C33.

AMERICAN SOCIETY FOR TESTING MATERIALS 1965a. Resistance to abrasion of large-size coarse aggregate by use of the Los Angeles machine. ASTM Designation C535.

AMERICAN SOCIETY FOR TESTING MATERIALS 1965b. Petrographic examination of aggregates for concrete. ASTM Designation C295.

AMERICAN SOCIETY FOR TESTING MATERIALS 1965c. Potential alkali reactivity of cement-aggregate combination (mortar bar method). ASTM Designation C227.

AMERICAN SOCIETY FOR TESTING MATERIALS 1965d. Potential reactivity of aggregates (chemical method). ASTM Designation C289.

AMERICAN SOCIETY FOR TESTING MATERIALS 1979c. Standard test method for potential reactivity of aggregates (chemical method). ASTM C289.

AMERICAN SOCIETY FOR TESTING MATERIALS 1979d. Standard test method for potential alkali reactivity of cement-aggregate combinations (mortar-bar method). ASTM C227.

AMERICAN SOCIETY FOR TESTING MATERIALS 1979e. Standard test method for potential alkali reactivity of carbonate rocks for concrete aggregate (rock cylinder method). ASTM C586.

AMERICAN SOCIETY FOR TESTING MATERIALS 1979f. RILEM Technical Committee 17 BM. Problem of the bond between bituminous binders and aggregates. Paris.

BRITISH GRANITE AND WHINSTONE FEDERATION 1966a. Sampling. Pamphlet INF/9.

BRITISH STANDARDS INSTITUTION BS 812:1975. Methods for sampling and testing of mineral aggregates and fillers.

BRITISH STANDARDS INSTITUTION BS 882:1975:1201. Aggregates from natural sources for concrete.

BRITISH STANDARDS INSTITUTION BS 1377:1975. Methods of test for soil for civil engineering purposes.

BRITISH STANDARDS INSTITUTION BS 5328:1976. Methods for specifying concrete.

BRITISH STANDARDS INSTITUTION BS 5835:1981. Alkali-aggregate reaction in Great Britain—the present position. *Concrete* **15** (3), 24–7.

BRITISH STANDARD INSTITUTION BS 5835:1980. Recommendations for testing of aggregates. Part 1 Compactability test for graded aggregates.

BROCK, E. & FRANKLIN, J. A. 1972. The point-load strength test. *Int. J. rock Mech. Ming Sci. Geomech.* **9,** 669–97.

CHAYES, F. 1949. A simple point counter for thin-section analysis. *Am. Mineralogist* **34,** 1–11.

COX, E. A. 1973. Roadstone assessment—an art or a science? *Quarry Managers J. London* **57,** 169–77.

DHIR, R. K., RAMSAY, D. M. & BALFOUR, N. 1971. A study of the Aggregate Impact and Crushing Value Tests. *J. Inst. Highway Eng.* **18,** 17–27.

DUNCAN, N. 1969. Engineering geology and rock mechanics. Vol. 1 and 2. International Textbook Co.

EDWARDS, A. G. 1963. Shrinkage of natural aggregates in concrete. *Building Res. Stn Digest* **35**.

EDWARDS, A. G. 1967. Properties of concrete made with Scottish crushed rock aggregates. *UK Build. Res. Stn.* Garston, *England. Pap.* **42**.

EDWARDS, A. G. 1970a. Scottish aggregates: rock constituents and suitability for concrete. *Building Res. Stn CP* **28/70**.

EDWARDS, A. G. 1970b. Shrinkable aggregates. *Build. Res. Stn Scottish Lab.,* East Kilbride. *SL* **1** (1) 70.

FIGG, J. W. & LEES, T. P. 1975. Field testing the chloride content of sea-dredged aggregates. *Concrete* September 1975.

FOOKES, P. G., DEARMAN, W. & FRANKLIN, J. A. 1971. Some engineering aspects of rock weathering with field examples from Dartmoor and elsewhere. *Q. J. eng. Geol. London* **4,** 139–85.

FOOKES, P. G. & HIGGINBOTTOM, I. E. 1980. Some problems of construction aggregates in desert areas, with particular reference to the Arabian peninsula. 2: Investigation, production and quality control. *Proc. Instn Civ. Engrs* **68,** 69–90.

FRANKLIN, J. A. 1970. Observations and tests for engineering description and mapping of rocks. *Proc. Second. Cong. Int. Soc. for Rock Mechs. Belgrade,* Vol **1,** paper 1–3.

GEOLOGICAL SOCIETY ENGINEERING GROUP REPORT 1970. The logging of rock cores for engineering purposes. *Q. J. eng. Geol. London* **3,** 1–24.

GEOLOGICAL SOCIETY ENGINEERING GROUP REPORT 1977. The description of rock masses for engineering purposes. Report by the Geological Soc. Eng. Gp. Working Party. *Q. J. eng. Geol. London* **10,** 355–88.

GILLOTT, J. E. 1975. Alkali-aggregate reactions in concrete, Review. *Eng. Geol.* **9,** 303–326.

GOODSALL, G. D. & MATHEWS, D. H. 1970. Sampling of road surfacing materials. *J. Appl. Chem.* 361–6.

GRANT, E. L. 1952. Statistical quality control. McGraw-Hill Inc. New York. 557 pp.

GRIFFITHS, J. C. 1967. Scientific method in analysis of sediments. McGraw-Hill Book Co. New York. 508 p.

HAWKES, J. R. & HOSKING, J. R. 1972. British arenaceous rocks for skid-resistant road surfacings. Report LR 488, Road Research Laboratory, Crowthorne.

HOSKING, J. R. & TUBEY, L. W. 1969. Research on low grade and unsound aggregates. Report LR 293, Road Research Laboratory, Crowthorne.

HARR, M. E. 1977. Mechanics of particulate media. McGraw-Hill Inc. New York, 543 p.

HARTLEY, A. 1970. The influence of geological factors upon the mechanical properties of road surfacing aggregates (with particular reference to British conditions and practice). proc. 21st. Sympos. Highway Geol. Univ. of Kansas.

HARTLEY, A. 1971. The development and testing of artificial gritty aggregate. Proc. Sympos., The influence of the road surface on skidding. University of Salford.

HURLBUT, C. S. & KLEIN, C. 1977. Manual of mineralogy. J. Wiley & Sons, New York 532 p.

IRFAN, T. Y. & DEARMIN, W. R. 1978. The engineering petrography of a weathered granite in Cornwall, England. *Q. J. eng. Geol. London* **11,** 233–44.

JONES, F. E. & TARLETON, R. D. 1958. National Building Studies Research Paper No. 25. Reactions between aggregates and cement H.M.S.O., London.

KRUMBEIN, W. C. 1941. Measurement and geological significance of shape and roundness of sedimentary particles. *J. Sediment. Petrol.* **11,** 64–72.

KRUMBEIN, W. C. & PETTIJOHN, E. J. 1938. Manual of sedimentary petrography. Appleton-Century Crofts Inc. New York 549 pp.

LEES, G. 1963. A new method for determining the angularity of particles. *Sedimentology* **3,** 2–21.

LEES, G. 1964. The measurement of particle shape and its influence in engineering materials. *J. Brit. Granite and Whinstone Fed.* **4,** 1–22.

LOMAS, K. J. & JONES, R. H. 1981. An evaluation of a self refrigerated unit for frost heave testing. *Trans. Res. Rec.* **809,** 6–13.

MENDES, F., ARIES-BARROS, L. & PERES RODRIGUEZ, F. 1966. The use of modal analysis in the mechanical character of rock masses. *Proc. 1st Cong. Int. Soc. Rock Mech., Lisbon,* **1,** 217–23.

NORTHERN MINERAL PRESS LTD. 1955. Mining explained in simple terms. Toronto.

PALMER, D. 1981. Alkali-aggregate reaction in Great Britain—the present position. *Concrete* **15** (3), 24–7.

PIKE, D. C. 1978. Variability in grading results caused by standard sample-reduction techniques for aggregates. Supplementary Report SR 489. Tpt & Rd. Res. Lab. England.

PROPERTY SERVICES AGENCY 1979a. Standard specification clauses for airfield pavement and works, Pt. 4 Bituminous surfacing Appendix 4B. Soundness of aggregate. Property Services Agency, Department of the Environment.

PROPERTY SERVICES AGENCY 1979b. Standard specification clauses for airfield pavement and works. Pt. 3, Concrete. Appendix 3A. Property Services Agency, Department of the Environment.

RAMSAY, D. M. 1965. Factors influencing Aggregate Impact Value in rock aggregate. *Quarry Managers J. London* **49,** 129–34.

RAMSAY, DHIR, R. K. & SPENCE, J. M. 1973. Non-geological factors influencing the reproducibility of results in the Aggregate Impact Test. *Quarry Managers J. London* **57,** 179–81.

RAMSAY, DHIR, R. K. & SPENCE, J. M. 1974. The role of rock and clast fabric in the physical performance of crushed-rock aggregate. *Eng. Geol.* **8,** 267–85.

RAMSAY, DHIR, R. K. & SPENCE 1977. The practical and theoretical merits of the Aggregate Impact Value in the study of crushed rock aggregate. Proc. Conf. on Rock Engineering, Newcastle upon Tyne, England.

ROAD RESEARCH LABORATORY, 1959. Roadstone tests data presented in tabular form. Road Note 24 H.M.S.O., London.

ROSSLEIN, D. 1941. Steinbrecheruntersuchungen unter besonderer Berucksichtigung der kornform. Forsch a.d. Strassenwesen Band 32, Berlin.

SHERGOLD, F. A. 1950. The classification and mechanical testing of road making aggregates. *Quarry Managers J. London* **34,** 27–35.

SHERGOLD, F. A. 1959. A study of the granulators used in the production of road making aggregates. Road Research Tech. Paper No. 44.

SHERGOLD, F. A. 1960. The classification, production and testing of road making aggregates. *Quarry Managers J. London* **44,** 47–54.

SHERGOLD, F. A. 1963. A study of the variability of roadstones in relation to sampling procedures. *Quarry Managers J. London* **47,** 3–8.

SHERGOLD, F. A. & HOSKING, J. R. 1963. Investigations of test procedures for argillaceous and gritty rocks in relation to breakdown under traffic source. *Roads and Road Construction,* **41,** 376–8.

SHERGOLD, F. A. & GREYSMITH, M. G. 1947. Factors governing the grading and shape of crushed rocks—a survey of the literature. *Quarry Managers J. London* **30,** 703–12.

SHERWOOD, P. T. 1981. Research at T.R.R.L. on the Frost-Susceptibility of Roadmaking Materials. Proc. Symposium on Unbound Aggregates in Roads. Dept. of Civil Eng., Univ. of Nottingham, 151–160.

SIMS, I. & POOLE, A. B. 1980. Potentially alkali-reactive aggregates for the Middle East. *Concrete* **14** (5) 27–30.

TRANSPORT & ROAD RESEARCH LABORATORY 1981. Specification for the T.R.R.L. frost heave test. Materials Memo. No. MM/64, Materials Division, T.R.R.L. Crowthorne.

TUBEY, L. W. & SZAFRAN, W. 1970. Aggregate Abrasion Test: an improved technique for preparing test specimens. *Quarry Managers J. London* **54,** 185–7.

WADELL, H. 1933. Sphericity and roundness of rock particles. *J. Geol.,* **41,** 310–31.

WEINERT, H. H. 1964. Basic igneous rocks in road foundations. *C.S.I.R. Research Report* **218,** 1–47.

WOODRUFF, S. D. 1966. Methods of working coal and metal mines Vol. 1. Pergamon Press, Oxford 538 pp.

WRIGHT, P. J. F. 1955. A method of measuring the surface texture of aggregate. *Mag. Concrete Research,* **7,** 151–60.

ZINGG, T. 1935. Beitrag zur schotteranalyse. *Schweiz. mineral. Petrog. Mitt.,* **15,** 39–140.

Additional Recommended References

LAWSON, J. E. & DELTON, J. P. 1978. Sampling and evaluating potential aggregate deposits. Ann. Convt. of National Sand and Gravel Assoc. Las Vegas, Nevada. 1–8.

LEES, G. 1969. The influence of binder and aggregate in bituminous mixes. *J. Inst. Highway Eng.* **16,** 7–19.

LEES & KENNEDY, C. K. 1975. Quality, shape and degradation of aggregates. *Q. J. eng. Geol. London* **8,** 193–209.

PERONIUS, A. N. 1979. Statistical applications for road soil surveys. *J. Inst. Highway Eng.* **26,** 15–18.

RITTENHOUSE, G. 1943. The relation of shape to the passage of grains through sieves. *Ind. Eng. Chem. Anal. Ed.* **15,** 153–5.

SHERGOLD, F. A. 1953. The percentage voids in compacted gravel as a measure of its angularity. *Mag. Concrete Res.* **5,** 3–11.

SHERGOLD, F. A. 1955. Results of tests on single-sized road-making aggregates. *Quarry Managers J. London* **39,** 2–8.

SHERWOOD, P. T. & RYLEY, M. D. 1970. Sulphates in colliery shale: effect on cemented materials. *Surveyor* **1–7**.

SPENCE, J. M., RAMSAY, D. M. & DHIR, K. R. 1974. A conspectus of aggregate strength and the relevance of this factor as the basis for a physical classification of crushed rock aggregate. *Int. Cong. Rock Mechs.* **2A**, 79–84.

TUBEY, L. W. & JORDAN, P. G. 1973. The reproducibility and repeatability of the PSV determination. TRRL Report LR 552.

General Reference

RILEM Technical Committee 17 BM, 1979. Problem of the bond between bituminous binders and aggregates. Paris.

Chapter 8. Aggregates for concrete

8.1 Introduction

The desirable and undesirable properties of aggregates for concrete have been thoroughly reviewed in a number of texts on concrete technology (Orchard, 1976, American Society for Testing and Materials, 1978) Murdock & Brook, 1979, Fookes, 1980, Neville, 1981). In practice, difficulties are frequently encountered in translating these properties into specification requirements for aggregates, or in assessing aggregate test results to determine compliance or otherwise with already specified parameters. This chapter, therefore, will consider the principal properties of concrete aggregates with the aim of assisting in the selection of appropriate specification requirements.

The essential requirements of aggregates for concrete are that they remain stable within a concrete throughout its design life and in the particular environment involved, and that their characteristics do not affect adversely the performance or cost of the concrete in either the fresh or hardened states.

For both technical and contractual reasons, these requirements have to be defined quantitatively. This involves the selection of relevant tests and assessment procedures and the specification of appropriate acceptance criteria. However, despite the guidance given in Chapter 7, it can be difficult to assure the long-term satisfactory performance of an aggregate combination solely by means of a number of British or other standard tests and limits. In such cases there is a need for experienced judgement, but the specifier's aim should always be to minimise this need. For example, contract specifications often include the generalised requirements that aggregates "... be hard, durable and clean, and of approved quality". These are qualitative requirements, without specific acceptance limits, and which therefore necessitate subjective judgments sometimes leading, in turn, to contractual disputes.

It is important to avoid specifying aggregate qualities which are contradictory; thus creating a potential for dispute, or which are not relevant or are unnecessarily superior to the needs of the particular works, thereby restricting use to unduly expensive materials. Thus, the specifier bears considerable responsibilities in these respects under a contract, hence the need for continued vigilance despite the rare incidence of arbitrations and litigations resulting from disputes about aggregate quality.

In the United Kingdom, the identification of any potential aggregate problems is primarily the concern of aggregate producers, upon whom specifiers and others traditionally have tended to rely. This reliance is generally justified, in view of the well established characteristics for most UK aggregate sources and the long-standing of the UK aggregate production industry, but should nevertheless be confirmed by independent testing and periodic rechecks.

However, in many locations overseas this situation does not obtain, when extensive aggregate source and product investigations may be appropriate. Another important factor overseas, but often less well recognised, is that National Standards and Codes of Practice inevitably relate to the climatic regimes and other influences which prevail in the country of origin. For this reason Standards and Codes should be studied with some care before their recommendations are specified for works in climatic regimes appreciably different from those of the country in which they were published.

8.2 Specification

The specification requirements for aggregates should, ideally, be considered separately for each project, in order that the locally available materials can be used to maximum advantage at the least cost. In British practice, specifications commonly only require compliance with the requirements of BS 882 which has recently been substantially revised (BS 882:1983). This Standard Specification now contains essentially only three quantitative compliance requirements: grading, cleanliness and a mechanical property. The all-embracing but qualitative "general requirements" clause in the previous edition (BS 882:1973) has been omitted from the new edition.

The widespread use of this long-established British Standard presupposes a substantial weight of experience to justify the view that this limited number of requirements has proved adequate for most concrete construction work in the United Kingdom. However, experience in recent years has shown that problems can arise on overseas contracts when specifications are based solely upon BS 882 (Fookes & Collis 1975a, b). For example, an unsound aggregate which is mechanically strong in the dry state may comply with this specification, but could result in concrete failure in certain environments.

In contrast with the limited number of quantitative requirements given in BS 882, the related British Standard covering the testing of aggregates, BS 812, provides test methods for a wide range of properties. However, BS 812 does not provide any guidance for interpretation of the test results. Hence there is a need to appreciate the relevance of the various tests and to select appropriate tests and limits on the basis of experience of use in particular contract locations and environments; for example, aggregate shrinkage testing and Scottish aggregates (Building Research Establishment, 1968; Edwards, 1970).

The specification of concrete aggregates for overseas projects, especially where there are no local Standards or Codes, necessitates considerable care in order to impose appropriate requirements to minimise the risk of under- or over-specification with the resultant risk of claims and cost difficulties. When preparing specifications for such works it is essential to acquire detailed up-to-date knowledge about current aggregate resources, including their potentials for adverse variations, and also about their mineralogical, physical and chemical characteristics. This effort should enable the majority of undesirable features of the local aggregates to be identified. If it is considered that such features can be eliminated or reduced, by measures which are both practical and economic, then appropriate qualitative requirements, test methods and quantitative limits must be selected and specified. In this way the contractor can make any necessary allowances for the cost of implementing such measures and thereby meeting the specified requirements.

A good aggregate specification must be relevant to the available resources and to the characteristics required of the concrete in which the aggregates will be used. Sufficient flexibility for judgement should be provided, in order to prevent the unintended exclusion of adequate material, or the need to employ unnecessarily superior and usually more costly material. However, as explained earlier, qualitative requirements necessitating subjective judgements should be avoided whenever possible.

8.3 Classification

The provisions in BS 882 do not require a supplier to furnish a petrological classification unless it is requested by the specifier or purchaser. In practice, this information is only rarely requested. However, even when provided, the brevity of the BS 812 classification provides little information of real value, either for engineering or materials control purposes. This aspect is discussed fully in Chapter 6.

Detailed information on the petrography of an aggregate can be of considerable value when assessing its potential suitability in terms of rock type and the state of geological alteration or sub-aerial weathering. Also, simple petrographic examination is particularly useful for monitoring the variability of aggregates during the course of construction. Petrographic examinations can identify, fairly rapidly and relatively inexpensively, the presence of detrimental features or potentially deleterious materials, which may be present either naturally or as contaminants (Mielenz, 1955). These examinations should ideally be carried out no later than the Tender stage and, although the results are normally only semi-quantitative, they can indicate whether further investigations and quantitative tests are desirable.

A further benefit of petrographic analyses is that, in conjunction with other physical and chemical tests, they can help to "fingerprint" an aggregate or parent rock (Fookes & Collis, 1975b). This can be of value as a means of defining an approved material and can also be helpful subsequently should any disputes arise as to the nature of the originally approved material.

8.4 Grading

The overall grading of combined coarse and fine aggregates should provide a material having the lowest possible surface area and void content per unit volume, thus requiring the lowest possible water and cement contents to achieve the desired workability, strength and other concrete properties. The extent to which these aims can be achieved is dependent *inter alia*, upon the actual grading of the available coarse and the aggregates and, most importantly, upon their uniformity within and between deliveries to site. The use of appropriately proportioned coarse and fine aggregate gradings, each within the relevant envelopes given for example in BS 882, normally provides concretes having adequate properties in the fresh and hardened states.

Although infrequently appreciated by aggregate users, the four fine aggregate grading 'zones' given in earlier editions of BS 882 (1954, 1965 & 1973) simply represented a classification of fine aggregate gradings readily available in the UK, not envelopes based on experience or research (Foreword to BS 882: 1954). In the current edition (BS 882:1983), the overall fine aggregate grading limits have been widened, on the evidence of an exhaustive survey (BSI News, 1983), to permit use of a finer material and three overlapping sets of additional grading limits (termed coarse, medium and fine) have replaced the previous four 'zones'. The BS 882:1983 fine aggregate grading limits are given in Table 8.1.

The overlapping character of the grading limits frequently allows a fine aggregate to comply with two of the three categories (coarse and medium, or fine and medium), which may be found to be confusing. However, satisfactory materials are less likely to be needlessly excluded for failing to comply with an artificially rigid grading envelope than was sometimes the case previously with the four zone system. Moreover, BS 882:1983 also permits the use of fine aggregates which do not comply with the grading limits

TABLE 8.1. *BS 882:1983 fine aggregate grading limits*

Sieve size	Percentage by mass passing BS Sieve			
	Overall limits	Additional limits for grading		
		C	*M*	*F*
10.00 mm	100	—	—	—
5.00 mm	89–100	—	—	—
2.36 mm	60–100	60–100	65–100	80–100
1.18 mm	30–100	30–90	45–100	70–100
600 μm	15–100	15–54	25–80	55–100
300 μm	5–70	5–40	5–48	5–70
150 μm	0–15*	—	—	—

* Increased to 20% for crushed rock fines, except when they are used for heavy duty floors.

provided that "the supplier can satisfy the purchaser that such materials can produce concrete of the required quality".

However, it should be understood that more care is generally necessary when designing concrete mixes using either very coarse or very fine sands. The important consideration is that the grading of a sand is suitable for the aggregate blend in a particular concrete and that repeated gradings for the same source do not vary excessively. In respect of the latter considerations, BS 882:1983 requires that not more than one in ten consecutive samples should fall outside the specified grading limits.

In the case of very coarse sands, the concrete mixes can tend to be 'harsh' and to suffer 'bleeding' (for concrete terminology, see Barker 1983), especially at lower cement contents. Very fine sands instead present a conflict between the amounts required to ensure that the concrete is sufficiently cohesive and the need to minimise the effect on water demand of the high aggregate surface area (Gaynor, 1964). However, these problems can be reduced or even eliminated by careful design of the concrete mixes, also by the judicious selection and use of air entrainment for coarse sands and water-reducing admixtures for fine sands (Concrete Society, 1980; Higgins, 1984).

In certain areas of the United Kingdom, for example South Wales and the South East Channel Coast, the gradings of the locally available natural fine aggregates may not lie within specified 'zones' or grading limits, although being acceptable in all other relevant aspects and on the basis of their long-term use in the area. In such situations it is usually possible to design concrete mixes to obtain adequate strength and durability potentials.

Where concrete of particularly high strength, uniformity or aesthetic quality are required, it is usually beneficial to specify that single-sized coarse aggregates are used; further that they are separately batched to provide continuous or non-continuous overall gradings as appropriate. In general, continuously graded aggregates tend to provide relatively straightforward and easy-to-place concretes, whereas gap-graded concretes usually require appreciably more expertise in mix design, production, handling and placing (Shacklock, 1959; Schaffler, 1979). For these reasons the labour costs of placing gap-graded concretes may sometimes be greater, but this may be justified where their special properties can be used to advantage; for example, where high mechanical stability is required immediately after compaction, or increased resistance to scour is required in inter-tidal work.

Coarse aggregates of nominal maximum size greater than 40 mm are sometimes specified for massive placements, the aims usually being to minimise the cement content and to reduce heat generation. To specify, for example a 75 mm or 150 mm maximum size, presumes that the physical characteristics of the larger size fractions are not inferior to those of the 40 mm and smaller sizes, in which case small but possibly significant reductions in cement content and cost may be achieved, but the possibility of increased risks of segregation should not be overlooked. However, certain rock types may produce crushed aggregates whose shape characteristics deteriorate with increasing particle size, and this could negate the technical and cost benefits that might otherwise be obtained. Such an effect will occur with rock materials that normally benefit from an improvement in particle shape during secondary and tertiary crushing to produce smaller aggregate sizes, or with some thickly-bedded or coarsely-cleaved rock sources. Thus, some investigation into the likely shape characteristics should be undertaken before specifying exceptionally large nominal maximum sized aggregates for mass concrete placements (Bloem, 1959).

In the United Kingdom, scant attention tends to be given to the question of gradings, probably because of the abundant supplies of materials whose gradings have not given cause for any general concern. However, this situation does not always apply overseas, and the specifier should investigate, *inter alia,* the grading characteristics of potential aggregate resources, in order to minimise the risk of imposing impracticable and uneconomic specification requirements.

The selection of aggregate types and gradings for architectural concrete involves special considerations in view of the aesthetic requirements involved, also those of aggregate and concrete durability, and reference should be made to specialist literature on this subject (Greater London Council, 1962; Blake, 1967).

8.5 Particle shape

The shape characteristics of aggregate particles are classified in qualitative terms in BS 812, but this Standard only provides test methods for quantifying the proportions of 'flaky' or 'elongated' particles in

TABLE 8.2. *BS 882:1983 flakiness index limits for coarse aggregate*

	Limits on flakiness index	
Grade of concrete*	Uncrushed gravel	Crushed rock or crushed gravel
C7.5 to C15	no requirement	no requirement
C20 to C35	50	35
over C35	35	35

* See BS 5328 for definition of grade of concrete.

coarse aggregates within a size range from 6.3 mm to 63 mm. BS 882:1983 provides limits in respect of the percentages of 'flaky' particles (the 'flakiness index') in coarse aggregates, and these vary according to the crushed or uncrushed nature of the aggregate and the grade (based upon strength) of the concrete. These limits are given in Table 8.2.

No limits are provided in BS 882:1983 for the percentage of 'elongated' particles and, indeed, limits are rarely specified for the 'elongation index'. Flakiness and elongation are not usually specified for fine aggregates, although these characteristics are easily determined qualitatively by visual examination under a low-power microscope.

Experience suggests that the widespread absence, until recently, of quantitative limits on the shape characteristics of concrete aggregates generally has not resulted in any undue difficulties with structural concretes in the UK. The probable reasons for this include the natural adequacy in this respect of the majority of UK gravels, and also the success with which aggregate producers have been able to control the shape characteristics of those materials which have greater tendencies on crushing to produce flaky and/or elongated particles.

Nevertheless, the shape characteristics of coarse and fine aggregates can have marked effects on the properties of both fresh and hardened concrete (Lees, 1964), these effects tending to be beneficial where the predominant particle shape is generally equi-dimensional, or detrimental where it is flaky and/or elongate (Kaplan, 1958, 1959; Neville, 1981).

Flakiness of coarse aggregate can have an adverse influence on the workability of concrete. In one set of experiments, with other concrete parameters held constant, an increase in 'angularity number' (see Chapter 7) from 1 to 10 caused a 10% reduction in the compacting factor (Kaplan, 1958). The strength of concrete also tends to be reduced by increasing aggregate flakiness, with flexural strength being more affected than compressive strength (Kaplan, 1959). Flakiness in fine aggregate also affects workability and increases the water demand of the concrete mix, which can lead to problems of 'bleeding' (Barker, 1983) and constituent segregation in the fresh mix and reduced strength and durability of the resultant hardened concrete.

Particle shape limits for concrete aggregates can be a major consideration when developing and designing aggregate production plant for new sources. In the case of overseas works, the producer or the specifiers should seek to select target maxima for flaky and elongated particles in coarse aggregates. The cost-benefit of complying with relatively low target maxima is largely a matter for investigation by the aggregate producer and, for overseas work particularly, by the consulting engineer no later than the Tender review stage.

Improvement of aggregate particle shape usually increases production costs, but may result in an enhancement of the durability potential of the concrete, possibly at a lower overall cost.

8.6 Particle surface texture

The main influence of particle surface texture is the effect on bond between the aggregate and the cement paste in hardened concrete. The classification in BS 812 is generalised; it is based on subjective assessment and does not therefore lend itself to specified limitations. Indeed, the specific exclusion of material of any particular surface texture is virtually unknown.

Surface texture is generally only considered in relation to concrete flexural strengths, which are frequently found to reduce with increasing particle smoothness. However, surface texture can adversely affect compressive strength in high-strength concretes (say $>60\ N/mm^2$), when the bond with the cement matrix may not be sufficiently strong to enable the maximum strength of the concrete to be realised (Kaplan, 1959; Neville, 1981).

8.7 Clay, silt and dust

BS 812 defines 'clay, silt and dust' in aggregates as being all the material passing a 75 μm sieve. In practice, however, the nature of the $-75\ \mu$m material, if any, can have a considerable influence upon the properties of concrete. The relative proportions of clay-sized ($-2\ \mu$m) and silt-sized ($+2\ \mu$m) particles are important, and there will also be a significant difference between the influence of clay-sized material comprising 'rock flour' and that of clay-sized material actually comprising water-active clay minerals (9.5.1).

BS 882 currently limits the amounts of 'clay, silt and dust' material passing the 75 μm sieve and these limits are given in Table 8.3.

Coarse aggregates and natural sands produced in the UK only rarely fail to meet these limits, whereas compliance with the crushed rock fines limit tends to be less well assured. This is due sometimes to the

TABLE 8.3. *BS 882:1983 clay, silt and dust limits*

Aggregate type	Quantity of clay, silt and dust (max. % by mass)
Uncrushed, partially crushed or crushed gravel	1
Crushed rock	3
Uncrushed or partially crushed sand or crushed gravel fines	3
Crushed rock fines	15 (8 for use in heavy duty floor finishes)
Gravel all-in aggregate	2
Crushed rock all-in aggregate	10

TABLE 8.4. *Grading comparisons for crushed rock fines*

Sieve size	% by weight passing		
	A	*B*	*C*
10.00 mm	100	100	100
5.00 mm	99	100	100
2.36 mm	88	89	86
1.18 mm	59	64	57
600 μm	39	45	34
300 μm	26	32	18
150 μm	17	23	7
75 μm	11	17	0

where:
A—Apparent overall grading determined in accordance with BS 812: Part 1: 1975, Clause 7.1.4.
B—True overall grading determined and calculated, in accordance with BS 812: Part 1: 1975, Clause 7.1.6.
C—Grading of material excluding −75 μm fraction determined in accordance with BS 812: Part 1: 1975, Clause 7.1.6.

natural tendency of −75 μm material in crushed rock fines to segregate during production and handling and, more frequently, to the fact that such fine aggregates are often an uncontrolled by-product. 'Clays' and 'silts' tend to predominate in the fines of natural (uncrushed) coarse and fine aggregates, whilst relatively coarser 'dust of fracture' is usually predominant in crushed rock aggregates.

The principal effects which these fine materials have on concrete are to increase the water demand of a concrete to achieve a given workability and, where present as coatings which are not removed during mixing, to impair the aggregate-matrix bond (Neville, 1981). The higher limit given for crushed rock fines in BS 882 reflects the typically coarser character of these fines. A higher limit for crushed rock fines ('manufactured sand') is also to be found in the American Specification for concrete aggregates (ASTM C33), which limits the maximum to 7% (or 5% if the concrete is subject to abrasion), provided the fine material consists of the 'dust of fracture' with clay or shale being essentially absent. However, the much lower ASTM limits, in comparison with the BS 882 limits, are noteworthy and many limestones and softer or weathered rocks are likely to fail these requirements, unless they have been subjected to additional processing, which may not be possible for economic or other reasons. Even after extensive processing, certain types of rock, especially limestones, will generate further quantities of dust during subsequent handling and transportation.

Maximum water/cement ratios and/or minimum cement contents are the criteria normally specified to enhance concrete durability. However, the product of these, the water content, is the more important but, usually, less well appreciated parameter. Therefore, any component or feature of an aggregate which increases the water requirement of concrete can have an adverse effect on the durability potential of that concrete. Hence, the more stringent ASTM limits for crushed rock fines may be considered preferable where the durability requirements of the concrete are paramount, also for some construction overseas where the composition or batch-to-batch variability of the fines may be greater than those which prevail in similar types of aggregate from UK sources.

The presence of substantial amounts of material passing the 75 μm sieve can result in misleading overall gradings when the sample is subject to routine sieve analysis in the dry condition. This is particularly so for crushed rock fines, as is illustrated by the data given in Table 8.4. This example demonstrates that a routine dry sieve analysis can substantially underestimate the amount of material passing the 75 μm sieve, and also indicate a markedly coarser grading than is actually the case. Where such an effect is not appreciated errors can be made when assessing the nature and yield of both natural and crushed resources, also in preliminary mix designs, decisions regarding admixtures, in estimates of material demands, and in fines disposal.

Sometimes, higher amounts of fine material can be less harmful or even beneficial. For example, where the concrete strength and durability requirements are less critical and a particularly cohesive concrete is required, or perhaps when the cement used is exceptionally coarse. A high fines content in coarse aggregate sometimes may help to improve workability when the local natural sand and/or crusher-run sand are coarse and 'sharp'. In the case of some fine dune sands, the roundness and smoothness of the particles may help to compensate, in terms of workability, for the fineness (Fookes & Higginbottom, 1980). These examples illustrate the general importance of flexibility in approach when considering limits for particular situations.

Occasionally, and particularly overseas, it is necessary to consider whether British Standard or other limits could be relaxed for practical or economic reasons. In these circumstances comparative concrete trial mixes are essential, but, as explained earlier, it is also helpful to identify the composition of the fine material involved. This identification is desirable because BS 882 effectively defines 'clay, silt and dust' as being all material finer than 75 μm, whereas in soils terminology (BS 1377:1975) 'clay' and 'silt' materials are separated on the basis of particle size (clay being less than 2 μm and silt less than 63 μm) and also these materials can be characterised and defined mineralogically. Thus the −75 μm component of a fine aggregate may be found to be entirely clay-grade material consisting of clay minerals, when the BS 882 maximum limit of 3% (for crushed rock coarse aggregate or natural sand) might be considered too high. Conversely the −75 μm component might be entirely silt-grade material, when a higher maximum limit might be tolerated.

Where 'clay, silt and dust' are present, they may occur partially as particle coatings and this can have an adverse effect on aggregate-matrix bond, especially when present as clay-grade material. Clay may also occur in the form of discrete particles or 'clay lumps' which, due to normal dust coatings, are often visually indistinguishable in the field from the aggregate particles themselves. However, their presence normally becomes apparent if the aggregate is wetted or washed, when partial or total disintegration may occur, although some clay lumps may be hard enough to survive the concrete mixing process. The presence of clay lumps in an aggregate could, for example, affect adversely the soundness test results, the local effectiveness of cover to reinforcement, and also the durability or aesthetic appearance of a concrete surface. Certain forms of clay minerals, such as montmorillonites, exhibit unusual degrees of swelling when wetted and contraction when dried, which could exacerbate concrete drying shrinkage and moisture movement characteristics (Weinert, 1968).

Overall, it is generally more prudent to assume that the presence in aggregates of excessive amounts of 'clays', 'silts' and other very finely divided materials are potentially detrimental to strength, durability and cost, unless or until it can be established otherwise.

8.8 Bulk density

The bulk density of aggregate, or unit weight, reflects in part the void content of an aggregate at a given degree of compaction and is, therefore, an indirect measure of the grading and shape characteristics. This value is most commonly used to enable concrete mixes specified by volume to be converted into gravimetric proportions, and thus enable batch weights to be determined. The use of bulk density measurements has therefore declined with the increasing application of designed and prescribed gravimetric concrete mixes (BS 5328:1981).

More generally, the value of bulk density and void content determinations lies in providing a reference property for a material, although it is little used for this purpose. A particular use could be as a comparative measure of variations in the particle shape characteristics of fine aggregates since no British Standard tests are available to measure these directly. However, these are not properties which should be limited in contract specifications, although they may be used by concrete producers to assist in the comparison of mix designs using alternative materials.

8.9 Relative density

The relative density (specific gravity) of dense aggregates is a property of particular value in concrete mix design, concrete yield checks, and in the assessment of compaction and void content of hardened concrete (for example, in lean mix concrete; see Department of Transport, 1976). This is an aggregate property which is not usually subject to limits in contract specifications, except where, for design reasons, there is a need to ensure a minimum concrete density; for example in gravity structures, such as dams or marine armouring, or for radiation shielding.

In such cases, contract specifications often impose a minimum limit on "the specific gravity", but it is often not made clear whether this refers to relative density calculated on the oven dried basis, the saturated surface dry basis, or the apparent basis. The differences between these results can sometimes be substantial and contractual disputes can arise where the appropriate relative density is not specified, or is mis-stated. Some examples to illustrate this aspect are shown in Table 8.5.

As the relative density of an aggregate increases, there is an increased tendency for it to segregate in concrete during handling and compaction, also for any tendency of the concrete towards 'bleeding' and plastic

TABLE 8.5. *Relative density comparisons for various aggregates*

Aggregate ref:	*A*	*B*	*C*	*D*	*E*
Water absorption, %:*	9.5	3.7	1.6	0.6	0.2
Relative density: *					
–Oven dry	2.03	2.36	2.67	2.68	2.73
–Saturated surface dried	2.23	2.45	2.72	2.70	2.74
–Apparent	2.52	2.59	2.79	2.73	2.75
Numerical difference, (Oven dry—Apparent):	0.49	0.23	0.12	0.05	0.02

* As determined in accordance with BS 812: Part 2:1975.

settlement to be exacerbated. This is particularly important where durable top (as cast) concrete surfaces are required, since segregation and settlement during and after compaction can tend to produce a less durable surface.

8.10 Water absorption

Water absorption is an indirect measure of the permeability of an aggregate which, in turn, can relate to other physical characteristics such as mechanical strength, shrinkage, soundness and to its general durability potential. These relationships tend to be imprecise, although in general less absorptive aggregates often tend to be more resistant to mechanical forces and to weathering. A reliable classification of water absorption values for UK aggregates is not available, but a 'low' absorption value might reasonably be considered as less than 1%, with an acceptable range for UK aggregates being from less than 1% to about 5% (Newman, 1959). Absorption limits are rare in both British and other national standards, although BS 5337:1976 includes a recommendation that the aggregate absorption should not "generally" be greater than 3% and a maximum value of 2.5% has often been specified for overseas works.

Contract specifications do not normally include a maximum limit on the absorption of an aggregate. Indeed, limits should not be imposed unless it has been established, for a particular material, that water absorption relates closely to some other undesirable property, such as poor resistance to frost or other weathering agencies. In such cases it may be more expedient to adopt absorption limits for control and acceptance purposes, rather than specify other more time consuming and costly test procedures which cannot be carried out readily in a site laboratory.

8.11 Mechanical properties

BS 882:1983 includes quantitative requirements for the mechanical properties of aggregates based upon the 10% Fines Value test (see Chapter 7). As an alternative, BS 882:1983 also provides limits in respect of the Aggregate Impact Value test. These various criteria, which vary according to the intended application of the concrete, are given in Table 8.6.

BS 812 provides several other mechanical test procedures, variously to determine Aggregate Crushing Value, Aggregate Abrasion Value and Polished Stone Value. Different American test procedures are also available, including the Los Angeles test for resistance to abrasion (ASTM C131). Adoption of these tests, and others, which are not specified in BS 882:1983 is a matter for judgement by individual specifiers.

The common features of the various mechanical tests (Chapter 7) are that loads are applied to specified aggregate size fractions and that the results are generally related to the reduction in weight or size of the original test sample. The exception is the Polished Stone Value test where the effects of the applied forces are measured by a friction test. The methods of loading in the tests, which were selected to simulate particular performance situations, and of measuring losses or changes in the material under test, are summarised in Table 8.7. Modified procedures for some of these tests have been developed and these are discussed in Chapter 7.

TABLE 8.6. *BS 882:1983 mechanical property requirements for coarse aggregate*

Type of concrete	10% fines value (not less than)	Alternatively, aggregate impact value (not exceeding)
	kN	%
Heavy duty concrete floor finishes	150	25
Pavement wearing surfaces	100	30
Others	50	45

The mechanical properties of coarse aggregates should be such that the material does not disintegrate or degrade, which could occur during handling, transportation, concrete mixing or compaction, and that the compressive strength and subsequent performance of the concrete is not impaired. For structural concretes, except concrete road pavements and abrasion resistant floors, the extent to which it is generally necessary to specify mechanical property requirements in addition to those given in BS 882 is limited. Any significant degradation of an aggregate prior to the concrete mixing phase should be detected more directly by routine acceptance tests for grading and fines content. Furthermore, the more stringent of the BS 882 limits (for pavement wearing surfaces and heavy duty concrete floor finishes) in respect of 10% Fines Value should ensure that the characteristic strength of other than high strength concrete is not limited on this account (Orchard, 1976).

Relationships between the mechanical properties of aggregates and their subsequent effects on the performance of concretes are generally uncertain (Bloem and Gaynor, 1963). However, the selection of a suitably high characteristic concrete strength is alone normally taken to assure an adequate potential resistance to impact, abrasion and attrition, provided that adequate attention is paid to the concrete mix design, placement and curing. The mechanical strength of the aggregate is usually not considered apart from routine compliance with BS 882 or an equivalent specification.

Where the performance requirements for concrete surfaces are exceptionally rigorous and beyond what

TABLE 8.7. *Selected mechanical property tests for aggregates*

Test	Standard or country of origin	Measurement principle
Aggregate Impact Value	BS 812	Fines produced by specified shock or impact loading.
Aggregate Crushing Value	BS 812	Fines produced by slowly increasing load to a specified maximum.
10% Fines Value	BS 812	Slowly increasing maximum load required to produce a specified amount of fines.
Aggregate Abrasion Value	BS 812	Loss produced by specified amounts of abrasion
Dory Hardness	USA	Loss produced by specified amounts of abrasion
Polished Stone Value	BS 812	A measurement of friction after a specified amount of polishing.
Deval Abrasion (Rock)	ASTM D2 (discontinued)	Fines produced by self-attrition on tumbling of rock pieces for specified number of revolutions.
Deval Abrasion (Aggregate)	ASTM D289 (discontinued)	As ASTM D2 but using aggregates with steel spheres to introduce impact and increase attrition.
Los Angeles Abrasion	(ASTM C131) (ASTM C535)	As ASTM D289
Aggregate Crushing Strength	(BS 812)	Uniaxial compressive strength of rock cores (now deleted from BS 812).
Franklin Point Load Index	UK	Splitting strength (of cores or irregular lumps of rock) and indirect measure of compressive strength.

might be reasonably expected of a normal structural concrete, special finishing operations (such as the use of vacuum mats, power floating, metallic surface dressings, or other treatments) may have to be employed in addition to the careful selection of both coarse and fine aggregates (Fookes & Poole, 1981).

8.12 Aggregate durability

The overall stability of concrete aggregates may be defined as the ability of individual particles to retain their integrity and not to suffer any physical, mechanical or chemical changes, to extents which could adversely affect the properties or performance of concrete in either engineering or aesthetic respects. Aggregate stability cannot, therefore, be divorced from concrete durability which, in recent years, has become of increasing concern to the construction industry (Russell, 1976).

Furthermore, the consideration of aggregate stability also may be dependent upon the performance requirements for a concrete in the particular environmental conditions involved. For example, a low strength, high absorption aggregate may be acceptable for a low to medium grade concrete in a non-aggressive internal environment. Conversely, the same aggregate might be considered unacceptable for concretes exposed to severe climatic or industrial conditions. Additionally, both climate and industrial environments can vary greatly even within a single project or structure and may necessitate using a superior quality aggregates in particular locations.

The responsibility for determining the concrete durability criteria and, thereby, any aggregate performance requirements in a project normally rests with the Engineer or Consultant. Published reports of concrete durability failures in recent years have highlighted this responsibility (Ryell & Smith, 1968; Fookes & Collis, 1975a), which can be especially onerous in the field of industrial construction, where the long-term and, sometimes, even immediate in-service performance requirements may not reasonably be foreseeable, even to the Client. Post-construction investigations of concrete durability problems occasionally reveal that the likely in-service conditions, and the appropriate aggregate requirements, were not adequately investigated or appreciated (Fookes *et al.* 1982), in which cases both professional liability and contractual issues may arise.

The factors which govern aggregate stability are manifold and may merit detailed consideration in

TABLE 8.11. *Chloride contents permitted for concrete in British Standard Codes of Practice*

Standard	CP 114: Part 2: 1969	CP 110: Part 1: 1972		
Type or use of concrete:[1]		A	B	C
Date	Maximum chloride limits for concrete			
Publication	2% by weight of cement, calcium chloride.[2]	1.5% by weight of cement, anhydrous calcium chloride		
1977	Limit deleted without replacement, but Clause 205A(b) "strongly" recommends that "calcium chloride should never be added in prestressed concrete, reinforced concrete or concrete containing embedded metal".	0.35% chloride ion by weight of cement for 95% of test results with no result greater than 0.50%.	0.06% chloride ion by weight of cement	
1980	No further amendment	0.35% chloride ion by weight of cement for 95% of test restults with no greater than 0.50%.	0.2% chloride ion by weight of cement	0.1% chloride ion by weight of cement

[1] A = Concrete containing embedded metal (including reinforced concrete) and made with cement complying with BS 12.
B = Concrete (including reinforced concrete) made with cement complying with BS 4027 or BS 4248.
C = Prestressed concrete and steam-cured structural concrete.

[2] Although not specifically stated, this reference assumed the use of hydrated calcium chloride, $CaCl_2.2H_2O$, in flake form. 2% such flake calcium chloride, is approximately equivalent to 1.5% anhydrous calcium chloride.

pyrite (Midgley, 1958). However, this test is only applicable to individual particles separated from an aggregate sample and is therefore not usually helpful for overall suitability assessments. Also, pyrite concentrations, whether 'reactive', 'non-reactive' or both, usually show very considerable variability within an aggregate source, so that it would be difficult to control the pyrite content by specification without actually disallowing certain general types of aggregate for concrete works in which surface appearance was paramount.

8.21 Alkali-aggregate reactivity

Field investigations and research findings over many years have demonstrated that various forms of reaction can occur in concrete between certain aggregates and alkali hydroxides derived from the cement matrix (Gillott, 1975; Building Research Establishment, 1982b). The physico-chemical mechanisms involved are complex and research continues to extend understanding and to reveal new features.

TABLE 8.12. *BS 882:1983 (Appendix C) chloride content guidance values*

Type or use of concrete	Maximum total chloride content expressed as percentage of chloride ion by mass of combined aggregate
Pre-stressed concrete	0.02
Steam-cured structural concrete	
Concrete made with cement complying with BS 4027 or BS 4248	0.04
Concrete containing embedded metal and made with cement complying with BS 12	0.06 for 95% of test results, with no result greater than 0.08

The principal effects of alkali-aggregate reactivity are that affected concrete suffers expansion and cracking. The expansion causes misalignment of structures and can threaten structural integrity. Cracking can lead to reinforcement corrosion and other durability problems. A major consideration is that there are no proven means of arresting alkali-aggregate reaction and its consequences.

The main types of alkali-reactivity identified may be categorised as follows:

(i) Alkali-silica reaction:
Reaction between alkalies and the amorphous, poorly-crystalline or micro-crystalline forms of silica, such as opal, chalcedony, chert, natural glass and others. The mechanism of disruption involves the formation of an alkali-silicate gel that may swell on absorption of moisture. This is the most common of the alkali-reaction types.

(ii) Alkali-silicate reaction:
Reaction between alkalies and certain types of silicate rocks and minerals, such as phyllite, vermiculite, chlorite and others. These reaction mechanisms are complicated and difficult to characterise, but they are frequently expansive (Gillott *et al.* 1973).

(iii) Alkali-carbonate reaction:
Reactions involving carbonate rocks, which may be either expansive or non-expansive (French & Poole, 1974; Poole and Sotiripoulos, 1981; Sims & Sotiripoulos, 1983). The potentially deleterious carbonates appear to contain appreciable quantities of dolomite and usually also clay minerals; micro-textures may also be important. Mechanisms vary in detail, but one of the best known involves dedolomitisation by alkalies, followed by swelling due to moisture-activation of the interstitial clay.

The identification of potentially alkali-reactive rock types or aggregate constituents usually includes petrographic examination. However, these examinations can only indicate whether an aggregate has a reactivity potential, whereas engineers usually need to decide whether any reactivity is likely to be detrimental in a particular concrete structure in a specific environment, since not all potentially reactive aggregates are necessarily deleterious in service.

A number of test procedures are available to assist in the investigation of alkali-reactivity potential and some of the more common ones are listed in Table 8.13. The first stage of such an investigation should involve the petrographic examination of the materials in question, since this should enable known reactive constituents to be identified. On the basis of the

TABLE 8.13. *Selected testing procedures frequently used in assessments of potential alkali-aggregate reactivity*

ASTM = American Society for Testing and Materials

ASTM C294	Standard Descriptive Nomenclature of CONSTITUENTS OF NATURAL MINERAL AGGREGATES
ASTM C295	Standard Recommended Practice for PETROGRAPHIC EXAMINATION OF AGGREGATES FOR CONCRETE
ASTM C289	Standard Test Method for POTENTIAL REACTIVITY OF AGGREGATES (CHEMICAL METHOD)
ASTM C227	Standard Test Method for POTENTIAL ALKALI REACTIVITY OF CEMENT-AGGREGATE COMBINATIONS (MORTARBAR METHOD)
ASTM C586	Standard Test Method for POTENTIAL ALKALI REACTIVITY OF CARBONATE ROCK FOR CONCRETE AGGREGATES (ROCK CYLINDER METHOD)
Building Research Station	National Building Studies Research Paper No. 25, 1958. Reactions between aggregates and cement. Part V1. Alkali-aggregate interaction: Test procedures. HMSO, London.

findings it is then necessary to exercise judgement concerning the need for various supplementary tests. Knowledge of the potentially reactive components, identified by petrographic means, largely dictates the choice of subsequent test procedures, and will assist in the interpretation of the results (Sims, 1981; Grattan–Bellew, 1983). Long-term physical expansion tests, such as the mortar-bar method (ASTM C227) and the rock cylinder method (ASTM C585), sometimes may be considered necessary.

Thus various laboratory procedures are available for evaluating the alkali-reactivity of aggregate materials, but it should be appreciated that the results are not necessarily conclusive and that considerable reliance sometimes has to be placed on experience and expert opinion, especially when reliable and appropriate evidence of local concrete performance is not available.

No guidance is given at present in BS 882:1983 on the subject of alkali-reactivity (apart from a reference to a forthcoming revision of CP 110). However, an independent working party has recently provided recommendations for minimising the risk of alkali-silica reactivity in the United Kingdom (Hawkins *et al.* 1983).

8.22 Aggregate drying shrinkage

A limited number of United Kingdom aggregates suffer higher than normal volume changes on wetting and drying, which, in turn, can affect adversely the drying shrinkage and wetting expansion characteristics of concretes. This finding resulted from failure investigations and subsequent research by the Scottish Building Research Station (Snowdon & Edwards, 1962; Edwards, 1966; Building Research Establishment, 1968) which revealed that, with certain aggregates, concrete could suffer as much as a fourfold increase in drying shrinkage.

The rock types that exhibit excessive shrinkage characteristics are found to contain minerals with clay-like, or sometimes mica-like, structures. Hence altered dolerites and basalts ('whinstones') can be affected, also some greywacke sandstones, mudstones, shales and mica-schists. For some sources, it has been possible to derive a predictive relationship between water absorption of the aggregate and the aggregate drying shrinkage (Snowdon & Edwards, 1962). In South Africa, work with heavily-altered dolerites has shown a good relationship between the content of secondary minerals and the shrinkage potential (Weinert, 1968).

Shrinkage of aggregate in concrete, particularly the coarse aggregate, may lead to cracking or crazing of the concrete, thereby increasing the vulnerability of that concrete to deterioration by other agencies, such as frost action. In the case of reinforced concrete, apart from the increased risk of corrosion as a result of cracking of the concrete cover, excessive deflections of beams and other structural units can be caused by excessive drying shrinkage of the aggregates within the concrete.

An aggregate shrinkage test procedure has been developed by the Building Research Establishment (Digest 35, 1968) comprising a concrete drying shrinkage test employing a standardised concrete mix in which only the aggregate source varies. The results are classified to provide the designer and specifier with a basis for assessing whether or not particular aggregate combinations require the adoption of special design or concrete production measures. It is, therefore, essential for the test to be carried out at an early stage of a project, when consideration is being given to the potential suitability of local sources of aggregates, and subsequently for approval and monitoring purposes. The classification given in BRE Digest 35 are reproduced in Table 8.14.

The BRE Digest 35 test method, and implementation of the appropriate recommendations, can only provide assurance against the likelihood of concrete shrinkage problems due to the composition of the aggregates. However, cracking and other problems resulting from excessive concrete shrinkage can occur, due to a variety of factors which are wholly or largely independent of the nature of the aggregate employed (Concrete Society, 1983).

8.23 Chemical composition

The overall chemical analysis of an aggregate is not usually considered useful or necessary for the appraisal for suitability for use in concrete. Chemical analysis for certain specific constituents, such as sulphates, chlorides, organic materials and selected metals, has been discussed earlier. In general, a thorough petrographic analysis is more likely to be helpful in assessing a concrete aggregate than a complete chemical analysis. However, chemical analysis of some types of carbonate rock can yield valuable additional information.

An analytical aspect of special relevance to aggregates for concrete concerns the aggregate correction factors involved in the analysis of hardened concrete to determine the original cement content and aggregate/cement ratio. These determinations are very common in building and civil engineering work, usually for dispute purposes, and the results are relevant to the assessment of compliance with specifications and any liability issues. The aggregate correction factors can have a significant influence but representative samples of the aggregates originally used are often not available at the time of analysis. Thus, it could be of value to specify that the correction factors are determined for reference purposes at the time the materials are approved, using the procedures given in BS 1881: Part 6: 1971.

TABLE 8.14. *Aggregate shrinkage categories and recommendations given in Building Research Establishment Digest 35, 1968*

Drying shrinkage of concrete (in BRE test), %	Remarks	Use
Not exceeding 0.045	Possible with several types of very low shrinkage aggregates such as granite, limestone, unaltered felsite, blastfurnace slag and a few dolerites and gabbros. Quartz, flint gravel and marble which are very hard and non-shrinking will produce concrete have a shrinkage less than 0.025%.	Suitable for all concreting purposes. If very low values of drying shrinkage for precast products are required, it may be necessary to cure with high pressure steam but this would not apply to concrete made with quartz, flint gravel and marble.
0.046–0.065	Possible with hard dense aggregates having low shrinkage and with softer non-shrinkable aggregates. Most of the aggregates in Central Scotland are in this range.	Suitable for most applications. Special care should be taken in the design of such units as cladding panels and cast *in situ* floors, particularly if heated. Span/depth ratios should be reduced below the values given in CP 114 or of CP 116 or alternatively, additional reinforcement should be provided in the compression zone of the section. For prestressed concrete an extra allowance should be made for loss of prestress.
0.066–0.085	Possible with aggregates covering a wide geological range. At this and higher levels of drying shrinkage the durability of the concrete is likely to become affected.	Suitable for all general purposes but for unreinforced or thin reinforced members, exposed to the weather, air-entrained concrete should be used.
Exceeding 0.085	These are gravels consisting mainly of sedimentary rocks such as greywacke, shale and mudstone. They have produced concrete which has resulted in severe warping and widespread deterioration.	Suitable only in positions where complete drying out never occurs, for mass concrete surfaced with air-entrained concrete and for members that are symmetrically and heavily reinforced and are not exposed to the weather.

Note: 1. Different batches of ordinary Portland cement may be expected to give rise to variations of the order of 0.005% in the drying shrinkage of the concrete.
2. In general, although the coarse aggregate plays the major part in determining the shrinkage of the concrete, the type and grading of the fine aggregate also affect the result. When a low shrinkage fine aggregate is used with a high shrinkage coarse aggregate (in place of the fine aggregate from the same source) the shrinkage of the resulting concrete may be reduced by as much as 0.03%. Also, changing the grading of the fine aggregate from Zone 4 to Zone 1, BS 882:1973, may in extreme cases decrease the shrinkage by as much as 0.02%.
3. It is assumed that the normal recommendations for quality control given by the appropriate Code of Practice for the particular job will be observed.

References

AMERICAN CONCRETE INSTITUTE (ACI). 1979. Committee 515. A guide to the use of waterproofing dampproofing, protective, and decorating barrier systems for concrete. *Concrete International,* **1,** (11), 41–81.

AMERICAN SOCIETY FOR TESTING AND MATERIALS (ASTM). 1976. Living with marginal aggregates. ASTM STP-597.

AMERICAN SOCIETY FOR TESTING AND MATERIALS (ASTM). 1978. Significance of tests and properties of concrete-making materials. ASTM STP 169B.

BARKER, J. A. 1983. Dictionary of concrete. Construction Press, London.

BLAKE, L. S. 1967. Recommendations for the production of high quality concrete surfaces. C & CA Tech. Advisory Series.

BLOEM, D. L. 1959. Effect of maximum aggregate size on strength of concrete. *Cement, Lime and Gravel.* **34,** 170–174.

BLOEM, D. L. 1966. Soundness and deleterious substances. ASTM STP 169A, 497–512.

BLOEM, D. L. & GAYNOR, R. D. 1963. Effects of aggregate properties on strength concrete. *J. of the American Concrete Institute.* **60,** (10), 1429–1455.

BRITISH STANDARDS INSTITUTE NEWS, July 1983.

BUILDING RESEARCH ESTABLISHMENT (BRE). 1969. Shrinkage of natural aggregates in concrete. BRE Digest 35.

BUILDING RESEARCH ESTABLISHMENT (BRE). 1980. Deterioration due to corrosion in reinforced concrete. Information Paper IP12/80.

BUILDING RESEARCH ESTABLISHMENT (BRE). 1982a. The durability of steel in concrete: Part 1, Mechanism of protection and corrosion. BRE Digest 263.

BUILDING RESEARCH ESTABLISHMENT (BRE). 1982b. Alkali aggregate reactions in concrete. BRE Digest 258.

BUILDING RESEARCH STATION. 1958. Reactions between aggregates and cement. Part VI. Alkali-aggregate interaction: Test procedures. National Building Studies Research Paper No. 25, HMSO, London.

CHAPMAN, G. P. & ROEDER, A. R. 1970. The effects of sea shells in concrete aggregates. *Concrete,* Vol. **4,** No. 2, Feb. 1970, pp. 71–79.

CONCRETE SOCIETY, THE. 1980. Guide to chemical admixtures for concrete, Technical Report 18. Ref. 52. 032.

CONCRETE SOCIETY, THE. 1982. Non-structural cracks in concrete, Technical Report 22. Ref. 53.038.

CRAMMOND, N. J. 1984. Examination of mortar bars containing varying percentages of coarsely crystalline gypsum and aggregates., *Cem. Concr. Res.,* **14,** 225–230.

DEPARTMENT OF THE ENVIRONMENT. 1979. Standard specification clauses for airfield pavement works. Part 3, Concrete. Property Services Agency.

DEPARTMENT OF TRANSPORT. 1976. Specification for road and bridge works, 5th Edition. HMSO, London. (also Notes for Guidance, 1976).

DEWAR, J. D. 1963. Effect of mica in the fine aggregate on the water requirement and strength of concrete. C & CA Technical Report TRA/370.

EDWARDS, A. G. 1966. Shrinkage and other properties of concrete made with crushed rock aggregates from Scottish sources. BGWF Journal, Autumn 1966, 23–41.

EDWARDS, A. G. 1970. Scottish aggregates: rock constituents and suitability for concrete. BRE Current Paper CP28/70.

FOOKES, P. G. 1980. An introduction to the influence of natural aggregates on the performance and durability of concrete. *Q. J. eng. Geol. Lond.,* **13,** 207–229.

FOOKES, P. G., COMBERBACH, C. D. & CANN, J., 1983. Field investigation of concrete structures in South West England. Parts 1 & 2. *Concrete,* **17,** (3) & (4), 54–56 & 60–65.

FOOKES, P. G. & COLLIS, L. 1975a. Problems in the Middle East. *Concrete,* **9,** (7), 12–17.

FOOKES, P. G. & COLLIS, L. 1975b. Aggregates and the Middle East. *Concrete,* **9,** (11), 14–19.

FOOKES, P. G. & COLLIS, L. 1976. Cracking and the Middle East. *Concrete,* **10,** (2), 14–19.

FOOKES, P. G. & HIGGINBOTTOM, I. E. 1980. Some problems of construction aggregates in desert areas, with particular reference to the Arabian Peninsula. 1. Occurrence and special characteristics. 2. Investigation, production and quality control. Proc. Instn. Civ. Engrs. Part 1, 68, February 39–67 and 69–90.

FOOKES, P. G., POLLOCK, D. J. & KAY, E. A. 1982. Concrete in the Middle East, Part 2, Rates of Deterioration, pp. 12–22. Viewpoint Publication by Eyre & Spottiswoode Publications Ltd.

FOOKES, P. G. & POOLE, A. B. 1981. Some preliminary considerations on the selection and durability of rock and concrete materials for breakwaters and coastal protection works. *Q. J. eng. Geol. Lond.,* **14,** 97–128.

FOOKES, P. G. & REVIE, WA. A 1982. Mica in concrete—a case history from Eastern Nepal. *Concrete,* **16,** (3), 12–16.

FRENCH, W. J. & CRAMMOND, N. J. 1980. The influence of serpentinite and other rocks on the stability of concretes in the Middle East. *Q. J. eng. Geol. Lond.,* **13,** (4), 255–280.

FRENCH, W. J. & POOLE, A. B. 1974. Deleterious reactions between dolomites from Bahrain and cement paste. *Cem. Concr. Res.,* **4,** 925–937.

GAYNOR, R. D. 1964. Effect of fine aggregate on concrete mixing water requirement. Cement, Lime & Gravel, January 1964.

GILLOTT, J. E. 1975. Alkali-aggregate reactions in concrete. *Eng. Geol.,* **9,** 303–326.

GILLOTT, J. E., DUNCAN, M. A. G. & SWENSON, E. G. 1973. Alkali-aggregate reaction in Nova Scotia. Part IV. Character of the reaction. *Cem. Concr. Res.* **3,** 521–535.

GRATTAN-BELLEW, P. E. 1983. Re-evaluation of standard mortar bar and concrete prism tests for alkali-aggregate reactivity. *Materiaux et Constructions,* **16,** (94), 243–250.

GREATER LONDON COUNCIL (GLC). 1962. Practice Notes on the architectural use of concrete. LCC (now GLC) Architect's Dept, Publication Ref. No. 4156.

GREATER LONDON COUNCIL (GLC). 1979. Marine aggregates for concrete—a revised GLC specification. GLC Bulletin No. 121 (2nd series), March 1979, 2/1–2/2.

HAWKINS, M. R. *et al.* 1983. Alkali-aggregate reaction—Minimising the risk of alkali-silica reaction—Guidance Notes—Report of a Working Party. Published for the Working Party by the C & CA, ref. 97-304.

HIGGINS, D. D. 1984. Admixtures for concrete. Cement and Concrete Association, ref. 45.041.

KAPLAN, M. F. 1958. The effects of the properties of coarse aggregates on the workability of concrete. *Mag. Concr. Res.* **10,** (29), 63–74.

KAPLAN, M. F. 1959. Flexural and compressive strength of concrete as affected by the properties of coarse aggregates. *J. Amer. Concr. Inst.,* **55,** 1193–1208.

KAY, E. A., FOOKES, P. G. & POLLOCK, D. J. 1982. Concrete in the Middle East, Part 2. Deterioration related to chloride ingress, pp. 23–32. Viewpoint Publication by Eyre & Spottiswoode Publications Ltd.

KEEN, R. A. 1979. Impurities in aggregates for concrete. C & CA Advisory Note No. 18.

LEA, F. M. 1970. The chemistry of cement and concrete. 3rd Edition, Edward Arnold, London.

LEEMING, M. B. 1983. Corrosion of steel reinforcement in offshore concrete—Experience from the Concrete-in-the-Oceans Programme. Chapter 4, Corrosion of reinforcement in concrete construction, Society of Chemical Industry, London.

LEES, G. 1964. The measurement of particle shape and its influence in engineering materials. *J. of the British Granite and Whinstone Federation.* **4,** (2).

MIELENZ, R. C. 1955. Petrographic examination. ASTM STP 169, 253–273.

MIDGLEY, H. G. 1958. The staining of concrete by pyrite. *Mag. Concr. Res.* **10,** (29), 75–78.

MIDGLEY, H. G. 1970. The effect of lead compounds in aggregate upon the setting of Portland cement. *Mag. Concr. Res.,* **22,** (70), 42–44.

MILLER, E. W. 1983. High density concrete: Part 1. (Concrete Society Current Practice Sheet No. 90), *Concrete* **17,** (12), 33–34.

MURDOCK, L. J. & BROOK, K. M. 1979. Concrete materials and practice. 5th Ed., Edward Arnold, London.

NEVILLE, A. M. 1981. Properties of concrete. 3rd Edition, Pitman, London.

NEWMAN, K. 1959. The effect of water absorption by aggregates on the water–cement ratio of concrete. *Mag. Concr. Res.* **11,** (33), 135–142.

ORCHARD, D. F. 1976. Concrete technology. Volume 3, Properties and testing of aggregates. Applied Science Publishers Ltd., London.

POOLE, A. B., SOTIROPOULOS, P. 1980. Reactions between dolomitic aggregate and alkali pore fluids in concrete. *Q. J. eng. Geol. London.* **13,** (4), 281–287.

ROEDER, A. R. 1963. The detection and determination of organic impurities. Cement, Lime and Gravel, 283–286.

ROEDER, A. R. 1977. Some properties of flint particles and their behaviour in concrete. *Mag. Concr. Res.,* **29,** (99), 92–99.

RUSSELL, P. 1976. Effect of aggregates on the durability of concrete. C & CA Advisory Note, Publication ref. 45.021, 2nd Edition.

RYELL, J. & SMITH, P. 1968. Case histories of poor concrete durability in Ontario highway structures. Performance of Concrete, University of Toronto Press, Chapter 11, 181–204.

SAMARAI, M. A. 1976. The disintegration of concrete containing sulphate-contaminated aggregates. *Mag. Concr. Res.,* **28,** (96), 130–142. (Corrigenda, 29, (100), 155).

SAND AND GRAVEL ASSOCIATION. 1968. Sea-dredged aggregates for concrete. Proceedings of a symposium held at Fulmer Grange, Slough, 9th December 1968.

SCHAFFLER, H. 1979. Concrete with gap-graded aggregates. *Betonwerk + Fertigteil-Technik, Part 1.* **45,** (6), 341–345.

SHACKLOCK, B. W. 1959. Comparison of gap- and continuously graded concrete mixes. C & CA Tech. Rep. TRA/240.

SHERWOOD, P. T. & ROEDER, A. R. 1965. Organic impurities in sand that interfere with the hydration of Portland cement. The Surveyor and Municipal Engineer, July 1965.

SHIRLEY, D. E. 1981. Impurities in concreting aggregates, C & CA Construction Guide, ref. 45.016, 2nd Edition.

SIMS, I. 1981. Application of standard testing procedures for alkali-resistivity, Parts 1 and 2. *Concrete,* **15,** (10), 27–29 and 15 (11), 29–32.

SIMS, I. & SOTIROPOULOS, P. 1983. Standard alkali-reactivity testing of carbonate rocks from the Middle East and North Africa. Proceedings of the 6th International Conference, Alkalis in Concrete, Research and Practice, Copenhagen, June 1983.

SNOWDON, L. C. & EDWARDS, A. G. 1962. The moisture movement of natural aggregate and its effect on concrete. *Mag. Concr. Res.,* **14,** (41).

WEINERT, H. J. 1968. Engineering petrology for roads in South Africa. *Eng. Geol.,* **2,** (5), 359–362.

British and American Standards

BRITISH STANDARDS INSTITUTION BS 12:1978. Specification for ordinary and rapid-hardening Portland cement.

BRITISH STANDARDS INSTITUTION BS 812:Parts 1 to 4:1975. Methods for sampling and testing of mineral aggregates, sands and fillers.

BRITISH STANDARDS INSTITUTION BS 882, 1201:1954. Concrete aggregates from natural sources.

BRITISH STANDARDS INSTITUTION BS 882, 1201:Part 2:1973. Aggregates from natural sources for concrete 1973 (including granolithic).

BRITISH STANDARDS INSTITUTION BS 882:1983. Specification for aggregates from natural sources for concrete.

BRITISH STANDARDS INSTITUTION BS 1377:1975. Methods of test for soil for civil engineering purposes.

BRITISH STANDARDS INSTITUTION BS 1881:Part 6:1971. Methods of testing concrete. Part 6, Analysis of hardened concrete.

BRITISH STANDARDS INSTITUTION BS 4027:1980. Specification for sulphate-resisting Portland cement.

BRITISH STANDARDS INSTITUTION BS 4248:1974. Supersulphated cement.

BRITISH STANDARDS INSTITUTION BS 5328:1981. Methods for specifying concrete, including ready-mixed concrete.

BRITISH STANDARDS INSTITUTION BS 5337:1976. Code of Practice for the structural use of concrete for retaining aqueous liquids.

BRITISH STANDARDS INSTITUTION CODES OF PRACTICE CP 110:Part 1:1972. The structural use of concrete. Part 1. Design, materials and workmanship.

BRITISH STANDARDS INSTITUTION CODES OF PRACTICE CP 114:1969. Structural use of reinforced concrete in buildings.

BRITISH STANDARDS INSTITUTION CODES OF PRACTICE CP 116:1969. The structural use of precast concrete.

AMERICAN SOCIETY FOR TESTING MATERIALS ASTM C33–81 (1981). Specification for concrete aggregates.

AMERICAN SOCIETY FOR TESTING MATERIALS ASTM C40–79 (1979). Test for organic impurities in fine aggregates for concrete.

AMERICAN SOCIETY FOR TESTING MATERIALS ASTM C88–76 (1976). Test for soundness of aggregates by use of sodium sulphate or magnesium sulphate.

AMERICAN SOCIETY FOR TESTING MATERIALS ASTM C131–81 (1981). Test for resistance to abrasion of small size coarse aggregate by use of the Los Angeles machine.

AMERICAN SOCIETY FOR TESTING MATERIALS ASTM C227–81 (1981). Test for potential alkali-reactivity of cement: aggregate combinations (mortar-bar method).

AMERICAN SOCIETY FOR TESTING MATERIALS ASTM C289–81 (1981). Test for potential reactivity of aggregates (chemical method).

AMERICAN SOCIETY FOR TESTING MATERIALS ASTM C294–69 (1981). Descriptive nomenclature of constituents of natural mineral aggregates.

AMERICAN SOCIETY FOR TESTING MATERIALS ASTM C295–79 (1979). Recommended practice for petrographic examination of aggregates for concrete.

AMERICAN SOCIETY FOR TESTING MATERIALS ASTM C535–81 (1981). Test for resistance to degradation of large-size coarse aggregate by abrasion and impact in the Los Angeles machine.

AMERICAN SOCIETY FOR TESTING MATERIALS ASTM C586–69 (1975). Test for potential alkali-reactivity of carbonate rocks for concrete aggregates (rock cylinder method).

Chapter 9. Aggregates for mortar

9.1 Introduction

This chapter, which is concerned with aggregates for mortars, necessarily introduces some degree of overlap with Chapter 8. However, there are major differences between mortars and concretes, arising mainly from the manner of their use and these are treated in some detail in respect of their relevance to mortar aggregates. It is therefore strongly recommended that the text appearing in 8.1, and 8.2 be read in conjunction with this section.

The term mortar is used in the building industry to denote a mixture of natural sand or other fine aggregate and some binding agent, used as a jointing or as a surface rendering material. In the United Kingdom the demand for building sand has grown steadily since the early 1960's and at its peak in 1973 represented some 18% of the total sand and gravel production (HMSO, 1979) for the country. (Fig. 9.1). Until about 20 years ago, the choice of finely divided binders in the UK was limited to lime or cement or cement/lime mixes. Now the choice has been widened to include masonry cements (specially blended mixtures of Portland cement with an inert finely divided filler and an air-entraining agent to provide plasticity) and ready-mixed mortars.

Mortars are commonly used for the bedding and jointing of building units such as bricks and blocks (masonry) and also for surface finishes (rendering, plastering and screeding). Rendering is a term usually applied to external finishes, plastering being used for internal finishes, whilst floor screed refers to types of regularised floor surfaces to which a finishing coat may be applied.

Historically, the bricklayer and plasterer made mortar mixes from the materials at hand, giving what skill taught to be the most suitable properties for the job. Consequently specifications have been based largely on the average sands used in practice, the skilled craftsman often adjusting the specified proportions according to the materials available (Bessey, 1966).

Natural sand as defined in BS 882, (Part 2, 1975), (i.e. fine aggregate resulting from the natural disintegration of rock), forms the greater proportion of the fine aggregates used for mortars, their sources in the UK ranging in age from the Permian to Recent. They are excavated from sand and gravel pits in unconsolidated clastic deposits derived from the weathering of existing rock types, the sands being predominantly quartzose in composition. They are often used in their natural form or processed by screening and/or washing and classifying. Building mortars are comprised principally of sand (or crushed stone), lime, cement, air and water together with special additives as appropriate. Sand is the major constituent and so any variations in its quality will have a considerable effect on the properties of the resulting mortar. Bessey (1966) has indicated that the main factors affecting sand quality are:

(i) Average particle size
(ii) The range of the particle size
(iii) The particle shape
(iv) Impurities, particularly the clay content

9.2 Types of mortars

Various common types of mix depend on the fact that, a quantity of fine material (lime and cement) equal to about 1/3 of the volume of an average sand is required in order to fill the voids between the sand grains and thus give the wet mix sufficient plasticity. To increase the workability an inert finely divided filler, in addition to Portland cement, may be added. This also reduces the excessive strength likely to develop from the use of too much cement. In recent years there has been a rapid growth in the ready-mixed mortar sector of the building industry, where considerable advantages are achieved as a result of the greater degree of control over the production processes as compared with that generally obtainable on site. Current good practice is set out in BS 4721 (1981).

9.2.1 Masonry (brickwork or general purpose) mortars. To function as a bonding material for bricks or other masonry units, some strength is necessary, but the mortar must also be able to yield sufficiently to accommodate the various differential moisture or thermal movements which occur in a structure. If these movements are small, the stresses set up are usually within the elastic limits of the materials used. Beyond these limits the stresses are relieved by yielding along lines of relative weakness in the structure. In a composite construction of bricks and mortar the cracking will penetrate the bricks themselves if the mortar is stronger than the bricks; an unsatisfactory state of affairs. Where the mortar is weaker than the bricks the relief of stresses result in a series of fine cracks distributed inconspicuously among

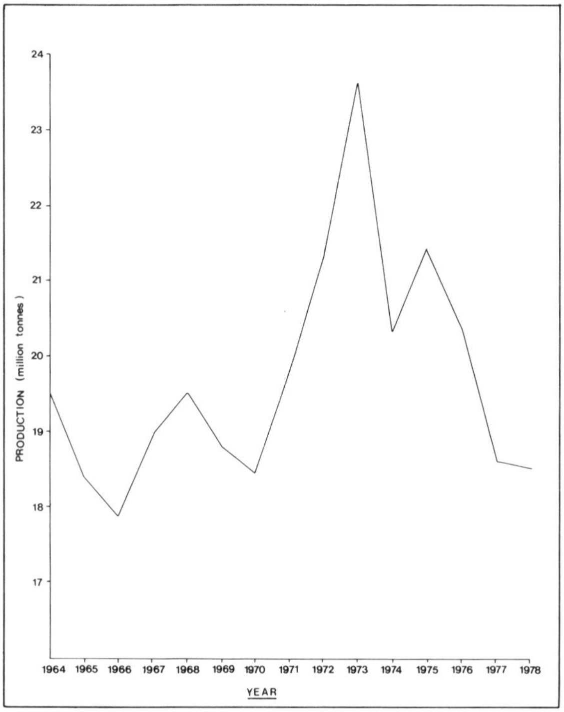

FIG. 9.1. UK Production of building sands for mortars 1964–1978 (HMSO 1979).

the vertical joints in the brickwork, there being only small relative movement between the bricks and the mortar in the horizontal joints. (McIntosh, 1970; BRE Digest 160, 1973).

9.9.2 Rendering and plastering. External rendering has been practised for many hundreds of years; the applications known as wattle and daub, used on primitive buildings as a weather resistant coating can still be seen in use in some Mediterranean countries. Mixtures of lime and pozzolan with hydraulic properties (i.e. which harden with water) were used extensively by the Romans. Calcined argillaceous limestones mixed with sand were misleadingly termed, Roman cement, from about AD 1800 onwards and were in common use as an external rendering up to the beginning of the 20th Century. The "Roman" variety, was subsequently replaced by Portland cement, a material which when mixed with lime produces rendering mixes having good application properties together with adequate strength and resistance to weathering. Of all the materials used in building construction, rendering and plastering mortar mixes are most subject to variability, due largely to differing local practice and lack of strict site control.

Internal plastering mortars serve two purposes, (a) they provide a continuous smooth, plane surface to the walls and ceilings and (b) they provide a layer suitable for receiving a decorative finish. Plaster can vary considerably in strength and porosity, but in practice those with the ability to absorb and release moisture and moisture vapour have certain advantages over the relatively impervious finishes. Cement/sand and cement/lime/sand plastering mixes are commonly in use today while the gypsum plaster/lime (or sand) blends are occasionally employed.

Some misunderstanding has arisen in the past due to the term plaster being used as synonomous with gypsum. However, recent specifications for plastering (BRS Digest, 1971) to distinguish between cement based mixes and gypsum based mixes. The latter have some advantage in that they are capable of drying out and hardening more quickly than the former, and the drying shrinkage which occurs is much less than with cement based mixes.

9.2.3 Floor screeds. Cement/sand screeds are generally used for levelling a concrete sub-floor to provide both a level and a smooth surface, usually to receive a decorative floor finish. It has been suggested that floor screed mixes bonded directly to a concrete sub-floor and subsequently covered with a decorative finish should have a mix ratio of $1:3$ to $1:4\frac{1}{2}$ by weight of Portland cement to sand. (BRE Digest 160, 1973). Mixes richer than $1:3$ are not recommended because of the higher risk of drying shrinkage which can result in cracking and curling of the screeds; nor are mixes weaker than $1:4\frac{1}{2}$ acceptable since they tend to be damaged either by abrasion or impact. Floor screeds should not be used as a wearing surface except as a temporary expedient.

9.3 Description, classification and testing of mortar sands and fillers

9.3.1 Description and classification. The present British Standard, BS 812 (1975), requires the provision of a simple petrological classification and whilst most British sands are quartzose, Ragsdale and Birt (1976) have recommended that a petrological examination be carried out so that the constituents of the sand and in particular the presence of any deleterious material can be identified. They also recommend the use of a binocular microscope to enable a detailed description of the material to be given. This can be of particular importance in the investigation of possible new sources of material.

Particle shape and surface texture, whilst easily ascertained for coarse aggregates, are mose difficult to determine for finer grained particles. Microscopic study will show particle shape (such as rounded and angular (Fig. 9.2) but the surface texture (i.e. glassy, smooth etc.), as defined in BS 812:1975 may not be so easy to describe. The geological occurrence of the deposits in which the sands are found will usually give an indication of the petrology of the fine aggregate to be expected. Natural sands have rounder, smoother

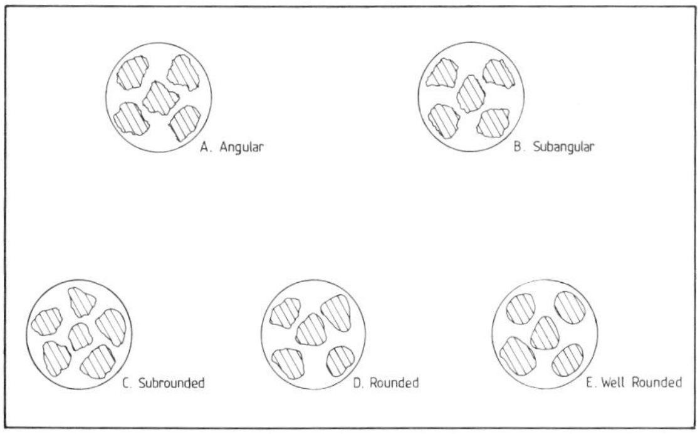

FIG. 9.2. Degree of particle roundness (after Pettijohn 1949).

particles than manufactured sands. The latter material with its characteristic sharp and angular particle shape is likely to produce mortars with workability properties quite different from those made with naturally occurring sands (9.4).

BS 812 (1975) states that it is convenient to classify and group all aggregates according to their petrological characteristics, but difficulties occur when the "trade group" method is applied to sands. The simplified method proposed in Chapter 6 offers a more satisfactory solution to fine aggregate classification in the broader sense.

In considering the quality of sands, BS 1198, 1199 and 1200 (1976) indicate that "sands shall consist of natural sand, crushed stone sand or crushed gravel sand, or a combination of any of these. They shall be hard, durable, clean and free from adherent coatings, such as clay, and from any appreciable amount of clay in pellet form". However all three standards allow for a proportion of clay, silt and dust, the limits for these fine fractions being greater for crushed stone sources than for sand or crushed gravel sources, it being established that the fine powder resulting from the artificial crushing of rocks is different in type and more acceptable than clay minerals (8.7).

Current British Standards indicate that a guide to the uniformity of materials supplied in respect of the silt and clay content of natural sand and crushed gravel sand can be assessed by the field settling test (BS 812:1975). As mentioned below the 'decantation' or 'sedimentation' methods are at present recommended for the accurate determination of clay, silt and fine dust in sand and crushed gravel sand or in crushed rock sand respectively. These may well be superseded in due course by the adoption of a wet sieving method which includes a 38 μm sieve.

9.3.2 Sampling and testing of fine aggregates. In the UK sampling and testing of sands are carried out in accordance with the requirements of the appropriate sections of BS 812 (1975), a standard which is currently under revision. Gradings of sands for particular use (according to BS 1198, 1199, and 1200) are determined by sieve analyses, the sizes used being as follows: 5.00 mm, 2.36 mm, 1.18 mm, 600 μm, 300 μm, 150 μm.

As the sand portion of mortar generally forms a large part of the total volume of a mix and its characteristics will influence the performance and properties of the fresh or hardened mortars in many ways. The inherent properties of the sand, such as chemical inertness, thermal coefficient of expansion, and durability are important, but in particular the shape and grading of the fine aggregate particles are likely to have a major influence on the properties of the mortar mix.

9.3.3 Grading of sands. Early work by Connor (1948) on the performance of sands concentrated on the grain size and the range of sieve fractions, detailed investigations giving strong indication that the graded quality of the sand in the mortar was very important. Further investigation by Connor (1953) on sieve analyses of sands which produced good mortars confirmed that the standard grading requirements recommended in the USA were not entirely reliable and should only be considered as tentative.

A well known method of representing the results of sieve analyses is by plotting the data cumulatively on log-probability paper to give a particle size distribution chart. By using this method of plotting, the line, or grading curve, obtained can be described in terms of its median size and slope, the former being the grain size in microns corresponding to 50 per cent passing fraction and the latter being indicated by the size interval between 16 and 50 per cent passing fractions, which is the standard deviation of the log-normal distribution. Swallow (1964) investigated several sands by this method and concluded that 'good' sands have a median value of 440 μm to 570 μm and standard deviation of the log-normal distribution between 200 μm and 275 μm. By determining these particle size characteristics for a mortar sand and comparing them with Swallow's 'good' sand values, it is possible to indicate whether a sand is too fine or coarse and whether the particle size distribution is too wide or too narrow. By plotting particle size limits for sands recommended for general-purpose mortars and for reinforced brickwork as given in BS 1200 (1955) the results shown in Table 9.1, are obtained. Swallow (1964) noted that the limits set for the grading of sands in BS 1200 (1955) are far too wide when compared with this scheme of median particle size and standard deviation, and indicated that although a sand may comply with the specifications this is no indication that it will be satisfactory when used to produce masonry mortars.

It is widely accepted that the present British Standards BS 1198 (1976) 1199 (1976) and 1200 (1976) (Tables 9.5, 9.4, and 9.2) do not cover the full range of fine aggregate types that are currently used for masonry, plastering and for floor screeds in the UK. In 9.5 reference is made to the proposed concept of "ideal" sands. This is suggested as an

TABLE 9.1. *Grade size variations for mortars referred to BS 1200 gradings for sands for general purpose and reinforced mortars. (after Swallow 1964)*

	Low limits for both types	High limits for reinforced mortars	High limits for general purpose mortars
Medium particle size	900 μm	340 μm	230 μm
Standard deviation	470 μm	160 μm	80 μm

alternative to the BS grading limits specification which would give a better description of the materials that have and still are being used successfully in practice. The debate on the revision of the British Standards for building sands continues and representative sources of materials are being considered which may provide evidence for more acceptable grading limits.

9.3.4 Tests for impurities. There has long been concern with impurities in building sands. British Standards 1198, 1199 and 1200 (1976) all refer to the possible occurrence of "harmful materials such as iron pyrites, salts, coal or other organic impurities, mica, shale and similar laminated materials, or flaky or elongated particles".

Prior to the revision of BS 812 in 1975, the presence of organic impurities was assessed by using caustic soda, a method which was quick, simple and widely used. It was, however, subject to some defects such as, for example, the failure to detect a serious contaminant like coal, where a single particle would not produce a colour change sufficient to detect the impurity. "Loss on ignition determination" has similar limitations in that the presence of calcium carbonate could also obscure the presence of coal (Westbrook, 1960). American standards do not provide an adequate test for organics either and currently there is a particular need for a test for identifying lignite, coal and similar organic impurities likely to cause blemishes in the mortar.

Although iron compounds are present in most sands, which generally provide the colouration, pyrites can occur as discrete particles which can then be considered as an impurity. The identification of pyrites is carried out by visual inspection. At the present time there is a need for a more satisfactory means of determining the quantitative effects of pyrites in sands used for building mortars.

Mica is considered as a deleterious material according to BS 812 (1975) and yet when used in renderings it can provide a very acceptable sparkle. In the constructions where it has been used, the quantity present in such mixes does not appear to have caused problems (Westbrook, 1960). The use of mortar sands resulting as a by-product of china clay production has been common in the past in the South West of England. They have a mica content varying from 0.5% to 15% but this apparently has no noticeable effect on the durability of the finished work and may be due to the type of mica mineral in the mix. Being a by-product, the sands are not specially processed to conform to British Standards. However in recent years it is accepted, at least by the major contractors, that the mica increases the water requirement and problems can occur with surface finishes related to "shelling out" (particles popping out). One contractor in the Plymouth area will only use Permian sands transported from the pits to the east of Exeter, a distance of some 50 miles, thereby avoiding using the local china clay sands. This contrasts with plain-faced rendering to pre-war houses in Middlesex, where 2–3% mica grains were incorporated in the mix for decorative effect, with no noticeable problems resulting (Ragsdale and Birt, 1976).

In some overseas countries natural micaceous sands used in the production of mortar for masonry work are known to have given rise to a number of problems such as poor strength, poor bond, and porosity of mortars, as for example that reported by Fookes and Revie (1982) in Eastern Nepal. Tests carried out to determine optimum mixes for a series of mica contents in sands conforming to a fine BS 882 Zone 3/Zone 4 grading gave the following limits:

Mica Content	Mix
1 to 2%	3:1
2 to 6%	$2\frac{1}{2}$:1
6 to 8%	2:1
Above 8%	Not suitable

These mixes gave the desired initial strength and bonding properties for low grade mortar, but strict control of the water, sand and mica contents was needed to ensure success.

Mica increases the water requirement of the mortar mix and this is thought to be due to water absorption taking place by the mica crystals in 3 ways:

(1) Absorption between the atomic layers of the crystals
(2) Absorption on the surface of the crystal layers
(3) Additional absorption by the parts of the crystals which have suffered from chemical alteration (e.g. Biotite to Chlorite)

In plastering, rendering or similar work the presence of soluble salts in excessive quantities is likely to influence the movement of water within the materials. Thus, since sea salts are deliquescent, dampness can occur as a result of their presence and danger can then arise from the corrosion of metal work, conduits etc. The movement of salts to the surface as a result of the water present at the time of construction, or from exposure or condensation, can cause efflorescence and damage to finishes. Sea water contains 2 per cent of chloride ion and Chapman and Roeder (1969) have stated that for sands having up to

10 per cent moisture an acceptable limit for chloride content would appear to be 0.2 per cent.

9.3.5 Testing for clays. In the particle size testing of building sands BS 812 (1975) accepts that dry sieving will not give the same result as wet sieving (Table 8.4), the differences in the results occurring because of possible aggregation of particles. The existing standard suggests that a visual examination of the residue of each test sieve will usually confirm when an appreciable aggregation of the particles occurs. If the latter situation exists pretreatment of the sample by decantation prior to dry sieving is suggested.

As the material finer than 75 μm increases so does the difference between the wet and dry test results. This discrepancy could be overcome and consistency in results achieved if wet sieving were adopted as the standard method. Pike and Limbrick (1981) propose that a 38 μm test sieve be included in the range to be used, as a result of which a wet-sieve test could be more rapidly and reliably deployed. Currently two different methods are used to specify the clay, silt and dust content for sands and crushed rock fines in BS 1198, 1199 and 1200 (1976). For sands, the decantation method giving a separation at 75 μm is specified, whereas for crushed rock fines the sedimentation test which has a 20 μm demarcation is recommended. Supporters of the introduction of a 38 μm sieve indicate that it produces similar results to those obtained by the sedimentation test and suggest that it can be used for the determination of the clay, silt and dust content for both sands and crushed rock fines by wet sieving.

9.4 Influence of sand and fines characteristics on mortar properties

The distribution of particle sizes have a major influence on the main properties of mortars, such as workability, water retentivity, strength and durability.

9.4.1 Workability. The working properties of mortar have two essential aspects (a) the consistence, which is the characteristic affected by the amount of water added to the solid and active constituents and (b) the plastic characteristics referred to simply as 'cohesiveness' at a given consistence. Although cohesiveness is improved by the presence of increased fine material including cement, lime and air-entraining additives, the presence of some silt and clay, has the effect of making the mix more workable, and also provides the 'fattiness' so typically required by the artisan (McIntosh, 1970).

The shape and grading of the sand and silt particles also has a marked effect on the workability, or handling properties, of a mortar mix, in the same way that aggregate shape and grading is known to influence concrete workability.

Apart from hydrating the cement, water also lubricates the cement and sand in a mortar mix and so makes the mix workable. The lubricating water subsequently evaporates leaving voids in the set mortar, and thus the ideal mix is that which has an optimum water content providing both good workability together with the minimum number and volume of voids.

Two main characteristics of the aggregates which affect the water content of a mix are the total surface area and the particle interference. Both according to McIntosh (1970) are functions of grading, who states that the greater the surface area of the mix constituents, the larger will be the amount of water required to lubricate the surface of the particles. A large surface area can be either the result of a fine grading or the presence of a large proportion of sharp, angular particles or any combination of both.

In trying to achieve consistent workability of mortars the shape of the particles must influence the grading to be employed if a particular surface area is to be maintained, and it follows that aggregates having a wide variety of shapes and gradings can possess the same total surface area and so result in the same water content requirement or workability. However the surface area concept is apparently less reliable with very fine particles, because these appear to demonstrate their own lubricating mechanism which requires less water than that needed for the relatively larger particles.

These factors of size and shape of fine aggregates could benefit from further investigation and research to provide more information on the influence that they have on mortar properties. Fig. 9.7 shows some of the more common shapes that occur in fine grained particles.

9.4.2 Water retentivity. The water retention characteristic of a mortar is an important property closely related to workability. It is evidenced by the mortar's ability to retain the mix water and still remain workable when brought into contact with an absorptive brick. The water retentivity of a mortar can be increased by the use of air-entraining agents or finely ground particles, and the presence of clay has also been shown to influence retentivity. The use of a well graded sand helps reduce the void content and also the separation of materials in a mortar mix, which in turn reduce bleeding and improve water retentivity.

The water requirement, and therefore the strength, of a mortar is influenced by the shape and texture of the aggregate particles and these factors are expressed indirectly by the packing of the aggregate i.e. the percentage of voids when the sand is in a loose condition (Currie and Sinha, 1981).

9.4.3 Strength. The effect of sand grain size on mortar strength has been the subject of research, the characteristics of the sand, together with the proportion of cement to sand, being found to be a significant factor (McIntosh, 1970). In some instances mortar strength is improved by the presence of a few voids, indicating that a completely dense mortar is not always suitable. A well graded mix improves compressive strength more than transverse strength.

Mortar strength as related to brickwork strength has been considered by many investigators, the common conclusion being that excessive mortar strength is to be avoided. It is generally considered that the adhesion between bricks and mortar basically depends upon the flow of moisture from the body of the mortar carrying cementitious particles to the brick-mortar interface. However, the mechanism of adhesion involved at this interface is not fully understood.

Tests have shown that compressive strengths of mortars made with two different sands can differ more than would be expected from the effects of grading alone (McIntosh, 1970), the particle shape, proportion of material finer than 150 μm, together with the presence of chemical impurities all being possible causes of this strength difference.

Fookes and Marsh (1981) noted a significant reduction in workability and strength due to the presence of discrete white mica particles in mortars in East Nepal. Though the longer-term effects of free mica in cement sand mixes are not well understood it is generally thought to be related to the mineralogy of the micas present.

9.4.4 Durability. Mortar specifications normally take into account the need for adequate durability and the protection of any embedded reinforcement. The grading of the sand is important when considering frost resistance, although normal protection from frost can be achieved by sufficient minimum cement content being used in relation to the grading. With well graded sands a lower cement content is needed than for a poorly graded material which has an increased water demand (Beningfield, 1980). Thus the grading of fine aggregates is important in the production of a mix in which the packing of the particles is such as to make it possible for the cement paste to fill the voids.

Particle shape will affect the particle interference whereby the distance between the larger particles is such that it is not possible to allow free passage of the smaller ones, and the voids ratio is affected. Particle interference usually occurs as a result of the fine aggregate containing an excess of larger sizes.

Although some general observations are available on fine aggregate shapes, their packing and association in mortar mixes to produce acceptable strengths and durability requires more detailed information. Thus, this area of knowledge offers considerable scope for research.

9.5 UK specifications and practice

Currently, the specifications for building sands (BS 1198, 1199, 1200) (1976) are being revised as a result of changes occasioned by the revision of BS 812 in 1975 and of BS 410 (Test Sieves) in 1969, together with the introduction of metrication. Reference to these standards is made below, but it should be noted that it is common practice in the UK to use sands which vary considerably in their properties and characteristics from those specified in British Standards, simply because local sources in many parts of the country provide the cheapest available form of fine aggregate.

9.5.1 Masonry mortars. Sand gradings are an important factor in determining the strength of a mix (Bessey, 1966; Lee and Pyle, 1969; Lee, 1975). Early cement based mortars were commonly produced, probably unknowingly, by filling the sand voids with cement, in the proportion of approximately 1:3 cement to sand. This could vary, however, from about 1:4 if a fine sand was used to about 1:2.5 or less with a coarse or poorly graded sand (Beningfield, 1980).

In some cases however, sands may not give the expected strengths, even though they comply with the wide limits set by BS 1200 (1976), because they need a high water content to obtain the necessary workability. Water in the mix has been referred to as a "necessary evil" (Pyle, 1967). The "evil" results from the fact that only 25–30% of the water is required to hydrate the cement, the remainder is necessary to provide a mix of workable consistency and subsequently evaporates during the early life of the building with a consequent lowering of strength.

Sands having gradings which comply with BS 1200 (1976) have been widely accepted for general purpose mortars. These grading limits are shown in Fig. 9.3 and the tabulated values are given in Table 9.2.

Where reinforced brickwork mortars are required, sands complying with BS 1200 can be used, and it has been found that even sand at the fine end of BS 1200 is capable of producing satisfactory results.

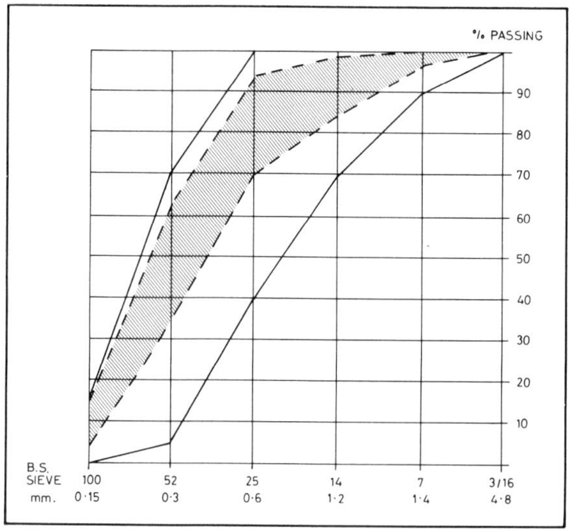

FIG. 9.3. BS Grading Limits of Sand for use in general purpose mortars (BS 1200:1976) "Ideal" sand limits shown hatched (see Table 9.2 and Table 9.3).

TABLE 9.2. *Sands for general purpose mortars (BS 1200:1976)*

B.S. sieve size	% By weight passing B.S. sieves
5 mm	100
2.36 mm	90–100
1.18 mm	70–100
600 μm	40–100
300 μm	5–70
150 μm	0–15

Practice has shown that the BS limits may be too wide for some mortar applications, and it has been suggested that a moderation of the limits of the fraction from the 1.18 mm size downwards (Smith and Plowman, 1972) would provide higher strengths generally. Some producers however are of the opinion that British Standard grading limits are so wide as to require considerable manipulation in order to produce the "ideal" sand. (Table 9.3).

Recent laboratory work on mortar sands suggests that the clay size particles, less than 20 μm in size, cause shrinkage problems, whereas silt sized material, ranging from 150 μm to 75 μm, benefits the mortar by giving better workability. The existing sizes used in British Standard specifications do not refer to particle sizes below 75 μm although it would appear that the 20–30 μm range could be critical to the performance of the mortar mix. Current proposals are that the Specification should be revised and should permit up to 8% of the material in the 150 μm–38 μm size range but not more than 3% of material less than 38 μm in size.

BS 1200 (1976) states that sands containing deleterious materials such as pyrite, salt, coal, mica, shale or similar laminated materials or flaky or elongated particles which may adversely affect the hardening, strength and durability of the mortar are not acceptable. In practice however, it is usually very difficult to assess the degree to which these materials are really deleterious, and several sources are known in the UK producing "contaminated" sands which are being used to produce satisfactory mortar. (9.7 below).

TABLE 9.3. *Grading limits of a suggested "ideal" sand for general purpose mortars (Benningfield, pers comm.)*

B.S. sieve	% by weight passing B.S. sieves
5 mm	100
2.36 mm	97–100
1.18 mm	85–99
600 μm	70–95
300 μm	35–65
150 μm	5–15

TABLE 9.4. *Sands for external renderings, internal cement plastering internal lime undercoat and floor screeds (BS 1199:1976)*

B.S. sieve size	% by weight passing B.S. sieve
5 mm	100
2.36 mm	90–100
1.18 mm	70–100
600 μm	40–100
300 μm	5–50
150 μm	0–10

9.5.2 External renderings. The materials used for external rendering mixes are frequently the same as those used for brickwork mortars, but sand grading becomes more critical where the finished appearance and durability is important. (Ragsdale and Birt, 1976). A relatively weak rendering made with an acceptably graded sand is likely to perform much better than one which is rich in cement but made from poorly graded sand. Sands used for rendering are considered acceptable if they conform to BS 1199 (1976) as shown in Table 9.4.

It will be noted that the range of particle size permitted on each sieve is wide when compared with BS 1200 (1976) but sands complying with the coarser mix would be harsh and are likely to be difficult to apply to a vertical surface. In some cases, unwashed dry-screened sands have been used and although sometimes complying with BS 1199 (1976) they require high water contents and so produce mortars which are relatively weaker and more prone to shrinkage than those made with well-washed sands. These defects may be due to particles of agglomerated clay passing as coarse sand grains which in normal wet processing would be removed.

9.5.3 Internal plastering. Plastering is intended to conceal uneveness in the background, and to provide a finish that is smooth, crack-free, hygienic, resistant to damage and easily decorated. Mixes used for internal plastering rarely include specifications incorporating cement as a binding agent, the lime/sand mixes being commonly preferred, their slow rate of hardening being an advantage. Sands used for cement/sand and cement/lime/sand plasters have been shown by experience to be acceptable when complying with BS 1198 (1976) as set out in Table 9.5. In the more usual case where gypsum is used as the cementing agent the more fine grained sands within the limits of BS 1198 (1976) are also generally acceptable.

In the UK it is not uncommon for sands to be used for rendering and plastering which do not comply with

TABLE 9.5. *Sands for gypsum plastering (BS 1198: 1976)*

	% by weight passing B.S. sieve		
B.S. Sieve	Undercoats type 1	Undercoats type 2	Finishing coats type 3
5 mm	100	100	100
2.36 mm	90–100	90–100	100
1.18 mm	70–100	70–100	90–100
600 μm	40–80	40–100	55–100
300 μm	5–40	5–50	5–50
150 μm	0–10	0–10	0–10

British Standards (Ragsdale and Birt, 1976). The quality of the sand is known to have an appreciable effect on the strength of the plaster/sand mixes, poorly graded sands not normally being recommended. However, there is less risk of defects in their use with gypsum mixes than when they are used in cement/sand mixes. Thus, when poorly graded sands (i.e. those with a higher proportion passing the 600 μm and 300 μm) only are available, gypsum mixes are preferable.

9.5.4 Floor screeds. Until recently, little published guidance was available for floor screeds. The development of special mixes and of ready-mixed materials, together with the widening range of special applications such as insulated and heated floors, has resulted in more attention being given to the technical requirements of such mortars.

The BRE Digest 104 (1973) proposed that the most suitable sands for such mixes are those complying with BS 882, but that the finer sands conforming with BS 1199 (Table 9.4) can also be used. According to Bessey (1966) the decision to include floor screeds in BS 1199 was taken on the basis that the application of such screeds is similar to that of a cement/sand rendering on a wall, and that furthermore, the specification is considered to be unduly restrictive at the coarser end, since a few per cent of material between 10 mm and 5 mm can be tolerated and may indeed be beneficial. In BS 882 (1973) Zone 3 sand (Table 9.6) is the nearest in grading to BS 1199 requirements, but Bessey suggests that BS 882 permits sand which would give too harsh a product in 1 : 3 and 1 : 4½ cement/sand mortar mixes. In recent years, the mortar industry has shown a tendency towards sands coarser than those required by BS 1199, one suggested "ideal" grading being that shown in Table 9.7.

TABLE 9.6. *Specifications for sands complying with BS 882 Zone 3 (1974)*

Sieve size	Percentage by weight passing B.S. sieve
mm	
10.0	100
5.0	90–100
2.36	85–100
1–18	75–100
μm	
600	60–79
300	12–40
150	0–10

TABLE 9.7. *An ideal sand grading for floor screeds (Benningfield, pers. comm.)*

B.S. sieve size	Ideal	Ideal range
5 mm	100	96–100
2.36 mm	90	86–95
1.18 mm	70	60–80
600 μm	55	45–60
300 μm	20	15–25
150 μm	3	0–5

9.6 Specifications and practice outside the UK

A review of standards applicable in other countries has been carried out by Ragsdale and Birt (1976) in which comparisons were made of specifications in Belgium, Bulgaria, Canada and the USA (Table 9.8). Particular note was made of the wide range of grading requirements used to meet these various specifications. Thus in the USA and Belgium a limitation on the fineness modulus is included (i.e. the sum of the total percentage retained on the 150 μm, 300 μm, 600 μm, 1.18 mm, 2.36 mm and 5 mm sieves divided by 100), and acceptable limits are suggested which range from 1.88 for fine to 3.30 for coarse mixes: this criterion is not incorporated into British Standards. The grading limits for plastering sand in the USA and Canada are identical and allow less fine material then the corresponding UK standards for sands when used with gypsum plaster. However, there is no standard in either of these countries for sands to be used in external rendering. They also noted that whilst Belgium has established standards for building sand for various purposes the Netherlands has only one standard for sand, which is applicable solely to reinforced and pre-stressed concrete. In contrast, Bulgaria has two categories of sand "for building purposes" which are termed (a) coarse and (b) fine. While the finer limits of category (b) are close to those of BS 1200, the coarse limits of category (a) are much higher than those permitted by the corresponding British Standards.

Ragsdale and Birt (1976) also provided comparative graphs applicable to the standards used in Canada,

TABLE 9.8. *Some relevant overseas Standards (after Ragsdale and Birt 1976) with additions*

Country	Reference	Year of issue	Use for which standard applies
U.S.A.	ASTM C114—70	1970	Masonry mortar
U.S.A.	ASTM C35—70	1970	Gypsum plaster
Canada	A82.56	1950	Masonry mortar
Canada	A82.57	1954	Interior plaster
Belgium	NBN 589—101	1969	General requirements for building sand
Belgium	NBN 589—106	1969	Masonry mortar
Belgium	NBN 589—108	1969	Facing (rendering)
Bulgaria	2271—67	1967	Building
Germany	DIN 1053	1974	Masonry mortars
Germany	DIN 18550	1967	Plastering mortars

USA, Belgium and the UK (Fig. 9.4) which again shows the wide difference in standards from country to country. In view of this divergence of specifications the suggestion has been put forward by the UK industry that the only real criteria are price and the ability of the material to perform satisfactorily in a variety of mixes. Thus in Hong Kong crushed rock sands are used together with supplies from marine dredged material as fine aggregates in mortar, the blending of the two sources producing a sand which is acceptable and complies with BS 1200 (1976).

Investigation of the particle size distribution of the building sands used in Germany (Albrecht and Wisotzky, 1968) confirmed that the variations in grading observed within the mortar industry could be attributed to the differences in their geological and geographical distribution. The sands are derived from river and glacial deposits as washed and unwashed pit products, some sources being dredged materials from rivers and lakes. Where better quality sources are not available, crushed rock sand, or even sands washed out of pumice gravel are used (e.g. as in the Coblenz region). This accounts for the fact that the specification and recommendation of the German Standards are only met by some of the sands used.

9.7 Occurrence of mortar sands in the UK

The current use of local sands for all building purposes is a matter of their commercial availability and the fact they often perform satisfactorily, although they may not comply with existing standards.

In Scotland particularly, sands conforming to British Standards are difficult to obtain (Ragsdale and Birt, 1976) and many non-standard sands from glacial and fluvial sources are accepted provided they are washed and have an acceptable performance. Sands used for mortars in Scotland were shown to exhibit a wide range of lithology in a study by Currie and Sinha (1981), the lithology including basalts, limestones, mudstones but with quartzite predominating. An exception to the glacial or fluvio-glacial source materials is a soft white friable sandstone near Kilwinning in Strathclyde which has been worked as a source of foundry sand in the past but is now often used for plastering and external rendering mortar mixes (IGS, 1977). Lignite is a common impurity in some Scottish sands and the occurrence of shrinkage problems is thought to be related to the presence of high percentages of this contaminant in some sources of greywacke (8.22).

The major sources of building sands for mortars in Northern Ireland are obtained from the fluvio-glacial deposits which are well distributed in the south-east part of the country. In addition some sands are dredged from Loch Neagh to meet the needs of the Belfast area, the source deposits probably being pre-glacial in origin.

In England and Wales mortar sands are obtained from a wide variety of sources. Thus in Durham, non-standard Permian sands have commonly been used, the coarse washed varieties being applied for rendering work to avoid possible shrinkage problems. More recently, these sands have been used in their dry-screen form and in consequence the mixes have been found to be subject to shrinkage problems.

The building industry in the Midlands utilises a high proportion of fine sand derived from gravel processing, which falls outise the limits for BS 1199 Table 1 (1976). One Local Authority has insisted on the use of a non-local, non-standard sand of this type, because it has an established reputation as a good rendering sand (Ragsdale and Birt, 1976).

Sea dredged sands are used extensively in the Merseyside area (the local name of 'Mersey grit' or 'Mersey sand' covering a wide variety of marine sands taken from the Mersey estuary) and one such fine washed gravel sand has been successfully used for rough cast and stucco work without any apparent defects. The finer graded Mersey sands are used mainly for brickwork mortars and rendering. They tend to be single sized, between 300 μm and 150 μm, and so fall outside the grading requirements for BS 1200 and BS 1199. Although the shell content is fairly high in these sands, with up to 8% of fragmented shell present, no apparent defects have been reported and neither does the combined salts content of up to 0.1% appear to have caused any damage.

In West Cheshire, where most of the sands used are of glacial origin, only one sand pit produces material within the limits of BS 1200. One producer in the area has found it necessary to import sand from North Wales to blend with the local sands and so produce a satisfactory grading for "floor screed" use.

Mortar sand sources in Wales are usually in the extensive deposits of glacial sands as gravels which occur well distributed over North and South Wales. In addition beach sands from the Porthcawl, nr. Swansea have found local uses as general mortar aggregates.

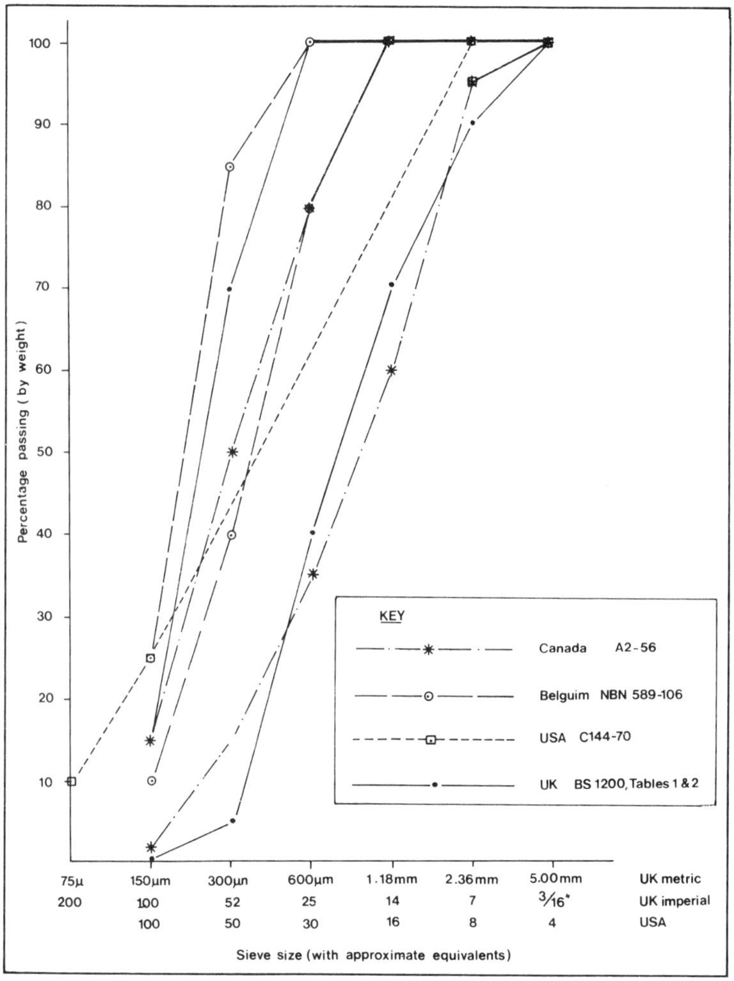

FIG. 9.4. Specified grading limits for masonry mortar sands (after Ragsdale and Birt, 1976).

In recent years it has been accepted that sea-dredged sands can be used successfully to produce mortar mixes. However, it is usual to add a plasticiser, such as limestone dust filler to increase the workability due to the single sized nature of these fined grained aggregates. In the Bristol and Cardiff areas, marine sands taken from the Bristol Channel are commonly utilised.

The extensive outcrops of sand in the Folkestone Beds within the Lower Greensand Series in SE England provide some well graded mortar sands, particularly after washing. Considerable lateral and vertical variation in lithology is common, and many quarries excavate up to 40 metres depth of sand from these deposits, some from below the water table (Humphries, 1964). Other horizons within the Lower Greensand of the Weald which yield building sands are the sandy facies of the Hythe Beds and the Sandgate Beds in the Redhill area although the latter are usually very fine-grained and limited in their use because of the high clay fraction. Although economic factors still dictate the use of local sands which are outside specifications, the building industry is usually able to obtain the sands it needs from indigenous sources without blending. For example, the local "Runfold" sands, taken from the Folkestone Beds near Farnham in Surrey, produce sand material complying with BS 1200, 1198 and 1199, in addition to BS 882 Zone 3 and Zone 4 concreting sands.

In the western part of the Hampshire Basin, the sands in the Reading Beds and Bagshot Beds are common sources of building sands, the popular "Sherfield English" pits located in the Reading Beds near Romsey have been noted for many years for the excellent general mortar sand supplied to local markets.

9.8 Engineering performance and problems of mortars with particular reference to UK case histories

It should be noted that when the proportioning of cement to sand for a mix is by volume, the weight of sand in a given volume can vary by as much as 30% between the dry and damp condition, as a result of which the true proportions used will not be known. (Bessey, 1966). This, together with the fact that the bulk density of cement can also alter, can lead to disputes between contractor and client should defects occur. It follows then, that the use of the BS 812 (1975) methods for assessing relative density and water absorption are of particular importance, especially with crushed rock sands where the water absorption can vary considerably in hot conditions. In recent years, a better understanding of the materials used for mortars has enabled more reliable guidance to be given as to the type of mix suitable for various uses and conditions.

The design of mortar mixes is normally based on the premise that the voids in the sand, which generally amount to 25 to 40% (typically about 33%) should be filled with binder. Inability to completely fill the voids does not necessarily result in failure however, but an increasing proportion of unfilled voids in a mix can increase the probability of trouble arising (Beningfield, 1980). With external rendering mortars, the main factors which indirectly contribute to their reliability and the weather resistance (i.e. durability) of the wall are those which affect the characteristics of drying shrinkage, frost resistance, adhesion, and flexural and compressive strength.

9.8.1 Effect of grading. Early research (Furnas, 1931) applied to mortars which considered close packing of solid particles, concluded that the best workability and packing was obtained from continuously graded aggregates.

A more recent publication (Bloem, 1962) dealing essentially with 1:3 cement/sand mortars reported that by employing air entrainment coarser sands than those now permitted by ASTM (1976) specifications could be used.

It is generally accepted that finer grained sands require more water to obtain reasonable workability and this causes reduction in strength and increased drying shrinkage. Investigations by Bessey, (1966), attempted to rationalise methods for specifying natural sands for mortars using the median diameter and a sorting coefficient but these have generally lacked the definition that can be achieved by the traditional gradings. Lee and Pyle (1968) and Lee (1975) have shown that for some sands used in mortars the strength of a mortar can be related to the grading of the aggregate. On the assumption that the aggregate complies with the requirements of BS 1200, Lee (1975) proposed that proportions of material retained on a 600 μm sieve, passing the 600 μm but retained on the

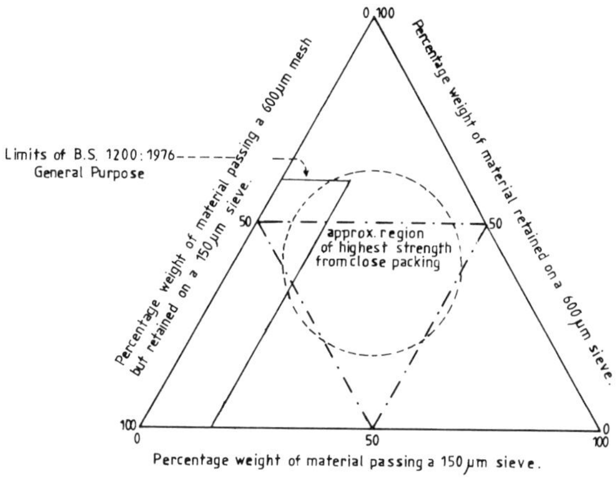

FIG. 9.5. Use of triangular graph paper for showing sand gradings (after Lee 1975).

150 μm sieve, and that passing a 150 μm sieve can be plotted on triangular graph as shown in Fig. 9.5, as a result of which a prediction of mortar strength is possible. It can be seen that as the central region is approached so the particle packing characteristics improve until an ideal packing is obtained at the centre point, having a blend of 1/3 of each fraction (sand, cement, lime and filler, if present) which would be expected to produce maximum packing and therefore optimum strength for the mortar. In theory the ideal mortar sand with 66.7% passing a 600 μm sieve and 33.3% passing a 150 μm sieve, when plotted on log normal paper, would show the grain size distribution given in Fig. 9.6.

In practice, it is suggested that the ideal sand is generally finer grained than that shown in Fig. 9.5, the reason being that this is necessary to obtain good

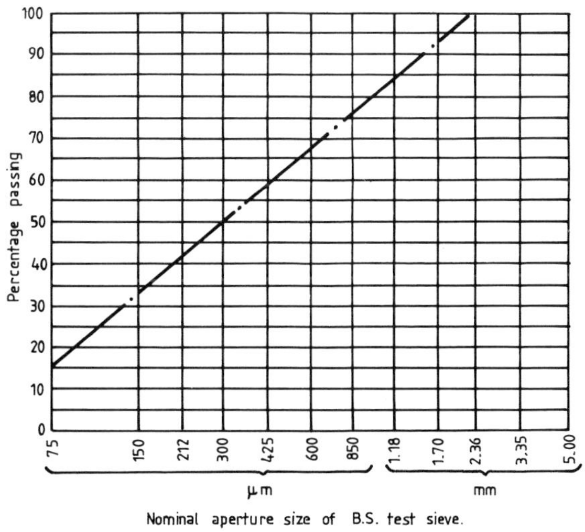

FIG. 9.6. Ideal grading curve of a building mortar if no voids (after Lee, 1975).

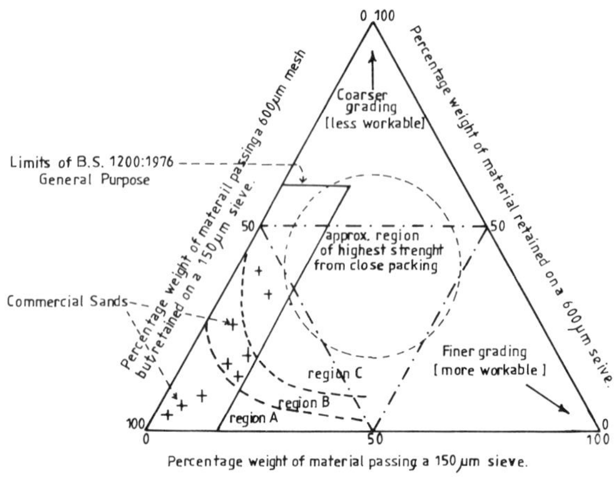

FIG. 9.7. Gradings of some commercial sands (after Lee 1975).

workability. Usually the ideal gradings are rarely available. Some of the more common gradings of sands used by the industry are shown diagramatically in Fig. 9.7. Sands in the regions B and C can be used without further treatment to produce satisfactory cement/lime/sand and gypsum/lime/sand mortar. Sands in region A will require blending, and in this case the experience of the bricklayers and plasterers is usually necessary to establish the mix required.

The packing of sand influences its void content (Lee, 1975) and the dry compacted bulk density may be considered as a true indication of the voids present. Silica has a relative density of 2.635 i.e. one cubic metre weighs 2635 kg (Table 9.9). However in practice, it is found that dry natural sands have compacted dry bulk densities of between 1400 and 1800 kg/m^3 and thus the void content represents 32 and 47 per cent respectively.

A poorly graded sand i.e. having a low dry compacted density requires a high proportion of material to fill the voids (Lee, 1975). If this material is cement or water, then high drying shrinkage will result. Poorly graded sands can be improved by blending with other sands but when this is not possible, the upgrading can be achieved by adjusting

TABLE 9.9. *Void contents of Sands (Lee 1975)*

Sand dry bulk density Kg/m^3	Void %
2635	0
1800	32
1700	35
1600	39
1500	43
1400	47

the lime content or adding an air entraining agent so that discrete gas pockets are formed.

A study of the factors affecting the packing and porosity of particles has been made by Lees (1970) on aggregate gradings for dense asphaltic compositions which may have similar applications for the mortar mix design. Although a general association between strength properties and density can be assessed Lees' studies indicate that there is no unique ideal maximum grading curve. The correct proportions for minimum voids are inevitably affected by changes in aggregate particle shape both from source to source and from size to size. Later investigations on the structural arrangements of aggregate particles in concrete were used by Lees (1973) to develop a method by which the packing properties can be designed such that maximum-density gradings can be derived which take into account the changes in aggregate packing properties that may occur at any size level. In addition the method recognises the importance of any lubricating or adhesive coatings present, the surface static effects on small particles and the degree or type of compactive effect as well as the effect of external boundaries. These findings may have a use in the design of mortar mixes and further research into the relationship between density and the engineering properties of mortars could prove well worthwhile.

9.8.2 Current experience. Mortar has in the past been mainly a site product, dependent for its properties upon the mixing skill of the bricklayer and plasterer, rather than a material which could be precisely specified and manufactured. However, with the much greater use of factory made ready-mixed mortars this position has changed and the emphasis is now on the use of known sources of sand and close control of the mix proportions with a consequent reduction in the site labour. The move from site mixing towards the ready-mixed product means that the use of local sands will become less common, and there will be a greater imput of technical expertise from the ready-mixed industry.

In a case study analysis of some 63 sands in the UK by Ragsdale and Birt (1976), it was found that only 8% complied in all respects with the British Standards against which they were supplied. Even so, the customers were generally prepared to accept these sands without complaint. The sands involved were mostly supplied for external rendering and brick work mortar. The grading tests showed that only 2 (4.5%) of the sands for rendering complied with BS 1199 Table 1 (1976) whilst some 26 (59%) of the sands for mortar complied with BS 1200 (1976).

9.8.3 Future trends in the use of sands. It has long been recognised that sands containing appreciable amounts of material passing the 150 μm sieve are needed to achieve adequate workability of the freshly-mixed mortar, although this material is unnecessary in terms of performance. It has been suggested that for external rendering and dubbing out as much as 20% to 25% of the material should pass the 150 μm sieve, while still maintaining a good continuous grading for the coarser fraction. This would have the effect of flattening the present grading envelope of BS 1199 Table 1 (1976) thereby reducing the maximum amount passing the 2.36 mm, 1.18 mm, and 600 μm sieves by about 10%, 15% and 35% respectively. As a better understanding of the technology of sands in mortars is developed the trend towards blending of sands from different sources may become more practical and economical to ensure consistency of product.

9.9 Discussion

Much of the sand used for the production of mortar in the UK, particularly that used for external rendering, does not conform to the appropriate limits prescribed by the current British Standards. Nor is reference made to the quantitative limits on 'impurities' in the existing UK standards, there being a considerable difference of opinion by both producers and users on the acceptability of sands containing impurities such as mica, lignite or high salts content.

Although the use of non-compliance sands is a common practice, both in the UK and elsewhere in the world, the most important criteria for performance appears to be the quality of cleanliness. Unwashed dry-screened sands, including crushed rock sands, whilst sometimes complying with British Standard requirements, tend to produce mortars having a high water demand as a result of which they are relatively weaker and more prone to shrinkage than mortars made with well-washed sand, where the clay content has been considerably reduced. Standards differ from country to country, largely due to the fact that they are based on the availability of natural resources and the manner of their use. Hence sands which are acceptable in one place may well be rejected elsewhere. The use of different cements and construction techniques will also influence whether particular sands can be accepted as suitable for construction mortars.

There is a considerable need for further investigation work to be carried out on aggregates for mortars in order to close the gaps in our present knowledge. Although much technical data is available for concrete aggregates, it is clearly apparent that the accumulation of comparable information relating to the fine-grained aggregates used in the production of mortars has been much neglected.

References

ALBRECHT, W. & WISOTSKY, I. T. 1968. Survey of the particle size composition of sands for masonry and plastering mortars. *Bau und Bauindustrie,* **11,** 692–697.

AMERICAN SOCIETY FOR TESTING AND MATERIALS ASTM Standard Specification for Aggregate for Masonry Mortar C144–76.

BENNINGFIELD, N. 1980. Aspects of cement-based mortars for brickwork and blockwork. *Concrete*, (Jan.) 27–30.

BESSEY, G. E. 1966. Current developments affecting the design and use of mortars for building purpose. *Journal of Applied Chemistry,* **16,** 313.

BLOEM, D. L. 1962. Effects of aggregate grading on properties of masonry mortar. Symp. on Testing. ASTM Spec. Tech. Publ. No. 320. 67–92.

BUILDING RESEARCH STATION DIGEST. 1971. Choosing specification for plastering. 49. 1–7.

BUILDING RESEARCH ESTABLISHMENT DIGEST. 104. 1973. Floor screeds.

BUILDING RESEARCH ESTABLISHMENT DIGEST. 160. 1973. Mortars for Bricklaying. 1–4.

BRITISH STANDARDS INSTITUTION BS 882. 1973. Specification for aggregates from natural sources for concrete (including granolithic).

BRITISH STANDARDS INSTITUTION BS 812. Part 1. 1975. Methods for sampling and testing of mineral aggregate sands and filters.

BRITISH STANDARDS INSTITUTION BS 1200. 1976. Sands for mortar for plain and reinforced brickwork, blockwalling and masonry.

BRITISH STANDARDS INSTITUTION BS 1198. 1976. Sands for internal plasters with gypsum plasters.

BRITISH STANDARDS INSTITUTION BS 1199. 1976. Sands for external renderings, internal plastering with lime and Portland cement, and floor screeds.

BRITISH STANDARDS INSTITUTION BS 4721. 1981. Ready-mixed building mortars.

CHAPMAN, G. P. & ROEDER, A. R. 1969. Sea dredged sands and gravels. *Quarry Managers. J. London* 251–263.

CONNOR, C. C. 1948. Factors in the resistance of brick masonry walls to moisture penetration. *Proc. Am. Soc. Test. Mats.* (48).

CONNOR, C. C. 1953. Some effects of the grading of sand on masonry mortar. *Proc. Am. Soc. for Testing Materials*, (52) 933–948.

CURRIE, D. & SINHA, B. 1981. Survey of Scottish sands and their characteristics which affect mortar strength. *Chem. and Ind.* 639–645.

FOOKES, P. G. & MARSH, A. H. 1981. Some characteristics of construction materials in the low to moderate metamorphic grade rocks of the Lower Himalayas of East Napal. 2: Engineering Characteristics Proc. Instn. Civ. Engrs. Pt. 1, 70., Feb., 139–162.

FOOKES, P. G. & REVIE, W. A. 1982. Mica in Concrete—a case history from Eastern Napal—*Concrete,* **16,** 3, 12–16.

FORDER, E. E. 1971. Some effects of mica in concrete. Project report prepared for a C. & C.A. Advanced Concrete Technology Course.

FURNAS, C. C. 1931. *Ind. Eng. Chem.* 23.

H.M.S.O. 1979. Production of Aggregates in Great Britain. 1977 and 1978.

HUMPHRIES, D. W. 1964. The stratigraphy of the Lower Greensand of the south Weast Weald. *Proc. Geol. Assoc. London.* **75,** 39–59.

INSTITUTION OF GEOLOGICAL SCIENCES. 1977. Mineral resources Consultative Committee, Mineral Dosier No. 18., *Silica*. H.M.S.O., London.

LEE, H. N. 1975. The technology of mortar design. *Chem. and Ind.* 114–117.

LEE, H. N. & PYLE, M. A. 1968. The design of mortars for calculated load bearing brickwork. *Pro. Brick. Cera. Cos.* **11,** 143–161.

LEES, G. 1970. The rational design of aggregate gradings for dense asphaltic compositions. *Proc. of Assoc. of Asph. Paving Tech.,* 39–97.

MCINTOSH, J. D. 1970. Specifying the quality of bedding mortar. *Proc. Brit. Cer. Soc.* **17,** 65–82.

PETTIJOHN, F. J. 1949. Sedimentary Rocks, Harper & Row.

PIKE, D. C. & LIMBRICK, A. 1981. A study of sieve tests for building sands. *Chem. & Ind.* **19,** 626–630.

PYLE, M. 1967. Modern Mortars for brick and blockwork. Building Materials.

RAGSDALE, L. A. & BIRT, J. C. 1976. Building Sands, availability usage and compliance with specification requirements. Ciria Report 59., 1–30.

SMITH & PLOWMAN. 1972. Bricks and Mortars Ibstock Building Products Ltd. Publication No. 1.

SWALLOW, H. T. S. 1964. Building mortars and bricks. Brit. Ceram. Res. Assn. Tech. Note 57.

WESTBROOK, E. L. E. 1960. Sands for plastering, rendering and mortar work. Sand and Gravel Assn. of Gt. Britain.

Chapter 10. Aggregates in unbound pavement construction

10.1 Introduction

In highway and airfield pavements aggregates are used in various types of unbound or bound materials (standard nomenclature for pavements as used in the United Kingdom is illustrated in Fig. 10.1). This chapter is concerned with aggregates which are not bound by cementitious or bituminous binders i.e. with various forms of wet mix and dry mix macadam used in bases and sub bases and with other less norrowly defined granular sub base materials including "as dug" gravels etc. Cement bound aggregates are discussed in Chapter 8 and bitumen and tar bound aggregates are discussed in Chapter 11. The terms of reference of this working party exclude consideration of artificial aggregates. However it should be noted that several types of artificially produced aggregates are widely used as granular sub-base materials. Important in this respect are well burnt colliery shale and crushed slags.

The prime function of the base and sub-base layers of a pavement, whether composed of unbound or bound materials, is to distribute wheel load stresses over a sufficiently wide area of the subgrade to reduce the deformation (transient and permanent) of that subgrade to acceptably low levels. It has been demonstrated in several carefully controlled experiments (Highway Research Board, 1962; Croney and Loe, 1965) that unbound aggregates are less well able to perform this function than bound materials. Nevertheless the considerably lower cost of the former* means that very large quantities of unbound aggregate materials have been and are still being used in pavement construction. While this is true for construction in the United Kingdom and other developed countries it is even more so in the developing countries where unbound aggregate materials may comprise the whole of the structural component of the road construction, with or without a thin application of single or multiple coat surface dressing.

* This lower cost applies to the straight comparison of price per tonne. It has however frequently been pointed out that the better load distributing properties of the bound materials would enable them to be laid at a lower thickness than the unbound. While this is the case in certain circumstances, for example where the requirement for frost protection of the subgrade does not apply, the overall economic advantage in terms of price per square meter is still generally with the unbound materials.

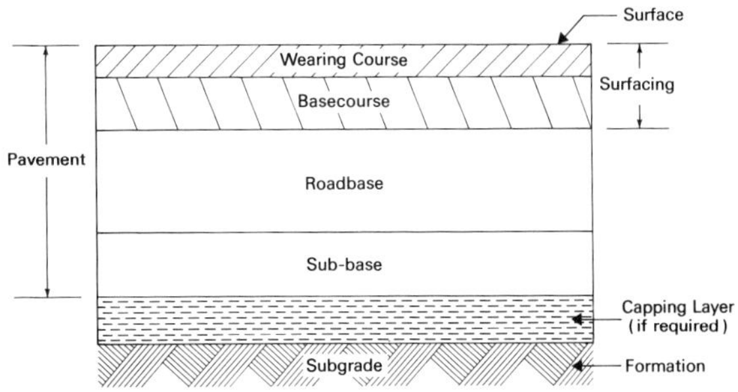

FIG. 10.1. Typical flexible pavement construction layers.

10.2 Desirable properties of aggregate for unbound layers in pavement construction

As with other uses of aggregates for construction materials it should not be supposed that an aggregate's suitability can be defined by its petrographic type alone. Aggregate petrography can only be considered together with particle shape, aggregate grading and degree of compaction, any of which can govern the degree of success in performance. It is therefore important to pay attention to those parts of a specification which deal with shape, grading and compaction requirements in addition to noting permitted aggregate types.

The strength properties (see Chapter 7 for test methods) required for the unbound layers of a composite pavement are:

(a) hardness, i.e. resistance to abrasion and attrition during mixing, handling and transporting,

(b) toughness, i.e. resistance to crushing and impact loads during compaction and under traffic.

In addition to possessing sufficient initial strength to resist the degradation mentioned above, the aggregate must also maintain an adequate strength during its service life, i.e. the aggregate must have an adequate resistance to weathering processes (10.2.4).

10.2.1 Resistance to degradation (crushing, abrasion, attrition)

10.2.1.1 Igneous rocks. In the unweathered state igneous rocks as a class are usually hard, due to the nature of the silicate materials from which they are formed, and strong, due to the interlocking nature of the constituent materials. Plutonic varieties are in general less tough than their hypabyssal and volcanic counter-parts (Knight and Knight, 1948). In addition the plutonic varieties rich in mica tend to show less resistance to abrasion than those containing appreciable quantities of hornblende and augite. (See also Chapter 11, sub-section 11.4 for further discussion on strength properties of igneous, sedimentary and metamorphic rock aggregate).

10.2.1.2 Sedimentary rocks and unconsolidated sediments. Sedimentary rocks to be used as aggregates fall generally into the two categories of sandstones and limestones. Of the former the quartzites tend to show the greatest strength due to the near complete cementation with secondary silica. As is well known the lower strengths of calcium carbonate and iron oxide cements combined with a tendency to less complete cementation leads to generally lower strength for calcareous and ferruginous sandstones.

Post Carboniferous limestones also tend to be lower in strength due to incomplete cementation but limestones of the Carboniferous and earlier periods frequently possess aggregate crushing values equal to many igneous rocks.

Excessive amounts of clay mineral in argillaceous limestones of the Aymestry period of the Silurian have been known to lead to base failure in a road site in Staffordshire. Here plastic fines from the decomposed base intruded the open textured bitumen macadam base course and resulted in disintegration of this layer under traffic.

The component particles of the unconsolidated sediments classified as gravels and sands possess the intrinsic strength and durability properties of the constituent rock and mineral types. In the mass, as in the case of all other rock types, the properties are governed by the shape (including ratio of crushed to uncrushed particles) and grading characteristics.

10.2.1.3 Metamorphic rocks. Metamorphic rocks of the foliated varieties, owing to the preferential orientation of flaky minerals (slaty, schistose and gneissose textures), are comparatively low in toughness and wearing resistance. Non-foliated varieties (e.g. hornfels and some gneisses) in which more or less complete recrystallisation of quartz and other minerals has occurred and which are often rich in hornblende, augite and garnet, are extremely hard and tough.

10.2.2 The degradation process. Whether occurring in the pavement or in laboratory testing, the degradation of particles proceeds in the following manner (Lees and, Kennedy 1975). Initially, as a crushing load is applied, the load is transmitted to the solid through its protruding points, if any, leading to the crushing of these points and the formation of larger contact areas. As the load is increased, either the crushing or flexural strength is exceeded and a crack is formed, leading to particle breakage.

10.2.2.1 Influence of particle shape and size. The strength of a particular rock type (as distinct from aggregate strength) is an intrinsic property of the rock itself but both shape and size of particle affect the strength of an aggregate as measured in laboratory tests and as encountered in service. Examples are the effect of flakiness index upon Aggregate Impact Value of a variety of petrological types (Ramsay, 1965), and the effect of size of particle upon Aggregate Crushing Value (Shergold, 1955).

In the former case the greater degradation of flaky (and of elongate) particles of a given size (i.e. volume) is associated with the greater bending moments that can be applied to given cross sectional areas. The latter case concerning the effect of size, demonstrates that the theory of statistical distribution of flaws, which states that the probability of a critical flaw occurring increases as the size of the member increases, applies to irregular rock particles in a similar way as it applies to concrete cubes and beams, metal test specimens etc.

10.2.2.2 Influence of moisture content. The effect of moisture content on aggregate strength is often neglected due to the failure of British and other standards (a) to specify precisely the state of dryness at which comparative standard tests should be performed and (b) to require that investigation should extend over a range of moisture contents. BS 812: 1975 states that "aggregate shall be tested in a surface-dry condition. If dried by heating the period of drying shall not exceed 4h, the temperature shall not exceed 110°C".

The looseness of the term surface dry and the unguided option left to the tester to oven dry for 4 hours or not as he considers necessary are such that aggregate could be tested from a zero moisture content to a nearly saturated state without recognition of or allowance in interpretation for the precise moisture content at time of test.

While the effects of this variable may generally not be significant in the range of commonly employed aggregates in the UK the effect may be considerable in certain cases in the UK and in the context of the wider range of aggregates required to be used in overseas countries. Testing aggregates only in the dry state if this were specified would conceal the extent to which aggregate strength falls with increasing moisture content, e.g. the loss in strength was found by Markwick and Shergold (1946) to be up to 40% for some igneous rocks. Work to investigate this tendency further has continued at the Transport and Road Research Laboratory in more recent studies. Irwin (1959) as reported by Hosking and Tubey (1969) showed that some low grade aggregate types from East and West Africa gave siginificantly poorer results when tested in a saturated state, with lateritic gravels losing nearly 20% of strength, some limestone losing nearly 80% of its strength and murram gravels losing over 70%. Coral limestone from the Bahamas showed a strength loss of some 38%.

Shergold and Hosking (1963) compared results of tests on dried and water saturated samples of various low grade agrillaceous and gritty rocks from the UK.

For aggregate judged on engineering experience to be acceptable, modified Aggregate Impact Tests showed about 17% loss in strength while modified 10% Fines value fell by 24%. Aggregates which had been judged unacceptable from practical experience showed much greater loss in strength in the saturated condition, namely 35% loss in the modified Aggregate Impact Test and 63% loss in the modified 10% Fines test. Shergold and Hosking concluded that in respect of aggregates for unsurfaced roads, bases on lightly trafficked roads and sub bases, useful tentative limits would be a maximum of 40 for the saturated modified Aggregate Impact Value and a minimum of 5 for the saturated 10% Fines Value.

These results indicate the vital importance of carrying out additional non-standard tests for ACV, 10% Fines, AIV, Modified AIV and/or Los Angeles at the anticipated maximum field moisture content if moist conditions and moisture susceptible aggregates are suspected.

10.2.2.3 Unstable aggregates. Unstable aggregates are defined by Hosking and Tubey as "apparently sound aggregates which rapidly decompose after being laid in the road". Examples are quoted from Australia, Basutoland, Fiji, Great Britain, Kenya, Mauritius, South Africa and the USA. All rocks subject to this decomposition in stock piling and/or in use are of basic composition, and all gave satisfactory results in initial routine laboratory strength testing. Methods to detect these unsound aggregates at the assessment stage have included:

1. Petrographic examination (Weinert, 1964).
2. Sulphate soundness tests (Chapter 7).
3. Washington Degradation Test (Chapter 7).
4. Modified Aggregate Impact Tests on samples that have either been soaked in water for 21 days or boiled in water for 24 hours.

Some indication of performance is given by all forms of testing listed, Weinert having, for South African rocks, established reasonable agreement between degree of alteration (determined by point count analysis) of mainly ferromagnesian minerals and risk of decomposition. Suspect materials in general showed a greater than 30% degree of alteration.

Tests for sulphate soundness or impact tests on soaked or boiled samples have shown promise but problems of either low reproducibility or that results may apply only to distinguish acceptable performance within a rock type, not across different categories of rock, prevent these tests being generally and reliably applicable. Useful information and guidance can however be obtained with tests performed under close control and for limited range of aggregate types.

10.2.2.4 Types of degradation. In respect of all the above the degradation process may be considered as:

1. The breakage of particles into approximately equal parts
2. The breakage or crushing of angular projections which may exist on the exterior surface of the particles
3. The grinding of small-scale asperities off the major faces or planes of the particles

The distinction between these processes is mainly in the different ratios of size of the broken product to the size of the original particle. Their relative importance differs with shape and petrology of the aggregate. For example, non-crushed high sphericity rounded gravels will only experience type 1 degradation as will parallel sided slabby particles. Angular cubical particles and tapering rod, disc and blade-like particles will often experience type 2 degradation before any type 1 breakage occurs. Weakly cemented sedimentary rock particles and surface weathered rock particles will be susceptible mainly to type 3 degradation. Repeated stressing at small strains, a fatigue process, will also tend to produce this kind of surface degradation. Attrition of this nature may eventually degrade the particles to their individual crystal or grain sizes.

Discussion of the mathematical solutions to the theory of degradation are contained in Rittinger (1867), Kick (1885), Gaudin (1926) Shelburne (1940), Bond (1952) and Yashima *et al.* (1970).

10.2.2.5 Influence of grading on aggregate degradation. Where aggregate of low crushing strength is considered for use it should be borne in mind that the amount of crushing to be expected is a function not only of the intrinsic strength for a given size and shape, but also of the aggregate grading. Clearly the standard Aggregate Crushing and Aggregate Impact tests cannot assess this factor since they are carried out upon 'single size' materials. Even the Los Angles test which permits the test to be carried out on a limited variety of gradings does not set out specifically to measure this effect.

The importance of grading on crushing is seen in Table 10.1 (after Ghani, 1969) which demonstrates that the denser gradings experience much less degradation than the more open gradings. The results encountered by Ghani in this series of tests prompted him to propose that modifications to the Aggregate Crushing Test and Aggregate Impact Test might be helpful in assessing aggregates susceptible to degradation. In these tests an AGIV (Aggregate Grading Impact Value) and an AGCV (Aggregate Grading Crushing Value) are determined by substituting the combined aggregate grading for the standard single size aggregate of BS 812. The values calculated are the increase in per cent fines, passing the 2.36 mm sieve, and the per cent increase in total surface area. In these tests the main advantage is that the extent of crushing of the aggregate is studied on the basis of the actual grading to be used in service, from which it also follows that the grading to produce least degradation can be chosen. The aggregate considered in this study was a Millstone Grit and the tests were carried out without a binder. Other work performed both at the University of Birmingham (same series of tests) and the USA (Shelburne, 1940; Moavenzadeh & Gotez,

TABLE 10.1. *Degradation of a Millstone Grit aggregate from Ambergate, Derbyshire, combined to 5 different gradings (after Ghani 1969)*

	Grading	Initial porosity (after vibration)	Aggregate Grading Impact Value AGIV	Marshall hammer compaction on dry aggregate 50 blows each 'face' % increase in fines (passing BS 2.36 mm sieve)	Marshall hammer compaction on dry aggregate 50 blows each 'face' % increase in surface area
BS 594	Base course 79% stone content			25.0	
BS 594	Base course 59% stone content			14.0	
BS 594	Wearing course 48% stone content			8.3	
Fuller (1907)	Grading curve	22.9	20.0	13.4	33.3
Lees (1970)	Grading method	1817	13.5	7.5	14.3

1963) has shown that employing the aggregates within the context of a bituminous mix, while generally reducing the degree of crushing, did not appear to affect the order of merit of the gradings or rock types used. It thus appears that an increasing use of low crushing strength aggregates and of flaky and/or elongate aggregates is possible provided dense gradings are designed to suit their special characteristics (Lees, 1970).

10.2.3 Engineering effects of degradation. The initial effect of degradation of a dry granular material, in service, is therefore likely to be a reduction in the two principal original components of stability namely rock particle interlock and interparticle surface friction between the original particles. However, densification, due to the smaller fragments produced filling the spaces within the aggregate structure is likely to occur, thus producing additional points of interparticle contact within the structure at which surface friction may be mobilised. In summary, the physical significance of disintegration of an aggregate structure is dependent upon the relative importance of these two changes. Stability will decrease if the effect of the additional number of contact points formed by degradation is less than the effect of reduced interlock and surface friction between the major particles but will increase if the reverse is true.

For example, it is by now generally accepted that in single size unbound or waterbound aggregates of large maximum size intended for road bases and sub-bases some crushing of the aggregate under the roller is desirable because of the greater density and further frictional contacts which result. It is emphasised that in order to be beneficial this process must take place under the roller i.e. during initial compaction. Post rolling degradation under traffic inevitably leads to a certain degree of rutting due to the further compaction which takes place, independently of whether the final aggregate product has a higher or lower shearing resistance than the original.

10.2.4 Weathering. In addition to the strength properties mentioned above a considerable degree of resistance to weathering during the service lifetime is required, including resistance to disintegration with cycles of wetting and drying or freezing and thawing. Some of the severest cases of failure of roadstone in unbound bases have been associated with physical breakdown of rock already suffering from varying degrees of chemical weathering e.g. Washington State (Turner & Wilson, 1956; Minor, 1959), Virginia (Melville, 1948), Oregon (Scott, 1955) and Idaho (Day, 1962).

The rock types identified in the majority of these studies were black basalts and gravels of Eocene age, although a number of other rock types including other Tertiary sandstones and shales and gravels of Mesozoic age have also been involved. In the failed areas a layer of saturated high plasticity fines has practically always been found between the surface layers and the underlying base. This slurry, which may be as much as 10 mm in severe cases, is formed when aggregate containing secondary minerals degrades into plastic fines under the combined action of water plus movement of rock against rock (attrition) as the pavement is flexed under moving loads. The plastic fines quickly permeate the entire base course. When this happens the stability of the base course is reduced sharply and failure of the pavement structure is inevitable.

Scott (1955), Day (1962) and Weinert (1968) all showed that rather than the primary minerals it is the secondary minerals formed from the primary minerals under the action of weathering and disintegration on a geological time scale, that are the principal causes of distress and failure of granular layers in pavements.

Irfan and Dearman (1978a) following Mendes *et al.* (1966) calculated a micropetrographical index I_p from the modal analysis as:

$$I_p = \frac{\%\ \text{sound constituents}}{\%\ \text{unsound constituents}}$$

for a weathered granite from Cornwall. Sound constituents were the primary minerals such as quartz, plagioclase, potash feldspar, biotite, muscovite and accessory minerals such as zircon and magnetite; unsound constituents were secondary minerals such as sericite, gibbsite, kaolinite (altered felspars), chlorite, secondary muscovite, iron-oxides, together with microcracks and voids resulting chiefly from weathering. The inclusion of microcracks and voids as "unsound constituents" is novel and may be justifiable here. However it might lead to a false conclusion of a highly weathered state if applied outside the context of this study e.g. to porous sandstones or vesicular unweathered basalts.

In a companion paper Irfan & Dearman (1978b) have carried out "laboratory and field determinations of a range of physical and mechanical properties of the various stages of weathering in granite materials. The results are correlatable with both micropetrographic and microfracture indices, often as a simple linear relationship with a high correlation coefficient".

Sedimentary rocks employed as aggregates, chiefly the sandstones and limestones, being themselves the product of weathering processes suffer comparatively little from mineral alteration. (Exceptions occur where feldspathic sandstones or arkoses formed in desert type climates are later exposed to wet temperate or tropical conditions). Sedimentary rocks do however suffer to varying degrees from softening and disintegration due to wetting and drying and freezing and thawing. Porosity and water absorption are the main factors influencing the susceptibility to these mechanical weathering effects.

As stated in Chapter 8, specifications do not normally limit the water absorption value of an aggregate. However a value below 1% would be considered low and generally associated with a more resistant aggregate, while 1% to 3% would be a typical range. If durability is suspect, petrological examination should be carried out together with tests for water absorption, related where necessary to standard freeze–thaw tests or soundness tests (Chapters 7 and 8).

10.2.5 Correlation between tests and service behaviour. It has been shown (Turner and Wilson, 1956; Scott, 1955; Collett *et al.* 1962) from investigations of failures of a number of American highways that aggregates which perform adequately under tests with the Deval Abrasion Machine, will not necessarily prove suitable in pavement construction. The failure of such tests to predict performance lies in their inability to reproduce the degradation pattern that occurs in the pavement, the fines produced in the road tending to be far more plastic and hence more harmful than those produced in conventional laboratory tests.

It appears improbable that the increased plasticity of fines produced in the pavement over those produced in laboratory tests is due to any significant production of clay minerals by chemical weathering during service. The more likely explanation is that softening and solution condition the aggregate in use to further mechanical breakdown and hence to translocation and concentration of already formed clay and associated minerals.

By contrast normal mechanical tests on aggregates carried out in the laboratory are so accelerated (and usually, though not always, carried out on dry aggregate) that the fines produced are merely ground down particles of the same average chemical composition as the parent aggregate and so show lower plasticity than the predominantly clay fines produced in the field by a softening, mechancial breakdown, solution and concentration process.

Sweet (ASTM, 1948) commented as follows on work performed by Goldbeck (1935). "The results from the Los Angeles, Deval and toughness tests were compared with those of a laboratory roller test (simulating the action of a roller in the field). Goldbeck concluded "In view of the fact that the roller test is really a service test, it is clear that neither the toughness test nor the Deval abrasion test can be relied upon to indicate service behaviour, but on the other hand, the Los Angeles rattler test does agree exceptionally well with the service behaviour". A further direct comparison of aggregate crushing resistance to the Los Angeles test results was obtained by Shelburne (1940). He determined the amount of degradation produced by 5 and 10 ton rollers on different aggregates and found good agreement between the change in surface area caused by fourteen passes of a 10 ton roller and 100 revolutions in the Los Angeles machine. Although there was considerably more degradation from 500 revolutions in the machine, (the standard number in ASTM 131–69) the relative resistance of the aggregates in the abrasion test remained in accord with their resistance to the roller.

Notwithstanding the good agreement shown in these cases, none of the results relate to the kind of long term degradation occurring in the field results referred to above. Thus although it has been shown that good agreement exists betweeen the Los Angeles test and short term crushing under the roller, somewhat lower correlation may occur between these laboratory tests and longer term degradation under traffic in wet climates where partially weathered rocks are concerned.

If laboratory tests are correctly to predict behaviour of certain types of already partially weathered rock in wet climates it is necessary that the solution and softening actions are correctly simulated and combined with the mechanical action normal in such tests.

Weinert (1968) has already stressed the importance of climate in assessing the potential risk of weathering of rocks used in road construction in South Africa. Amongst the several meteorological factors which comprise climate, Weinert regards availability of moisture as the most important, and has produced a map of climatic zones based on an analysis of precipitation and evaporation records. Further, Weinert has produced a classification of rock types based upon the percentage of secondary minerals present which is then related to the climatic factor. Thus by comparison of present degree of weathering and the climatic factor Weinert suggested that it would be possible to predict which rocks would be sound and which unsound in practice.

This approach has much to commend it. The importance however of recognising the precise climatic factors which would apply to regions different from South Africa is stressed. Also even for South African conditions a further factor relating to the potential stress environment and exposure environment of the aggregate would further improve the method. In other words it would appear that many rocks which would be unsuitable if used as unbound aggregates, because of the risk of softening and further weathering, would suffer little or not at all if used in a dense bituminous (or concrete) mix providing both an impermeable coating and structural support to the aggregate in question. Further discussion on the engineering properties of aggregates and the relation between tests and service behaviour is contained in Meininger (1978).

10.3 Existing Department of Transport (United Kingdom) Specifications

The Department of Transport Specification for Road and Bridge Works 1976 clause 801 refers to two tables, namely 8/1 for sub-base materials and 8/2 for road base materials (reproduced here as Table 10.2(a) and (b)). In respect of both it is mentioned that the contractor may also substitute for the specified materials any material which is permitted to be specified for any course of pavement to be superimposed on it. The implication is that any material which is inferior (note 'superior' and 'inferior' are used here in the general sense of strength and durability).

TABLE 10.2

(a) *Sub-base materials*

(after clause 801, table 8/1 and clauses 803, 804, 805, 806, 807, 815 of D.Tp. Specification)

Alternative permitted materials complying with Clause:

803 Type 1 granular sub-base: crushed rock, crushed slag, crushed concrete or well burnt non-plastic shale, all to grading of table 8/4 of D.Tp. specification as follows:

BS sieve size	Percentage by mass passing
75.0 mm	100
37.5 mm	85–100
10.0 mm	40–70
5.0 mm	25–45
600 μm	8–22
75 μm	0–10

804 Type 2 granular sub-base: natural sands, gravels, crushed rock, crushed slag, crushed concrete or well burnt non plastic shale, all to grading of table 8/5 of D.Tp. specification as follows:

BS sieve size	Percentage by mass passing
75.0 mm	100
37.5 mm	85–100
10.0 mm	45–100
5.0 mm	25–85
600 μm	8–45
75 μm	0–10

805 Soil Cement*
806 Cement-Bound Granular Material**
807 Lean Concrete***
815 Wet Lean Concrete**

(*b*) *Roadbase Materials*

(after clause 801, table 8/2 and clauses 805, 806, 807, 808, 809, 810, 811, 812 of D.Tp. Specification)

Alternative permitted materials complying with Clause:

805 Soil Cement*
806 Cement-Bound Granular Material**
807 Lean Concrete***
808 Wet-Mix Macadam
809 Dry-Bound Macadam
810 Dense Tarmacadam B.S. 4987
811 Dense Bitumen Macadam B.S. 4987
812 Rolled Asphalt B.S. 594

Any other approved roadbase material as described in the Contract.

* (soil, chalk pulverised fuel ash, washed or processed granular material, crushed rock or slag, well burnt shale or spent oil shale, all to grading of table 8/6 of D.Tp. specification).

** (naturally occurring gravel-sand, a washed or processed granular material crushed rock or slag, all to grading of table 8/7 of D.Tp. specification).

*** (naturally occurring concrete aggregates complying with the requirements of B.S. 882—see Chapter 7).

As can be seen from Table 10.2, some of the materials permitted for base and sub base are bitumen or tar bound and some are cement bound. The present chapter is concerned only with those aggregates which are unbound, namely 803 Granular sub base Type 1, 804 Granular sub base Type 2, 808 Wet-mix macadam roadbase and 809 Dry-bound macadam roadbase.

10.3.1 General requirements. All materials intended for use within 450 mm of the surface of the road are required to be non frost susceptible when tested in accordance with TRRL Supplementary Report SR 318 and Materials Memorandum MM64 (following the earlier TRRL Laboratory Report LR90). For further discussion see Chapter 7.5.4.2. The development from the earlier tests (Croney & Jacobs, 1967; TRRL, 1977) to the procedure specified in MM64 (TRRL, 1981) was discussed at a recent Symposium (Jones, 1982). It is hoped that the modifications will improve the reproducibility of the test.

Frost heave and associated thaw weakening is a function of the pavement system so that consideration needs to be given to subgrade properties and to the water table position as well as to the unbound aggregate. Within a given system, frost heave of the aggregate is controlled by both the between particle pores (the size distribution of which is governed by the initial grading, the compaction and the degradation on laying) and the within particle pores which are a function of the parent rock. Thus, knowledge of the grading, saturation moisture content and suction characteristics is a valuable aid in the preliminary assessment of frost susceptibility (Jones & Hurt, 1978; 1980). However, in view of the complex interactions and despite the difficulties with reproducibility, the direct frost test appears unlikely to be superseded as the definitive test for frost susceptibility of aggregates.

A further requirement is that aggregates which may be placed within 500 m of concrete structures or pavements, or cement bound paving materials shall have a soluble sulphate content (SO_3) not exceeding 2.5 g/l when tested according to BS 1377 (Test 10).

10.3.2 Granular sub-base. The specifications for both Types 1 and 2 granular sub-base permit the use of crushed rock, crushed slag, crushed concrete and well burnt non plastic shale. Type 1 materials are intended to withstand substantial trafficking by construction plant. Type 2 materials, for less demanding circumstances, permit in addition natural sands and gravels. Tech. Memo. H6/78 restricts use of these sands and gravels to roads carrying less than 6 msa (million standard axles); though implying that the restriction could be relaxed in the light of successful previous experience. As seen in Table 10.2 the grading envelope for Type 2 is widened to permit the use of materials of lower maximum size and finer grading than Type 1.

Restrictions on the fines content and the plasticity of the fines have appeared in several editions of the Ministry of Transport Specification. The 1963 edition restricted the fines passing 75 μm to a maximum of 2%, this percentage being permitted to rise to 10% if the fines *of this size* were non-plastic. The current specification has been changed to permit the 10% maximum of passing 75 μm material to operate provided that the fraction passing the 425 μm sieve is non-plastic. This relaxation, recommended by Dunn (1966) and others, allowed again the use of a great many crusher run aggregates which prior to 1963 had clearly been used successfully but which did not meet the 1963 restrictions on plasticity.

Dunn showed that the inclusion of a test for plasticity on material passing the 75 μm sieve was unjustified not merely because it had led to the exclusion of many materials but because it could be shown that maximum soaked and unsoaked CBR values often occurred with fines contents of 4 to 9% even when the fines (<425 μm) were plastic.

So far as aggregate strength is concerned the constant need to keep the costs of road construction within acceptable limits has led to minimal requirements being given for Type 1 and Type 2 aggregates to meet any specified limit of ACV, AIV etc. No limits were in fact included in any of the first four editions of the Specification for Road and Bridge Works. Presumably because of the increased traffic loadings and intensities encountered in recent years, the fifth edition (1976) included a minimum value of 5 kN for the 10% fines test for all aggregates except well burnt non-plastic colliery shale (Fry, 1978). Supplement No. 1 of the Specification, published in 1978 amended the minimum value to 50 kN.

10.3.3 Roadbase and sub-base macadams. Coarse and fine aggregates for both wet-mix and dry-bound macadams are required to conform to BS 882 (Aggregates for concrete). This standard requires the aggregate to be hard, durable and clean and not to contain significant quantities of deleterious materials. The standard gives examples of the latter as: clay, flaky or elongated particles, mica, shale and other laminated materials, coal and other organic impurities, iron pyrites and soluble sulphates. A Flakiness Index not exceeding 35 and a minimum 10% fines value of 50 kN or alternatively a maximum aggregate impact value of 30 are additionally specified.

10.3.3.1 Dry-bound macadam. Dry-bound macadam consists of a compacted layer of coarse aggregate

TABLE 10.3. *Range of grading for wet-mix macadam to D.Tp. specification clause 808*

BS sieve size	Percentage by mass passing
50.0 mm	100
37.5 mm	95–100
20.0 mm	60–80
10.0 mm	40–60
5.0 mm	25–40
2.36 mm	15–30
600 μm	8–22
75 μm	0–8

TABLE 10.4. *Summary of Department of Transport Specifications for unbound aggregates in highway pavements*

	Permitted materials	Specified test/composition values							
		Grading	% passing 75 μm BS 1377	Plasticity index of passing 425 μm BS 1377	10% fines (kN) BS 812	AIV	Flakiness Index	Frost suscept-ibility	Soluble sulphate content (SO_3)
Road base Wet mix macadams	BS 882: any aggregate which is hard, durable, clean and without significant quantities of deleterious materials clay, flaky or elongate particles, mica, shale (and similar), coal (and similar), iron pyrites & soluble sulphates	Clause 808	$\nless 8$		$\nless 50$	$\ngtr 30$	$\ngtr 35$	any material used within 450 mm of surface of road is to be non frost sus-ceptible when tested in accordance with TRRL LR 90	any aggregate placed within 500 mm of con-crete structures, pave-ments, or cement bound paving materials shall have a soluble sulphate content not greater than 2.5 g/l when tested in accord-ance with BS 1367 (test 10)
dry bound macadams		Clause 809	Nil		$\nless 50$	> 30	$\ngtr 35$		
Sub-base Wet mix macadams dry bound macadams	as above								
type 1	Crushed rock, crushed slag crushed concrete, well-burnt non plastic colliery shale	Clause 803	$\ngtr 10$	Non plastic	$\nless 50$ (except colliery shale)				
Type 2	as above, plus natural sands & gravels (for roads carrying <6 msa (unless otherwise relaxed))	Clause 804	$\ngtr 10$	< 6	$\nless 50$ except colliery shale)	CBR require-ments (See DTp Specifi-cation)			

(50 or 40 mm nominal size, BS 63) into the surface of which fine aggregate is vibrated. Its use in the UK has gradually declined in recent years in favour of wet-mix macadam which is quicker to lay and requires less labour.

10.3.3.2 Wet-mix macadam. The grading of wet-mix macadam is similar to that of a Type 1 sub base. However it has a smaller maximum size and a smaller range of tolerances on a greater number of sieves (Table 10.3), in keeping with the higher degree of control and quality demanded of a material permitted to be used at a higher level in the pavement structure.

10.4 Discussion

Natural aggregates employed as unbound material in pavement construction are referred to in Department of Transport and British Standard Specifications. These chiefly specify the desired aggregate in terms of grading parameters but include some reference to strength tests and to frost resistance. A summary of the Department of Transport requirements for unbound aggregates in highway pavements is given in Table 10.4.

10.4.1 Specifications obtaining in countries other than United Kingdom. It is not possible for reasons of space to consider in any detail the specifications of territories outside the United Kingdom. However attention is drawn to the existence of these national, state and local specifications in all continents. Many differ in detail from those of the United kingdom but in general the same factors apply subject to local climate, traffic and economic considerations.

In respect of the latter the large increases in transport costs in recent years is causing a fundamental re-thinking on the specification requirements of aggregates in road construction, with the accent being currently on seeking means to maximise the use of local aggregates. These local aggregates may include natural waste materials such as mine waste (from metalliferous mining) slate waste and sand from china clay extraction. All can be considered capable of providing technically satisfactory aggregates subject to confirmation of performance in laboratory and/or field test and may be economically attractive. Well planned stage construction is recommended, where appropriate, as a means towards this end, allowing structural deformations (adjustments) within the construction layers (and to some extent in the foundation) to take place before the placement of, and hence without excess risk to, high cost surfacing materials.

10.4.2 Effects of degradation in service. Particle degradation in service (i.e. post rolling) may, by alteration of the grading produce materials of either higher or lower shearing resistance. Whichever occurs will, in the course of the change, result in further compaction and consequent rutting of the surface and is hence undesirable (unless allowed for in a planned stage construction of the highway). The importance of designing dense gradings to reduce particle degradation, particularly where low crushing strength aggregate is employed, has been stressed. Where aggregates of low initial strength or of suspect long term strength (because of evident weathering) are required to be used for economic reasons the importance of considering the pavement structure as a whole is also stressed. In such cases it is highly desirable to employ strong, dense and impermeable surfacing materials to distribute the traffic stresses and to protect the unbound aggregate from surface water, and to provide adequate drainage systems.

10.4.3 Moisture movements and salt damage in road foundations. While the above will be generally true i.e. that an impermeable surfacing is desirable, an exception occurs in drier climates where evaporation from capillaries has hitherto kept the water table at a low level (Brakey, 1970). In such circumstances the provision of an impermeable cover may prevent the normal evaporation and cause a rise in the water table into the higher subgrade and even into the road foundations with possible disastrous consequences. A surfacing of adequate vapour permeability is called for in such cases.

The contrary case of soluble salt damage to surfaced roads has been discussed by Fookes and French (1977). In this case it is the continuous rise of capillary moisture and the precipitation of soluble salts carried to the surface thereby that causes the distress to the pavement structure. To alleviate this situation one of the recommendations made by these authors is to provide an impermeable surfacing layer. It is clearly of great importance to distinguish between these two types of potential failure due to capillary moisture movement since one requires a permeable surfacing and the other an impermeable surfacing. It is also of course possible for the risk of both processes to exist at the same site i.e. for crystallisation of salts to occur if capillary rise is allowed to occur and for the water table to rise if capillary rise and evaporation are prevented. In this case, very careful analysis of the conditions is required. One solution may be to risk the latter event i.e. the rise in water table and as recommended by Fookes and French to design the road on the basis of a low CBR value. Attention to roadside drainage conditions will very materially assist in reducing both risks.

References

AMERICAN SOCIETY FOR TESTING AND MATERIALS, ASTM D1075. (1978). Test for effect of water on cohesion of compact and bituminous mixtures.

BOND, F. C. 1952. The third theory of comminution. *Trans. Am. Inst. Ming Eng.* **193,** 484–94.

BRAKEY, B. 1970. Colorado Division of Highways Denver Colo. Personal communication.

BRITISH STANDARDS INSTITUTION BS 63 1951, 1971. Part 1 Imperial Units. Part 2 Metric Units.

BRITISH STANDARDS INSTITUTION, BS 812, 1975. Sampling and testing of mineral aggregates, sands and fillers.

British Standards Institution, BS 882, 1973. Aggregates from natural sources for concrete.

British Standards Institution, BS 1047, Part 2 1974. Air-cooled blast-furnace slag coarse aggregate for concrete.

British Standards Institution, BS 1377, 1975. Methods of testing soils for civil engineering purposes.

Collett, F. R., Warnock, C. C. & Hoffman, D. S. 1962. Prevention of degradation of basalt aggregates used in highway base construction. *Highway Res. Board Bull.* **334,** 1–7.

Croney, D. & Jacobs, J. C. 1967. The frost susceptibility of soils and road materials. *Road Res. Laboratory Rep. LR90.*

Croney, D. & Loe, J. A. 1965. Full-scale pavement design experiment on A1 at Alconbury Hill, Huntingdonshire. *Proc. Instn Civ. Eng.* **30,** 225–270.

Day, H. L. 1962. A progress report on studies of degrading basalt aggregate bases, *Highway Res. Board Bull.* **334,** 8–16.

Department of the Environment, 1966. Roads in urban areas. H.M.S.O. Metric Supplement (1974), Guide to Revisions (1979).

Department of the Environment, 1968. Layout of roads in rural areas. H.M.S.O.

Department of Transport, 1976. Specification for road and bridge works.

Department of Transport. Tech. Memo H6/78. Road Pavement Design.

Dunn, C. S. 1966. The stability of aggregates used in road sub-bases. *Roads and Road Const.* **44** (519), 77–81.

Fookes, P. G. & French, W. J. 1977. Soluble salt damage to surfaced roads in the Middle East. *Highway Eng.* **24** (12), 10–20.

Fry, J. W. 1978. in Developments in Highway Pavement Engineering—2. Ed. P. S. Pell. Applied Science Publishers.

Gaudin, A. M. 1926. An investigation of crushing phenomena. *Trans. Am. Inst. Ming Eng.* **73,** 253–316.

Ghani, I. 1969. The design and properties of dense asphaltic mixes with low crushing strength aggregates. Ph.D Thesis, Univ. of Birmingham.

Goldbeck, A. T. 1935. Discussion on the Los Angeles Abrasion Machine. Proc. ASTM 35 (2) 530.

Hosking, J. R. & Tubey, L. W. 1969. Research on low-grade and unsound aggregates. Road Research Laboratory Report LR 293. Crowthorne.

Highway Research Board, 1962. The AASHO Road Test. Rep. 5. Pavement Res., Spec. Rep. 61E, (National Res. Council).

Irfan, T. Y. & Dearman, W. R. 1978(a). Engineering petrography of a weathered granite. *Q. J. eng. Geol. London* **11** (3), 233–44.

Irfan, T. Y. & Dearman, W. R. 1978(b). Engineering classification and index properties of a weathered granite. *Bull. Int. Ass. eng. Geol.* **17.**

Jones, R. H. (Ed). 1982. Unbound Aggregates in Roads. Proc. Symposium Dept. of Civil Eng., Univ. of Nottingham. Session 3 "Frost Heave", 45–62, 151–181.

Jones, R. H. & Hurt, K. G. 1978. An osomotic method for determining rock and aggregate suction characteristics with applications to frost heave studies. *Q. J. eng. Geol. London* **11** (3), 245–52.

Jones, R. H. & Hurt, K. G. 1980. The prediction of the frost susceptibility of limestone and sub-base materials. *The Highway Engineer Vol.* **27,** (1), 8–14.

Kick, F. 1885. Das Gesetz der proportionalen Widerstande und seine Anwedung. Leipzig.

Knight, B. H. & Knight, R. G. 1948. Road aggregates, their uses and testing. Arnold, London.

Lees, G. 1970. The rational design of aggregate gradings for dense asphaltic compositions. *Proc. Assoc. Asph. pvg. Tech.* **39,** 60–97.

Lees, G. & Kennedy, C. K. 1975. Quality, shape and degradation of aggregates, *Q. J. eng. Geol. London* **8** (3), 193–209.

Marwick, A. H. D. & Shergold, F. A. 1945. The aggregate crushing test. *J. Instn Civ. Eng.* **24** (6), 125–33.

Meininger, R. C. 1978. Aggregate abrasion resistance, strength, toughness and related properties. ASTM Spec. Tech. Publ. 169B.

Melville, P. L. 1948. Weathering study of some aggregates. *Proc. Highway Res. Board.* **28.**

Mendes, F., Aire-Barros, L. & Pere Rodrigues, F. 1966. The use of modal analysis in the mechanical characterisation of rock masses. *Proc. 1st Cong. Int. Soc. Rock Mech., Lisbon,* **1,** 217–23.

Minor, C. E. 1959. Degradation of mineral aggregates. ASTM Spec. Tech. Publ. 277.

Moavenzadeh, F. & Goetz, W. H. 1963. Aggregate degradation in bituminous mixtures. *Highway Res. Record* **24.**

Ramsay, D. M. 1965. Factors influencing Aggregate Impact Value in rock aggregate. *Quarry Managers J. London.*

Rittinger, P. 1967. Lehrbuch der Aufbereitungskunde. Berlin.

Scott, L. E. 1955. Secondary minerals in rock as a cause of pavement and base failure. *Proc. Highway Res. Board* **34,** 412–17.

Shelburne, T. E. 1940. Crushing resistance of surface treatment aggregates. *Purdue Univ. Eng. Exp. Stn.* **24** (5), 7–67.

Shergold, F. A. 1955. Results of tests on single-sized roadmaking aggregates. *Quarry Managers J. London* **38** (10), 636–42.

Shergold, F. A. & Hosking, J. R. 1963. Investigation of test procedures for argillaceous and gritty rocks in relation to breakdown under traffic. *Roads and Road Constr.,* **41,** (492) 376–8.

Sweet, H. S. 1948. Physical and chemical tests of mineral aggregates and their significance. *ASTM Spec. Tech. Publ.* **83,** 49–73.

Transport & Road Research Laboratory, 1977. The "LR 90" frost heave test—interim specifications for use with granular materials. SR 318, T.R.R.L. Crowthorne.

Transport & Road Research Laboratory, 1981. Specification for the T.R.R.L. frost heave test. Material Memo No. MM/64, Materials Division, T.R.R.L. Crowthorne.

Turner, R. S. & Wilson, J. D. 1956. Degradation study of some Washington aggregates. *Washington State Inst. of Tech. Bull.* **232.**

Weinert, H. H. 1968. Engineering petrology for roads in South Africa. *Eng. Geol.* **2(b),** 363–95.

Yashima, S., Morohashi, S., Awano, O. & Kauda, Y. 1970. (Chem. Eng. Japan). 34. English Abst. Particulate Matter.

General Reference

Rilem Technical Committee 26GM. 1978. Aggregates and Fillers. International symposium Budapest, Hungary.

Chapter 11. Aggregates in bituminous bound construction materials

11.1 Introduction

Both natural and artificial aggregates of coarse and fine sizes are employed in bituminous mixtures for highway and airfield pavements, hydraulic and building applications. The artificial aggregates such as steel slags, blastfurnace slags and calcined bauxite play an important part in many applications but are excluded from discussion here by the terms of reference of the Working Party. Similarly fillers are not dealt with in this chapter due primarily to the fact that many fillers are not natural aggregate products. The role of fillers has been dealt with extensively by Heukelom (1965).

11.2 Petrological types of aggregate used in bituminous compositions

A wide range of rocks of igneous, sedimentary and metamorphic origins are used in bituminous mixes. The general requirement is for rocks which will be durable (i.e. resistant to abrasive wear), strong (i.e. resistant to slow crushing loads and to rapid impact loads) and where appropriate, i.e. in wearing course and surface dressing applications, resistant to the polishing action of traffic.

A further property required especially for aggregates for surface dressing and for macadam type mixes is that they should show good adhesion with bituminous binders.

Some early works on the subject of roadstone aggregates such as that by B. H. and R. G. Knight (1935) carefully described the desirable qualities of rocks and proceeded to relate mineralogical and textural details to strength (and to a slight extent to polishing characteristics). Considerable emphasis was laid in such works on defining the best type of rock for a particular purpose. While such search for exactness is normally to be commended in engineering, the requirement for the best economic use to be made of available raw materials has in current practice proved to be of far greater significance. Both the high costs of haulage of aggregates and the depletion of some natural resources are of importance in this respect. In other words, the present problem is not so much how to identify the best type of rock for a certain usage but to decide what is the boundary between acceptable and non-acceptable (a boundary which must be recognised as a shifting boundary as technology changes) and how best to use what is acceptable. The latter will involve the evolution and application of improved design and control methods for bituminous mixes in order to enable satisfactory use of the most economically available suitable aggregates.

11.3 Desirable properties of aggregates for bituminous mixes

It is generally recognised (Hosking, 1970; Hartley, 1974; Lees and Kennedy, 1975) that the main qualities required from a roadstone to be used in pavement surfacing materials are:

(i) Toughness
  a) resistance to slow crushing—assessed by the Aggregate Crushing Test and Ten Per Cent Fines Test (see Chapter 7 for discussion and detailed reference on all tests in this section, except for tests on adhesion (Appendix A11).
  b) resistance to rapid loading—assessed by the Aggregate Impact Test, modified Aggregate Impact Test and the Los Angeles Test.

(ii) Hardness—resistance to abrasion/attrition—assessed by the Deval Test, Aggregate Abrasion Test and, to a limited extent, by the Los Angeles Test.

(iii) Resistance to polishing (for wearing course materials and surface dressings)—assessed by the Accelerated Polishing Test.

(iv) Resistance to stripping (ability to maintain adhesion to any bituminous binder with which the aggregate is used)—assessed by various tests described by Mathews (1958), Bituminous Materials in Road Construction (1962) and in the standards of various other national and regional authorities. (See Chapter 7 for descriptions of these tests.)

(v) Resistance to weathering effects in the pavement (e.g. resistance to frost action, swelling and softening by water)—assessed by determination of water absorption, sodium sulphate/magnesium sulphate soundness tests.

(vi) Ability to contribute to strength and stiffness of

total mix by intrinsic aggregate strength and shape properties.

11.4 Influence of aggregate petrography on engineering properties

11.4.1 Crushing strength. Hartley (1974) provided an useful summary of the main petrographical factors affecting the toughness or crushing strength of aggregates. Referring to the engineering classification for intact rock material, devised by Deere and Miller (1966) he states:

"Of the rocks tested few have strengths in excess of 225 MN/m^2 and those which do are almost entirely non-porous rocks possessing an interlocking texture, i.e. quartzites, dolerites and dense basalts. The rocks classified as of high strength (110 MN/m^2—210 MN/m^2) include the majority of the igneous rocks, the stronger metamorphic rocks, well cemented sandstones and indurated shales and about 40 per cent of the limestones and dolomites tested. The medium strength range includes poorly bonded porous rocks and foliated rocks. Rocks in the low and very low strength ranges are generally very porous, weakly bonded sedimentary rocks, highly foliated igneous and metamorphic rocks or chemically altered rocks of any origin".

11.4.1.1 Igneous rocks. The main factors causing a reduction in the strength of igneous rocks are thus an increase in porosity, grain size and the proportion of soft minerals, and the development of flow structure. Chemical alteration in general increases the proportion of soft minerals e.g. clays and allied minerals (with the exception of epidotisation) and peripheral alteration may result in weakening of the intergranular bonding. Vesicular texture decreases the strength, while the effect of amygdales depends upon the nature of the infilling material.

11.4.1.2 Sedimentary rocks. With sedimentary rocks, the interlocking fabric and mineralogy of limestones and dolomites appear to be responsible for their relatively high strength and Young's modulus. The main factors influencing the strength of sandstones are the intergranular bonding and porosity. Chemical alteration effects, though much less common due to the secondary nature of the sedimentary rocks are, when they occur e.g. in the weathering of the felspars of a felspathic sandstone or arkose, similar to the effects encountered in igneous and metamorphic rocks. Anisotropy related to the fabric of the sediment results in variation of the modulus ratio according to whether the load axis is placed perpendicular or parallel to the plane of the laminations.

11.4.1.3 Metamorphic rocks. The principal factors causing a reduction in the strength of metamorphic rocks are increase in grain size, porosity, proportion of soft minerals and, in particular, an increase in schistosity or foliation. The extent of the intergranular bond is also highly important. The effect of chemical alteration is similar to that noted for igneous and sedimentary rocks.

Fundamental strength tests such as that described above are not generally used in the UK as a method of assessing the potential of roadstone aggregates due to their poor reproducibility (Shergold, 1948). The effects of variability of individually prepared specimens have been reduced by the introduction of aggregate tests using specimens composed of a number of pieces of the aggregate. Even so the petrographical characteristics listed above together with the particle shape do influence the behaviour of the individual aggregate particles (Dhir, Ramsay and Balfour, 1971).

11.4.2 Resistance to abrasion. Abrasion of road surfacing aggregate takes place by the action of tyres carrying detritus. This detritus is in part material worn off the road surface and in part naturally derived sand and clay materials and artificial debris carried or blown onto the road (Maclean & Shergold, 1958).

Hartley (1974) considers petrographical characteristics of importance are the degree of hardness and proportion of hard minerals, the proportion, orientation and distribution of cleaved minerals, the grain size, the nature of the intergranular bond and the degree of liability to chemical alteration of the mineral content.

11.4.2.1 Igneous rocks. Of the fresh igneous rocks, those containing a high free silica content tend to resist abrasion better than the basic rocks which have a high ferromagnesian content. This is due to the hardness and lack of cleavage of the quartz in the former. Chemical decomposition generally results in increased abrasion, particularly if a high degree of peripheral alteration exists since this can destroy the intergranular bond and allow the harder cores to be plucked from the rock surface. Vesicular texture generally reduces the resistance to abrasion while the nature of the mineral infilling determines the effect of amygdales.

11.4.2.2 Sedimentary rocks. Among the sedimentary rocks, the low hardness and cleavage facility of the minerals contained in the younger and more porous limestones and dolomites renders them liable to rapid wear while the abrasion resistance of siliceous rocks is almost entirely dependent upon the nature of the intergranular bond. Thus flint and quartzite are highly resistant whilst poorly cemented ferruginous and calcareous sandstones are soon abraded due to plucking of the grains. Mixed mineral sedimentary rocks such as greywacke tend to have poor abrasion resistance although this is not always the case, especially with Palaeozoic examples.

11.4.2.3 Metamorphic rocks. Of the metamorphic rock used in road surfacing, gneisses tend to behave similarly to igneous rocks of the same mineralogy while hornfels and quartzites have a high abrasion resistance due to their hard mineral content and dense interlocking texture.

11.4.3 Resistance to polishing. The importance of

TABLE 11.2 (a). *Concentrations of surfactant required with various aggregates to reduce stripping to an acceptable level in a full-scale experiment with Bitumen Macadam (Lee 1958)*

Aggregate	Concentration of agent required to give satisfactory results (i.e. not >2% stripping)				Results with 10 per cent of road tar in bitumen*
	Agent A (amine)	Agent B (amine)	Agent C (amine)	Agent D (Quaternary ammonium comp)	
Group 1 Flint gravel, Quartzite	No satisfactory results obtained				
Group 2 Granite A	0.5	1.0	1.0	0.56	NS
Porphyry A	0.5	1.0	0.75	0.56	NS
Porphyry B	0.5	1.0	0.75	0.56	
Porphyry C	0.25	1.0	0.5	0.42	NS
Basalt A	0.5	1.0	0.5	0.56	S
Basalt B	0.75	1.0	0.75	0.56	S
Group 3 Gritstone A	0.25	1.0	0.5	0.28	
Gritstone B	0.25	1.0	0.5	0.28	S
Hornfels A	0.25	0.5	0.5	0.28	S
Granite B	0.25	0.5	0.5	0.28	S
Basalt C	0.25	1.0	0.5	0.28	
Group 4 Slag A, Slag B, Basalt D, Limestone A, Limestone B	Performed satisfactorily without agents				

*NS = Not Satisfactory, S = Satisfactory

TABLE 11.2 (b).

Surface active agent	Aggregate				
	Granite A	Porphyry A	Flint gravel A	Basalt A	Basalt B
	Percentage of agent required for good adhesion at 20°C				
A	1.0	2.5	2.0	0.5	2.0
B	1.0	2.5	2.0	1.5	2.5
C	2.0	2.5	2.0	2.0	2.0

11.5 Detailed requirements and conflicting requirements for aggregates for bituminous pavement materials

Certain aggregate properties have greater significace in some contexts than others. For example high crushing strength is required of aggregates for surface dressing and heavily trafficked open texture macadams. Good adhesion is similarly required in these situations but is less important as a factor in dense impermeable asphalts. High resistance to polishing is required for wearing course materials, surface dressings and coated chippings in difficult sites of high traffic intensity and high risk rating (Salt and Szatkowski, 1973) but is of less significance in easier sites and regions of lower traffic intensity and clearly irrelevant in the selection of base and base course aggregates. Apart from the need to recognise the different significance of the various properties in different engineering contexts, several of these traditionally designated 'desirable' qualities are partially or completely conflicting e.g. hard minerals are frequently brittle, hence high resistance to abrasion may be accompanied by low resistance to crushing or impact e.g. quartz, flint, chert. Further these silica minerals, while highly abrasion resistant, are also known to have poor adhesion to bituminous binders in the presence of water (Lee, 1936; Mathews, 1961).

This conflict in desirable properties extends also into the province of particle shape since it can be shown that highly rounded particles would have the best anti-stripping tendencies (Mathews, 1958) while being inferior to angular particles in terms of shearing resistance (Herrin & Goetz, 1954; Lottman & Goetz, 1956). In practice the latter property would be considered by far the more important since poor adhesion can be attended to by the use of adhesion agents.

Another familiar example of conflicting properties in roadstone concerns the qualities of skid resistance and resistance to abrasion as mentioned above. Although abrasion resistance is generally considered a desirable quality in a roadstone, since in general it will be associated with durability of the pavement, a rock which is uniformly hard will also wear uniformly, albeit slowly. Hence a polished and slippery surface will result. Conversely a rock which exhibits some wear, presents by so doing fresh and unpolished surfaces to the tyre.

11.5.1 Detailed requirements—Strength. The basic requirement for crushing strength remains in the case of coated materials as in the case of uncoated stone (Chapter 10) and concrete aggregate (Chapter 8). Certain categories of use however will be less demanding than others. Still others, such as for example surface dressings compacted by steel wheeled rollers on hard surfaces, will be very demanding and require aggregates of the highest strength (see below). Also the strength generally expected for an aggregate situated higher up in the construction layers is greater than for one situated at lower levels because of the diminution of the traffic stresses with depth.

An example of the different suggested minimum strength values required from aggregate intended for different uses is seen in Table 11.3 (Shergold, 1948). This table is now several decades old, but has not been superseded. The strength values are stated in terms of one current BS 812 test (the Aggregate Crushing Test), one current ASTM test (the Los Angeles Test) and some older tests not in current British use (the 'Deval' Dry Attrition Test, the 'Page' Impact Test and the Crushing Strength Test). Shergold reported a good correlation between Los Angeles results, Aggregate Crushing Value and field performance in New South Wales, Australia. These results according to Shergold showed that an upper limit of 25 for the Aggregate Crushing Test and 35 for the Los Angeles Test would eliminate nearly all of the unsatisfactory aggregates for this particular form of construction.

The table illustrates the principle that aggregates of different performance can be suitable for different locations in the pavement construction, for different types of material composition and under different traffic stresses. For example it is suggested that an aggregate of ACV not greater than 13 would be required for a waterbound macadam (or unbound material, Chapter 10) to be used under heavy traffic while at the other extreme, aggregate for concrete (Chapter 8) to be used in the body of the road in light to medium traffic conditions might be permitted to have an ACV up to 40. These differences emphasise the point that aggregates receive different degrees of support in the context of different engineering compositions, as well as experiencing different stresses according to position in the construction and traffic loadings. Even within the bituminous materials (which are not in this Table identified as of any particular composition) it is suggested that for the least exacting condition the maximum ACV would be 30 and for the most exacting, 17. If the bituminous compositions were subdivided into open texture mixes (say voids equal to or greater than 7%) and denser mixes (say voids less than 7%) the latter would be able to utilise lower strength aggregate than the former. It is tempting to consider that a revision of Shergold's table should be attempted in the light of current experience. However, the warning contained in the note to the table against using the given values as rigid specification limits and the fact that no attempt has been made to revise the table in 30 years reinforce the view that only very general guidance can or should be given (with the exception of certain special cases which are detailed below).

The present report subscribes to this view since no matter how accurate a relationship is achieved between laboratory tests and service behaviour there can never be any justification for absolute prohibition until the economic factors of long term durability for high initial cost versus shorter life for low initial cost have been assessed. These factors will vary from year to year and with geographical location. For example in countries or localities where sources of high quality stone are scarce, higher values of ACV, AIV, and AAV may well be tolerated in aggregates for road construction.

Current British Standards for bituminous mixes state only that natural aggregate shall belong to one of the following groups (BS 812 1975): (2) Basalt, (4) Gabbro, (5) Granite, (6) Gritstone, (7) Hornfels, (8) Limestone, (9) Porphyry, (10) Quartzite, but add that certain aggregates not belonging to the above groups may be permitted by agreement with the purchaser. Thus the only 'group' excluded is the Schist Group (Group 11) comprising phyllites, schists, slates and all severely sheared rocks. In terms of the Basic Commercial Classification proposed in Chapter 6 this would imply that all types of natural rock may be considered except those in the Metamorphic foliated class. However the inclusion of the statement concerning the possible acceptance of aggregates not in the named groups presumably recognises that even in the Schist group, varieties can exist which still meet the physical/engineering property requirements for

TABLE 11.3. *Summary of tentative limits that have been suggested for the mechanical properties of roadstone (Shergold 1948)*

Type of construction	Dry attrition less (%) (max) DEVAL	Impact (no. of blows) (min) PAGE	Crushing strength (lb./sq. in) (min)	Los Angeles test (% fines) (max)	Aggregate crushing test (% fines) (max)
		Light or medium traffic			
Waterbound					
Light traffic	8.0	5	15000	35	30
Medium traffic	5.0	10	25000	25	23
Concrete					
Body of road	—	—	10000	50	40
Surface of road	—	—	12000	40	35
Bituminous					
Body of road	8.0	5	15000	35	30
Surface of road	5.0	10	25000	25	23
		Heavy traffic			
Waterbound	2.7	18	40000	13	13
Concrete					
Body of road	—	—	12000	40	35
Surface of road	—	—	15000	35	30
Bituminous					
Body of road	5.7	10	15000	35	30
Surface of road	4.0	13	30000	17	17

Note: This table has been drawn up from figures given in a number of the papers that have been studied.

It is important to remember that in every case the figures are intended for use as an approximate guide to the properties required in aggregates for use in different forms of construction. They are not intended to be applied as rigid specification limits.

certain categories of bituminous composition. Gravels are permitted provided they are from Groups 2, 4, 5 or 9 or are of flint (Group 3). BS 4987 (1973) prefaces the above by requiring that the rock shall be hard, clean and durable. BS 594 (1973) merely adds that it shall be of a type and quality approved by the purchaser or his representative.

11.5.2 Detailed requirements—polishing resistance and abrasion resistance: Department of Transport (UK). For several years the Department of Transport Specification for Road and Bridge Works gave the only further guidance as to desirable quality and then only in the following terms for aggregates for wearing course materials: "the minimum polished stone value and the maximum aggregate abrasion value shall be as described in the contract". In 1976, however, the Department's Technical Memorandum H16/76 laid down specific requirements for aggregate properties (and texture depth) for bituminous surfacings for new motorway and trunk road construction. This memorandum, based upon work of the Transport and Road Research Laboratory (Hosking & Tubey, 1972; Salt & Szatkowski, 1973) matches the desired skidding resistance with the different highway conditions and the anticipated traffic flows (Table 11.4). The latter are calculated as the commercial vehicles per lane per day anticipated to be using the lane at the end of the projected life of the surfacing. The same requirements are also recommended for new construction of other types of road. In all cases the requirements are to be applied (except where otherwise stated) only to roads constructed within the geometric standards of the Department's Publications 'Layout of Roads in Rural Areas' and 'Layout of Roads in Urban Areas' (as amended by subsequent Technical Memoranda). For other roads not to these standards the engineer is recommended to select his own requirement for minimum PSV based upon his own and local or related experience.

The types of bituminous surfacing which the specification is to be applied are (i) the chippings for surface dressing, (ii) coated chippings applied to the surface of rolled asphalt, dense tar surfacing, fine cold asphalt and mastic asphalt, (iii) the coarse aggregate in

TABLE 11.4. *Categories of sites and minimum Polished Stone Value for flexible roads*

1 Site	2 Approximate percentage of all roads in England	3 Definition	4 Minimum Polished Stone Value	5 Remarks
A1 (difficult)	Less than 0.1%	(i) Approaches to traffic signals on roads with 85%ile speed of traffic greater than 40 mile/h (64 km/h) (ii) Approaches to traffic signals, pedestrian crossings and similar hazards on main urban roads	Less than 250 cv/lane/day: 60 250 to 1000 cv/lane/day: 65 1000 to 1750 cv/lane/day: 70 More than 1750 cv/lane/day: 75	Risk rating 6 Values include +5 units for breaking/ turning
A2 (difficult)	Less than 4%	(i) Approaches to and across major priority junctions on roads carrying more than 250 commercial vehicles per lane per day* (ii) Roundabouts and their approaches. (iii) Bends with radius less than 150 m on roads with an 85% ile speed of traffic greater than 40 mile/h (64 km/h) (iv) Gradients of 5% or steeper, longer than 100 m	Less than 1750 cv/lane/day: 60 1750 to 2500 cv/lane/day: 65 2500 to 3250 cv/lane/day: 70 More than 3250 cv/lane/day: 75	Risk rating 4 Values include +5 units for braking/ turning * In i., the 350 ov/ lane/day applies to each approach
B (average)	Less than 15%	Generally straight sections of and large radius curves on: (i) Motorways (ii) Trunk and principal roads (iii) Other roads carrying more than 250 commercial vehicles per lane per day	Less than 1750 cv/lane/day: 55 1750 to 4000 cv/lane/day: 60 More than 4000 cv/lane/day: 65	Risk rating 2
C (easy)	Less than 81%	(i) Generally straight sections of lightly trafficked roads i.e. less than 250 cv/day (ii) Other roads where wet skidding accidents are unlikely to be a problem	45	No risk rating applied Many local aggregates have a PSV well above 45 and normally these should be used

Note: Values should be amended in accordance with paragraph 4.1.4 of this Memorandum where abnormal risks occur. (Tech. Memo. H16/76).

wearing courses of tarmacadam and bitumen macadam and of rolled asphalt and dense tar surfacing without coated chippings applied to the surface, (iv) the coarse aggregate in non-surface dressing base courses and roadbases of tarmacadam, bitumen macadam and rolled asphalt used as running surfaces.

Tables 11.5 and 11.6 are reproduced from the Laboratory Report by Salt and Szatkowski referred to above and give more fully the information upon which Technical Memorandum H16/76 is based. Table 11.5 gives the value of PSV required to achieve a given Sideway Force Coefficient at 50 km/h under given traffic densities. Table 11.6 gives guidance on the minimum SFC_{50} values required for certain highway situations and risk ratings.

In addition to specifying minimum PSV's for aggregates to be used in such surfacing materials the Technical Memorandum states minimum AAV requirements for different traffic loadings calculated as that using the lane at a time equal to two-thirds of the anticipated life of the surfacing (Table 11.7).

The materials to which this requirement is to be applied are those listed as (i), (ii), (iii) above. No specification for minimum AAV is deemed necessary for category (iv). Further recommendations on desired strength for aggregates is contained in Road Note 39 (1972) which in addition to normal surface dressing deals with the case of surface dressing on concrete and other extremely hard surfaces (see also TRRL LR908, 1979). In view of the latter's particularly exacting

TABLE 11.5. *PSV of aggregate necessary to achieve the required skidding resistance in bituminous surfacings under different traffic conditions (Salt & Sgetkowski 1973)*

Required mean summer SFC at 50 km h^{-1}	PSV of aggregate necessary Traffic in commercial vehicles per lane per day 250 or under	1000	1750	2500	3250	4000
0.30	30	35	40	45	50	55
0.35	35	40	45	50	55	60
0.40	40	45	50	55	60	65
0.45	45	50	55	60	65	70
0.50	50	55	60	65	70	75
0.55	55	60	65	70	75	[a]
0.60	60	65	70	75	[a]	[a]
0.65	65	70	75	[a]	[a]	[a]
0.70	70	75	[a]	[a]	[a]	[a]
0.75	75	[a]	[a]	[a]	[a]	[a]

[a] Aggregate not greater than 10

conditions a minimum ACV of 18 for lane traffic categories 1, 2 and 3 of the Road Note's table 1 is recommended for these sites. With these exceptions, which all concern wearing course or surface dressing materials, current practice in the UK (and those countries following UK practice) is not to specify or even recommend tentative limits in terms of strength or other physical or chemical properties.

It follows that much greater responsibility rests with the engineer to assess the local requirement and to judge the potential quality and life of the materials available for purchase. This implies the need for a high

TABLE 11.6. *Minimum values of skidding resistance for different sites (Salt & Szetkowski 1973)*

Site	*Definition*	*SFC (at 50 km h^{-1}) Risk rating* 1	2	3	4	5	6	7	8	9	10
A1 (v. difficult)	(i) Approaches to traffic signals on roads with a speed limit greater than 40 mile h^{-1} (64 km h^{-1})						0.55	0.60	0.65	0.70	0.75
	(ii) Approaches to traffic signals, pedestrian crossings and similar hazards on main urban roads*										
A2 (difficult)	(i) Approaches to major junctions on roads carrying more than 250 commercial vehicles per lane per day										
	(ii) Roundabouts and their approaches				0.45	0.50	0.55	0.60	0.65		
	(iii) Bends with radius less than 150 m on roads with a speed limit greater than 40 mile h^{-1} (64 km h^{-1})										
	(iv) Gradients of 5% or steeper, longer than 100 m										
B (average)	Generally straight sections of and large radius curves on:										
	(i) Motorways										
	(ii) Trunk and principal roads	0.30	0.35	0.40	0.45	0.50	0.55				
	(iii) Other roads carrying more than 250 commerical vehicles per land per day										
C (easy)	(i) Generally straight sections of lightly trafficked roads										
	(ii) Other roads where wet accidents are unlikely to be a problem	0.30	0.35	0.40	0.45						

* Main urban roads would generally be included in Marshall road categories, 1, 2 and 3

TABLE 11.7. *Traffic loadings and maximum aggregate abrasion values for flexible surfacings*

Traffic in commercial vehicles per lane per day	Under 250*	Up to 1000	Up to 1750	Up to 2500	Up to 3250	Over 3250
Maximum AAV for chippings	14	12	12	10	10	10
Maximum AAV for aggregate in coated macadam wearing courses	16	16	14	14	12	12

* For lightly trafficked roads carrying less than 250 commercial vehicles per lane per day aggregate of higher AAV may be used where experience has shown that a satisfactory performance is achieved by aggregate from a particular source. (Technical Memorandum H16/76 Dept. of Transport)

standard of relevant education and experience. For conventional premixed bituminous materials the absolute lower limit of aggregate strength that may be tolerated will be that for which the aggregate can be screened, mixed, transported and laid without undue degradation (as confirmed by analysis methods of BS 598 and similar overseas specifications). This will tend to exclude all aggregates with an AIV greater than about 35 (Hawkes & Hosking, 1972) and AAV greater than 16 (Technical memorandum H16/76), but it may be noted that the footnote to Table 11.7 suggests that aggregates of AAV > 16 may be used for lightly trafficked roads where experience has shown satisfactory performance.

11.5.3 Detailed requirements—Property Services Agency UK (Airfields Branch). For airfield pavements in the United Kingdom, (and certain overseas territories) the Property Services Agency (Airfields Branch) of the Department of the Environment summarise (Table 11.8) the aggregate types permitted for various types of bituminous surfacing. As can be noted from Table 11.8 the PSA go further than the DTp in specifying limiting values of Flakiness Index, ACV, Absorption, Soundness and susceptibility to stripping.

11.5.4 Relationship between mix composition and desirable aggregate properties. As implied in Table 11.8 and elsewhere in this chapter, lower crushing strength aggregates can be included in dense bituminous mixes than would be acceptable in open texture mixes, because of the support afforded in the former case by the bituminous mortar and by the increased impermeability to water and air. This applies not only to the intrinsic strength of the rock but also the influence of particle shape on strength. For example the traditional aversion to low crushing strength aggregates and for flaky and elongate particles in aggregates for bituminous construction materials arises chiefly out of the greater degradation they exhibit both in conventional aggregate tests and in service behaviour as open texture macadam aggregates such as friction course (and as surface dressings). In the two uses quoted, the breaking of aggregate particles is at once evident and harmful (by reason of producing uncoated cohesionless surfaces). The breakage occurs readily in these high void content or single layer mixes because the aggregate particles are in especially vulnerable situations, with little or no protection from the development of high stresses. It is for these reasons that a maximum permissible limit on Flakiness Index was first incorporated in BS 63 (1951) and re-stated in BS 63 (1971) Part 2 Metric Units (Table 11.9).

On the other hand it has been shown (Lees, 1964; Salehi, 1968) that the risk of deleterious crushing of flaky aggregate decreases, in some cases to negligible proportions, when the so-called 'suspect' aggregate is used in dense bituminous mixes where support from the bituminous mortar is mobilised.

A method for designing aggregate gradings for dense bituminous compositions which takes into account variations in aggregate shape and other relevant factors has been described by Lees (1970).

11.6 Conclusions

It is apparent that in regard to specifications concerning aggregates for bituminous materials the following factors have been considered relevant:

(a) the type of usage e.g. highway or airfield pavement, hydraulic engineering
(b) the type of bituminous material e.g. surface dressing, friction course, rolled asphalt, asphaltic concrete
(c) the intensity of traffic
(d) the location in the pavement e.g. wearing course or base course
(e) the relative risk rating for skid resistance purposes

TABLE 11.8. *Properties of Aggregates for various categories of Bituminous Surfacing (Property Services Agency)*

	Coarse aggregates—test properties										
	All Aggregates				Crushed Rock and Gravel (BS 812)		Slag (BS 1047)				All Aggregates
	Flakiness Index (% MAX)		Aggregate Crushing Value (% MAX)								
Surfacing Material	Crushed Rock or Slag	Gravel	Crushed Rock or Slag	Gravel	Absorbtion (% MAX)	Aggregate Soundness (% MAX)	Stability	Sulphur Content (% MAX)	Bulk Density (Kg/m^3) (MIN)	Absorbtion (% MAX)	Stripping Test (in accordance with Appendix 4(A))
Marshall Asphalt	30	30	30	25	2	18 (3)	—	—	—	—	Not greater than 6 particles from a 150 particle test sample shall indicate evidence of stripping
Marshall DTS	30	30	30	25	2	18 (3)	—	—	—	—	Not greater than 6 particles from a 150 particle test sample shall indicate evidence of stripping
Rolled Asphalt Wearing Course	30	30	30	25	2	18 (3)	Requirements to Appx (A) of BS 1047	2%	1120 (2)	4	Not greater than 6 particles from a 150 particle test sample shall indicate evidence of stripping
Macadam Base Course	30	30	30	25	2	—	Requirements to Appx (A) of BS 1047	2%	1120 (2)	4	Not greater than 6 particles from a 150 particle test sample shall indicate evidence of stripping
Dense Tar Surfacing	30	30	30	25	2	18 (3)	Requirements to Appx (A) of BS 1047	2%	1120 (2)	4	Not greater than 6 particles from a 150 particle test sample shall indicate evidence of stripping
Friction Course	25 (1)	—	16 (1)	—	1.5	18 (3)	—	—	—	—	Not greater than 3 particles from a 150 particle test sample shall show evidence of stripping

Notes (1) Crushed rock only of specified groups
(2) Min when aggregate sample used in test complies with Table I of BS 63 for 20 mm nominal single-sized aggregate.
(3) Overall percentage mass loss based on the mass losses of the various nominal sizes tested from any one supply source

TABLE 11.9. *Requirements for grading and particle shape*

	Specified size			Oversize		Undersize				Specified size and oversize		
1	2	3	4	5	6	7	8	9	10	11	12	13
Nominal size	Passing BS test* sieve	Retained on BS test sieve	Minimum proportion of speci-fied size	All to pass BS test sieve	Maximum retained on BS test sieve quoted in col. 2	Passing BS test sieve	Maximum proportion of undersize	Passing BS test sieve	Maximum fines	Passing BS test sieve	Maximum dust	Maximum permissible flakiness index
mm	mm	mm	%	mm	%	mm	%	mm	%	µm	%	
50	50.0	37.5	70	63.0	15	28.0	5	2.36	2	75	1.5	40
40	37.5	28.0	60	50.0	15	20.0	5	2.36	2	75	1.5	40
28	28.0	20.0	60	37.5	15	14.0	7	2.36	2	75	1.5	35
20	20.0	14.0	60	28.0	15	10.0	7	2.36	2	75	1.5	35
14	14.0	10.0	60	20.0	15	6.30	7	2.36	2	75	1.5	35
10	10.0	6.30	65	14.0	15	5.00	10	2	2	75	1.5	35
								microns†				
6	6.30	3.35	65	10.0	15	2.36	10	600	2	75	1.5	—
3	3.35	1.70	65	6.30	15	1.18	10	600	2	75	1.5	—

* BS 410, 'Test sieves'

† One micron = 0.001 mm (1 µm)

Note 1. Aggregate which complies with this table will fall within the grading requirements for single sized aggregate of the same nominal size given in Table 1 of BS 882, 'Coarse and fine aggregates from natural sources'.

Note 2. Table does not include requirements for the immediate under size material, but this fraction is controlled by the remaining requirements of this standard (BS 63 Part 2 1971).

(f) the reputation and reliability of traditional sources of aggregate supply

Degrees of guidance varying from vague advice to precise specification limits have been given by various authorities in the United Kingdom and overseas in accordance not only with variation in usage, but also in accordance with history.

The wide variation in availability of suitable types of aggregate and in economic circumstance throughout the world makes impractical any attempt to establish a single uniform range of specifications applicable to all countries.

Emphasis is therefore placed on the following as aids to good engineering practice:

(i) good education and experience in the field of the components and the design of bituminous materials
(ii) the use of stage construction especially where economics dictates that known lower quality materials have to be used
(iii) a move towards end product specification for bituminous mixes based upon defined strengths, stiffness, and durability criteria (and skid resistance where appropriate) allowing the designer/manufacturer the freedom to design mixes in the most effective way to maximise the qualities of good aggregates and minimise the risks where aggregates of inferior quality need to be used (ASTM, 1976).

References

American Society for Testing and Materials, ASTM Standards Designation Part 15 (1978) D1075.

British Standards Institution BS 63. 1951. 1971. Part 1 Imperial Units. Part 2 Metric Units.

British Standards Institution, BS 594. 1973. Rolled asphalt (hot process) for roads and other paved areas.

British Standards Institution, BS 598. 1974. Part 2 Sampling and examination of bituminous mixtures for roads and other areas.

British Standards Institution, BS 812. 1975. Sampling and testing of mineral aggregates, sands and fillers.

British Standards Institution, BS 4987. 1973. Coated macadam for roads and other paved areas.

Deere, D. U. & Miller, R. P. 1966. Engineering classification and index properties for intact rock. Tech. Report AFWL-TR-65-116 Air Force Weapons Lab., Kirtland.

Department of Transport. 1966. Layout of roads in urban areas. Metric supplement 1974. A guide to revisions 1979.

Department of Transport. 1968. Layout of roads in rural areas. Metric supplement 1974.

Department of Transport. 1976. Specification for road and bridgeworks and Technical Memorandum H16/76. Specification requirements for aggregate properties and texture depth for bituminous surfacings to new roads.

Dhir, R. K., Ramsay, D. M. & Balfour, N. 1971. A study of the aggregate impact and crushing value tests. *J. Inst. High. Eng.* **XVIII,** 1 17–27.

Gray, J. E. & Renninger, F. A. 1963. Limestone and dolomite sands in skid resistant Portland cement mortar surfaces. Res. Report of National Crushed Stone Assoc. Washington.

Gray, J. E. & Renninger, F. A. 1965. Skid resistant properties of carbonate aggregates. Res. Report of National Crushed Stone Assoc., Washington.

Hallberg, S. 1950. The adhesion of bituminous binders and aggregates in the presence of water. Meddelande 78. Statens Vaginstitut, Stockholm.

Hallberg, S. 1958. Tests with oiled gravel roads. Meddelande 90. Statens Vaginstitut, Stockholm.

Hartley, A. 1968a. Petrographic aspects of the polishing of roadstones. Proc. 1st Symp. Influence of the Road Surface on Skidding, Univ. of Salford, 2.1–2.34.

Hartley, A. 1974. Mechanical properties of road surface aggregates. *Q. J. eng. Geol. London,* **7** (1), 69–100.

Hawkes, J. R. & Hosking, J. R. 1972. British arenaceous rocks for skid-resistant road surfacings. Transport and Road Research Laboratory Report LR 488. Crowthorne.

Herrin, M. & Gotez, W. H. 1954. Effect of aggregate shape on stability of bituminous mixes. *Proc. Highway Research Board,* **33,** 293–308.

Heukelom, W. 1965. The role of filler in bituminous mixes. *Proc. Assoc. Asph. Pvg. Tech.* **34,** 396–429.

H.M.S.O. 1962. Bituminous materials in road construction. Her Majesty's Stationery Office, London.

Hosking, J. R. 1967. An experiment comparing the performance of roadstones. Road Research Laboratory Report LR 81. Crowthorne.

Hosking, J. R. 1970. Synthetic aggregates of high resistance to polishing: Part 1-gritty aggregates. Road Research Laboratory Report LR 350. Crowthorne.

Hosking, J. R. & Tubey, L. W. 1969. Research on low-grade and unsound aggregates. Road Research Laboratory Report LR 293. Crowthorne.

Hughes, R. I., Lamb, D. R. & Pordes, 0. 1960. Adhesion in bitumen macadam. *J. Appl. Chem. London,* **10,** 433.

Knight, B. H. & Knight, R. G. 1935. Road aggregates, their uses and testing. Arnold, London.

Knill, D. C. 1960. Petrographic aspects of the polishing of natural roadstones. *J. Appl. Chem. London,* **10,** 28–35.

Lee, A. R. 1936. Adhesion in relation to bituminious road materials. *Jl Soc. Chem. Ind. London,* **55,** 23T–29T.

Lee, A. R. 1958. Roads and Road Construction.

Lees, G. 1964. The measurement of particle shape and its influence in engineering materials. J. British Granite & Whinstone Federation 4(2), 17–38.

Lees, G. 1969. The influence of binder and aggregate in bituminous mixtures. *J. Inst. highway Eng.* **16** (1) 7–19.

Lees, G. 1970. The rational design of aggregate gradings for dense asphaltic compositions. *Proc. Assoc. Asph. Pvg. Tech.* **39,** 60–97.

Lees, G. 1978. in 'Developments in Highway Pavement Engineering' Ed. P. S. Pell, Applied Science Publishers.

Lees, G. & Kennedy, C. K. 1975. Quality, shape and degradation of aggregates. *Q. J. eng. Geol. London* **8** (3), 193–209.

Lottman, R. P. & Goetz, W. H. 1956. Effect of crushed gravel fine aggregate on the strength of asphaltic surfacing mixtures. *Ntl Sand Gravel Assoc. Wash. Circ.* **63,** 3–22.

Maclean, D. J. & Shergold, F. A. 1958. The polishing of roadstone in relation to the resistance to skidding of bituminous road surfacing. DSIR Road Research Technical Paper 43, H.M.S.O. London.

MATHEWS, D. H. 1958. Adhesion in bituminous road materials: a survey of present knowledge. *J. Inst. Pet. London.* **44** (420), 423–32.

MATHEWS, D. H. 1961. Surface-active agents in bituminous road materials. Dept of Scientific and Industrial Research. RRL Research Note 3926/DHM.

PROPERTY SERVICES AGENCY. 1979. Standard specification clauses for airfield pavement works. Part 4 Bituminous Surfacing.

SALEHI, M. 1968. A study into the internal structure and flexural strength properties of bituminous paving materials. PhD Thesis, Univ. of Birmingham.

SALT, G. F. & SZATKOWSKI, W. S. 1973. A guide to levels of skidding resistance for roads. Dept of the Environment, T.R.R.L. Report LR 510.

SHERGOLD, F. A. 1948. A review of available information on the significance of roadstone tests. Road Research Technical Paper 10. DSIR, H.M.S.O., London.

SHUPE, J. W. & LOUNSBURY, R. W. 1959. Polishing characteristics of mineral aggregates, Proc. of the First Int. Skid Prevention Conference, 509–37.

SZATKOWSKI, W. S. 1980. Rolled asphalt wearing courses with high resistance to deformation, in The Performance of Rolled Asphalt Road Surfacings, *Instn Civ. Eng.*, London.

SZATKOWSKI, W. S. & HOSKING, J. R. 1972. The effect of traffic and aggregate on the resistance to skidding of bituminous surfacings, Dept. of the Environment, T.R.R.L. Report LR 504.

TRANSPORT AND ROAD RESEARCH LABORATORY. 1962. Bituminous materials in road construction. H.M.S.O. London.

TRANSPORT AND ROAD RESEARCH LABORATORY. 1972. Road Note 39. Recommendations for road surface dressing.

TRANSPORT AND ROAD RESEARCH LABORATORY. 1979. Surface dressing: proposed amendments to Road Note 39. N Wright LR 908.

WILLIAMS, A. R. & LEES, G. 1970. Topographical and petrographical variation of road aggregates and the wet skidding resistance of tyres. *Q. J. eng. Geol. London.* **2** (3), 217–36.

Appendix A11. Adhesion Tests

A11.1 Static immersion tests—coated aggregate particles are immersed in jars of distilled water and the degree of stripping is assessed as the visually estimated percentage of uncoated surface area after periods of 1, 3, 24 and 48 hours at 25°C and 40°C respectively. This test has only low reproducibility due to the human factor. A variety of static immersion test appears in Queensland standards Q16—1966 where the aggregate is not completely coated before immersion. In this test 50 representative pieces of the aggregate sample are pressed into a thin layer of the bituminous binder, left therein in an oven at 60°C for 24 hours and then immersed for 4 days in a water bath at 50°C. The particles are then plucked by hand from the tray and the degree of retention of binder assessed visually. In this test and the immersion tray test described below under (vi) there is a strong risk of influencing the result (as the present author has found in his own laboratory) by the handling process i.e. dust coatings are removed to varying degrees by handling and varying amounts of oil/perspiration are deposited from the fingers of the handler. These factors contribute to the low reproducibility mentioned above.

A11.2 Chemical immersion tests—as for static immersion tests but a chemical solution is used in place of distilled water, generally Na_2CO_3 in varying concentrations.

A11.3 Dynamic immersion tests—similar to the static test but agitation or stirring is involved.

A11.4 Immersion mechanical tests—these follow a variety of types of tests on graded and compacted coated material. After soaking for a specified number of hours at a specified temperature the materials are subjected to tests for say compressive strength, flexural strength, tensile strength, cone penetration, abrasion resistance (e.g. modified Deval, modified Los Angeles). It has been commented (HMSO 1962) that little is known about correlation between the tests and performance in service and that with many comparatively dense and strong mixes which have been assessed by these means it is doubtful whether an important practical problem really exists.

A11.5 Immersion trafficking test—an attempt to introduce the known significant factor of trafficking into adhesion testing came with the development of circular track machines and reciprocating loaded wheels (the Immersion Wheel Tracking Test, HMSO 1962). The time to failure of the specimens (continuously 'trafficked' under water at 40°C) is taken as a measure of resistance to stripping. Fairly good correlation with service behaviour has been established and examples are given in the reference to show the effect of different fillers on laboratory and field resistance to stripping.

A11.6 Coating tests—differ from those above in that the aggregate is not coated initially and its resistance to stripping subsequently assessed or measured. In coating tests the ability of the aggregate to attain a coating through the medium of water is assessed. The immersion tray test (Lee 1936) involves the placing of a layer of binder in a tray which is then covered with water. The aggregate particles are then applied by hand to the binder film, left therein for 10 minutes, then removed and the percentage of binder adhering to the face of the stone which has been pressed on the binder surface assessed visually (HMSO 1962). The test while low in reproducibility has been found useful to assess the degree of effectiveness of adding certain types and proportions of adhesion agent to standard bituminous binders.

A11.7 Adhesion can also be assessed by more fundamental tests including measurement of contact angle (θ) and interfacial tension between binder and water (γ_{BW}) (Lee, 1936; Hallberg, 1950) and by direct measurement of adhesion tension ($\gamma_{BW} \cos \theta$) (Hallberg 1950).

Chapter 12. Railway track ballast

12.1 Introduction

Railway track formations in the United Kingdom consist essentially of a layer of coarse aggregate, or ballast, in which the sleepers are embedded. The ballast may rest directly on the subgrade or, depending on the bearing capacity, on a layer of blanketing sand. The layer of ballast is intended to provide a free draining base, which is stable enough to maintain the track alignment with the minimum of maintenance. The function of the blanketing sand is primarily to provide a filter to prevent contamination of the ballast by fine particles derived from ascending waters.

The fundamental engineering is well understood. It concerns the transmission of the load from the base of the sleeper onto an area of closely packed ballast beneath. The sleeper bears onto the points of the angular ballast which are, in consequence, liable to fracture or abrade under load, a problem more severe with concrete than with wooden sleepers. If the ballast is not close packed there is a tendency for the sleeper to pump up and down as the wheels of the traffic pass in succession which eventually can cause the sub-grade to be disturbed if the ballast layer is not thick enough.

Consequently the ballast layer must be thick enough to hold the track in position and prevent traffic loads from distorting the subgrade, while the aggregate under the sleepers must be tough enough to resist the abrasion and degradation caused by the intermittent traffic loads. The main requirement, confirmed by practice on many railway systems, seems to be for a strong angular aggregate with a very good resistance to abrasions. Many tough, dense, siliceous rocks seem to be suitable but limestones, as a general rule, seem to be distrusted because of their low resistance to abrasion.

Specifications for ballast normally involve size, shape and rock quality requirements, the aim being to obtain a coarse angular material, the particles of which will interlock to form a layer of good dimensional stability. In addition, individual particles are required to be strong enough to resist surface attrition and degradation caused by the hammering action of rail traffic. It is important to exclude very fine material as its presence tends to reduce the bulk stability and creep resistance of the ballast.

The thickness of the ballast layer can be as low as 150 mm on lightly trafficked lines increasing to 400 mm on lines with high speed or heavy traffic. When concrete sleepers are used the ballast thickness may be increased by at least 50 mm. The blanketing sand layer normally has a minimum thickness of 150 mm.

It will be clear from the descriptions of requirements in various countries that the selection of suitable materials for railway ballast involves similar considerations throughout the world. The main area of difficulty lies in the lack of data linking in-service performance with various types of ballast. Hence specifications are often based on experience and judgement rather than on experimental evidence, not that such a situation is unusual with aggregate specifications, but in the case of railway ballast especially, there appears to be a dearth of reliable information.

12.2 Rock types suitable for track ballast

Application of the Wet Attrition Test (BS 812:1951, Clause 27) to various aggregates in the UK, mainly by British Rail, suggests that rocks which give the best values are likely to have a high crushing strength, a high resistance to abrasion and often a relatively low resistance to polishing. Such rocks, which fall mainly in the igneous and metamorphic category, might be of little value in other markets if low resistance to polishing makes them unattractive as wearing coarse aggregate in roads. Hence, the provision of optimum material for railway ballast might well justify the operation of a special quarry.

Materials currently produced in the UK which appear to have outstanding properties as railway ballast, in terms of the Wet Attrition Test, include rock which has been subjected to thermal metamorphism (e.g. hornfels) and certain fine grained acidic intrusives (e.g. felsites or microgranites). Coarser grained igneous rock together with dolerites and basalts often produce good quality ballast, but sedimentary rocks are not widely used although some quartzites and dolomites are considered adequate.

12.3 UK Practice

12.3.1 British Rail currently require aggregates for ballast to be finer than 50 mm, coarser than 14 mm and at least 80 per cent coarser than 28 mm with an even size distribution between 50 and 28 mm. The material is otherwise required to be clean, free from dust and angular in shape with all dimensions nearly equal.

The major quality requirement is specified in terms

of a Wet Attrition Value, which is defined in BS 812:1951, but which is now only applied to track ballast although it is related to abrasion resistance (7.5.4.1). The Wet Attrition Value should be as low as possible and a maximum value of 6 has been specified for ballast used on UK main lines rising to 8 on secondary lines. In addition, although of lesser importance, the Aggregate Crushing Value should not exceed 30, the Aggregate Impact Value should not exceed 25, the Flakiness Index should not exceed 50 and the elongation index should not exceed 50.

Sand used in the blanketing layer is specified in terms of the following size distribition:

BS Test Sieve size (mm)	14	2.36	1.18	0.60	0.30	0.15	0.075
% Passing by Weight	100	80–100	70–90	48–76	24–60	5–42	0–10

If more than 4% of the material is finer then 75 μm it should not show any plasticity (i.e. the Plasticity Index must be zero (BS 1377)).

12.4 Practice in some other countries

The general principles controlling track ballast specifications apply throughout the world, although there are minor differences between countries. Usually the Wet Attrition and other tests are so rigorous that they exclude materials which might be affected by the extremes of temperature or humidity found in various climates.

Brief examples of local specifications and recommendations for rail ballast in a selection of countries are as follows:

12.4.1 USA. American specifications for rail ballast are based on ASTM methods. They require that the material consists of hard, strong, durable particles of crushed stone or slag. Deleterious substances are restricted to a maximum of 5% soft and friable pieces, 1% of material finer than No. 200 (ASTM) sieve size and 0.5% of clay lumps. The percentage of wear of processed ballast, tested in the Los Angeles machine, should not exceed 40% except as otherwise specified by the Engineer and the proportion of flat or elongated particles, defined as having a length to average thickness ratio of at least 5:1, should not exceed 5%. Processed ballast for use in regions where frost is expected should have a weighted average loss of not more than 7% after 5 cycles when subject to the sodium sulphate soundness test.

12.4.2 France. French specifications are quite detailed but basically they require ballast to be produced by crushing material from quarries in "hard stone", (defined as granite, microgranite, diorite, micro-diorite, rhyolite, quartzite, basalt, sandstone, gneiss, etc.). The main rock quality assessment is carried out in terms of a hardness coefficient which is derived from the results of Deval and Los Angeles tests. Three separate size ranges of ballast are specified (25–50 mm, 16–31.5 mm and 10–25 mm) and there are restrictions on the proportion of flat or elongated particles. Friablc or altered fragments are not to exceed 3% of the total and the proportion of sand, dust, earthy material and foreign matter must be less than 1%.

12.4.3 West Germany. Track ballast used in West Germany is made from rock or blast furnace slag, which must be free from earth, clay, loam, marl, plant residues and other inferior or harmful constituents. The ballast is required to be resistant to frost and weather, capable of withstanding a pressure of 1800 kg/cm^2, and resistant to impact; all defined by appropriate DIN test methods. Two sizes of ballast are specified (65–25 mm and 30–15 mm) and in both a precise size distribution is required. There is also a requirement that the ballast should consist of irregular, angular shaped fragments.

12.4.4 India. On Indian railways three separate sizes of ballast are recommended (50 mm, 40 mm and 35 mm) depending on the type of sleeper. The material is expected to be "hard, durable, tough and wear resistant, cuboid in shape with sharp edges, non porous and non-absorbent of water. On some Indian railways, in addition to being durable, resistant to impact and free from adherent coating, the ballast is not to contain more than 10% of dust, rubbish or decomposed particles that can pass through a 5 mm sieve. There are restrictions on the use of gravel in that the ballast must be derived from stones weighing no less than 10 kg or having a diameter of less than 150 mm; faces of the ballast should be produced by crushing and only one smooth face is allowed.

12.4.5 Middle East. As an example, a specification for ballast developed by a British consulting engineer for a Middle East rail project is included which was based largely on British practice and Standards, but also included some additional requirements derived from American Standards. Thus a Los Angeles Abrasion Value (7.5.4.1c) was specified as an additional measure of resistance to attrition and a sulphate soundness test (7.5.4.2a) was specified to obtain a rough guide to resistance to salt weathering. More specifically, the requirements imposed were as follows:

> The ballast material shall be crushed rock. Naturally occurring gravels shall not be used as track ballast. The rock for ballast shall be strong to very strong rock free from deleterious materials and shall be capable of being crushed to provide ballast as specified below. Calcareous rock (limestone, dolomite, etc.) shall not be used as ballast.
>
> The crushed rock for ballast shall be strong angular particles free from incipient fractures. All material in the ballast passing the 63 μm sieve shall be non-plastic.
>
> The crushed rock ballast shall conform to the

following particle size distribution:

Limits to frequency distribution of per cent passing sieve	Sieve size mm (square hole), mm				
	50	37.5	28	20	2.36
	Per cent passing sieve by weight				
Upper 1 percentile	100	90	60	5	2
Lower 1 percentile	100	60	20	0	0

The crushed rock ballast shall comply with the following quality requirements:

(i) The percentage of wear as measured by the Los Angeles Machine shall not exceed 20.
(ii) The wet attrition value shall not exceed 6.
(iii) The flakiness index shall not exceed 30.
(iv) The elongation index shall not exceed 30.
(v) The aggregate crushing value shall not exceed 25.
(vi) The aggregate impact value shall not exceed 20.
(vii) The loss after 5 cycles of the magnesium sulphate soundness test shall not exceed 8.

References

AMERICAN RAILWAY ENGINEERING ASSOCIATION (AREA) 1976. Report on Committee 22, Bulletin 656, Chicago, 312–317.

AMERICAN RAILWAY ENGINEERING ASSOCIATION, (AREA) 1976. Manual Recommendations, Special Committee on Concrete Ties, Bulletin 660, Chicago, 133–138.

AMERICAN RAILWAY RECOMMENDATIONS, COMMITTEE 1, 1976. Report on Assignment 2, Ballast, Bulletin 660, AREA, Chicago, 110–115.

DALTON, C. J. 1973. "Field Durability Tests on Ballast Samples as a Guide to Specification Requirements", Canadian National Railways Technical Research Centre, St. Laurent, Quebec.

GASKIN, P. N. & RAYMOND, G. P. 1976. "Contribution to Selections of Railroad Ballast", *Proc. ASCE,* **102,** (TE 2), 377–394.

HAY, W. W., BAUGHER, R. & REINSCHMIDT, A. J. 1977. "A Study of Railroad Ballast Economics", U.S. Department of Transportation Report No. FRA/ORD-77-64, Federal Railroad Administration, PB 275102.

KNUTSON, R. M. & THOMPSON, M. R. *et al.* 1977. "Materials Evaluation Study", U.S. Department of Transportation Report No. FRA/ORD-77-02, PB 264215.

MAJIDZADEH, K. 1976. "An Evaluation of Engineering and Physical Properties, and Use-Specifications for Locally Available Ballast Materials", report of Development Project 6017, Engineering Experiment Station, Ohio State University, Civil Engineering Materials Laboratory.

RAIL INTERNATIONAL 1978. No. 9 Sept.

THE PERMANENT WAY INSTITUTION 1971. British Rail Track, Design, Construction and Maintenance.

Chapter 13. Aggregates for use as filter media

13.1 Introduction

Although the total volume of aggregate used for filters may be relatively small, filters nevertheless play important and diverse roles in many projects and structures. Table 13.1 shows some of the main uses of filter aggregates and the first four parts of this chapter discuss some of the requirements and properties of these materials. The final three parts of the chapter describe the functions and specification of filter aggregates in their three main applications, namely in water and effluent treatment, and as drainage filters for earthworks and other civil engineering structures. These will normally be subject to different design criteria and may also call for distinct sets of physical and mechanical properties.

Filter aggregates generally consist of sand, gravel or crushed rock. Manufactured aggregates are also occasionally used and these often include blast-furnace slags.

Although filter materials for water and effluent treatment works are often used in relatively small quantities, the high quality of aggregate normally required is not always readily available from commercial production processing, which may be designed to yield a satisfactory general-purpose aggregate at least cost. On the other hand, drainage layers in major civil engineering works such as embankment dams, are usually designed to make the best use of the available natural materials with the minimum of processing.

13.2 Key properties of filter aggregates

13.2.1 General. Particle size grading, strength and durability are the key properties, as these determine permeability, ease of construction, stability and longevity.

Particle sizes in filter aggregates can range from fine sands to boulder-sized material. The aggregates are

TABLE 13.1. *Uses of filter aggregates*

Category of works	Filter application	Function: Drainage	Pressure	Filtration
Land drainage	Surface blanket drainage	*		
	Ditch/trench fill	*		
	Drain/soakaway fill	*		
	Pipe filters			*
	Pipe envelopes	*		
Structure drainage	Foundation drainage	*	*	
	Roadbed drainage	*		
	Runway drainage	*		
	Retaining structure drainage	*	*	
	Embankment dam drains	*	*	
Remedial & stabilisation works	Uplift pressure relief		*	
	Seepage force prevention		*	
	Saturation control	*	*	
	Piping prevention		*	
Water supply	Well filters			*
	Well envelopes	*		
Filtration & purification works	Fine & coarse filter beds			*
	Gravel underdrain envelope	*		*
Marine works	Bedding layers		*	
	Transition layers			*
	Pressure relief		*	

usually placed and compacted in an unbound condition.

The completed filter will usually need to have the following properties: structural stability (especially if placed in an unconfined situation); durability; high permeability combined with resistance to internal erosion of fines; low frost susceptibility; low susceptibility to salt aggression, chemical attack and solution loss. Physical and chemical breakdown in service may seriously impair the design grading of the filter.

The strength, shape, surface texture and composition of the individual particles will have an important influence on the above properties, as also will the abrasion resistance and crushing strength of the aggregate, since if these are deficient, it may degrade during placing and compaction.

13.2.2 Grading. This is fundamental to the design of the filter, detailed consideration of which is beyond the scope of this report. Filters in contact with natural soils or controlled fill material will usually be designed in accordance with the criteria given by Terzaghi, and multi-layered graded construction will tend to be used for the thicker filter layers. A very wide range of particle sizes, from fine sand up to boulders, might therefore be used in different parts of a filter zone. An important requirement is freedom from silt and clay, and it is commonly specified that the material passing the 425 μm BS sieve shall be non-plastic when tested in accordance with BS 1377.

13.2.3 Aggregate strength. Since in the absence of a cementitious binder, the stability and load bearing properties of a filter depend on the aggregate particles alone, these must be strong enough to prevent breakdown during construction and when in use. Fine-grained materials normally used for filters are generally those which nature has selected as the most durable. Thus many sands and gravels, and particularly the finer gravels, are composed mainly of the hard and stable mineral quartz. However, aggregate particles coarser than fine gravel are not usually monomineralic but are composed of rock fragments in which the strength is derived from the interlocking or cementation of individual mineral grains. Serious consideration must then be given to the assessment of the strength and abrasion resistance of these aggregates.

13.2.4 Particle durability. It is essential that filter aggregates are durable in the long term and this is particularly important where high strength is not a characteristic of the aggregate. Any breakdown of the constituent particles would increase the fines content of the filter and alter its design grading and efficiency. Durability can be defined as the resistance of the aggregate under its working conditions to cyclic variations in temperature, load, moisture content, freezing and thawing, and chemical environment.

13.2.5 Particle shape. This property is, after grading, as important as strength in a filter aggregate and can be discussed under two main headings. The first relates to whether a particular particle is basically equidimensional in shape or whether it is flaky or elongated. In general, it is desirable for filter aggregates to be essentially equidimensional as this aids the flow distribution in the medium and also facilitates packing of the coarser and finer constituents and improves the effectiveness of the filter. The other important shape characteristic is roundness and again it is better for a filter aggregate to have rounded, as opposed to angular, edges to most particles. However, rather different considerations apply to biological percolating filters, where the aggregate acts mainly as a supporting medium to a microbial film, and its specific surface is therefore important. This is discussed below in 13.5.

13.2.6 Particle texture. Surface texture affects to a lesser extent both the flow of liquid through the filter and the filtration characteristics of the material. A smooth glassy surface is useful from a drainage aspect but not so efficient from a filtration or fines ingress viewpoint, where a rougher surface texture is considered more advantageous.

13.2.7 Surface coatings. Some aggregates have a coating that may, or may not, be easily removed during initial processing. Coatings are commonly composed of clay, silt, calcium carbonate, iron oxides, silica, or gypsum, but other coatings can occur. Such coatings may vary in thickness and hardness and are, on the whole undesirable in filter aggregates. They should either be removed by scrubbing or be as strong and durable as the particle itself.

13.2.8 Porosity. The functions of drainage and filtration are not significantly affected by particle porosity, which can be expected to have a negative correlation with strength and durability. Therefore, the aggregate porosity, or a related property such as water absorption, should be measured in order to assess its possible effects on, for example, durability, chemical reaction, freeze/thaw or other breakdown mechanisms.

13.2.9 Chemical reactivity. There are generally no specified requirements for chemical inertness or solubility, but this should be taken into account in relation to the use of the filter media. It is important that the aggregate is not affected by the chemical properties of the liquid to be treated, and also that constituents are not leached from the medium into the filtrate. The latter is more likely to occur with artificial aggregates, such as slag, than with naturally occurring aggregates. Slag may also be aggressive to concrete and if used, it is generally required to comply with the stability and sulphur content requirements of BS 1047. Other materials when used in civil engineering structures should not have a soluble sulphate content exceeding 2 g of sulphate (as SO_3) per litre when tested as described in BS 1377, Test No. 10.

13.3 Testing of aggregates

Tests relevant to filter aggregates are essentially those used for general aggregate testing, and are listed in

TABLE 13.2. *Suggested tests for evaluation of filter aggregates*

Category of test	Test	Guideline acceptance criteria*
Physical properties	Grading	Design test only
	Shape {Flakiness, Elongation}	Index not more than 30
	Relative density (oven dried)	Not less than 2.55
	Water absorption	Not more than 3% by weight
Mechanical properties	10% Fines Value	Not less than 100 kN
	Aggregate Crushing Value	Not more than 30%
	Aggregate Impact Value	Not more than 30%
	Aggregate Abrasion Value	Not more than 20%
	Los Angeles Abrasion	Not more than 40%
Durability properties	Soundness ($MgSO_4$)	Not more than 12%
	Deleterious substances, Petrographic analysis	To be assessed after examination

* Dependent on conditions of use and to be evaluated collectively, not in isolation.

Table 13.2, which also shows some guideline acceptance criteria.

Additional criteria applicable to some special purposes are considered in the following sections, which describe the main engineering applications of filters.

13.4 Water treatment: filtration through sand

13.4.1 General. There are basically two types of sand filter:

(i) Filters characterised by slow filtration rates, using a fine sand and requiring infrequent cleaning.

(ii) Rapid filters, which are coarser, operate at higher filtration rates and require frequent cleaning.

The action of a filter bed is complex and not yet fully understood, being a combination of, (i) straining, (ii) physical forces, which determine the conveyance of particles towards and onto the sand grains, and (iii) surface forces which hold the particles to the grains. In slow sand filters the bed usually comprises from 0.6 m to 1.2 m of fine sand, supported by layers of graded gravel 0.2 m to 0.3 m thick. Rapid filters usually comprise from 0.6 m to 0.8 m of coarse sand supported on several layers of fine to coarse gravel to give a total bed thickness of 1.0 m to 1.3 m.

13.4.2 Filter bed sand. The sand should be hard, abrasion resistant and free from contaminants. It should not lose more than 5 per cent by weight after immersion for 24 hours in 40 per cent hydrochloric acid at 20°C. Both rounded and angular grains may be used, but they should not be flaky. Filter sand is normally obtained from natural sand deposits by sieving, although crushed fines produced from rocks composed predominantly of quartz is sometimes used.

13.4.3 Filter gravel. The gravel or coarse layer has several functions. It supports the sand, permits the filtered water to move freely towards the underdrains and, in the case of a rapid filter, facilitates a uniform flow distribution. The coarse material should be clean, hard, durable, and free from flat, thin or long particles.

13.5 Aggregates for biological percolating filters

The term 'filter' is a misnomer as the process involved is not truly filtering but a biological contact process. The filter medium is primarily required to support colonies of organisms, including bacteria and fungi, which exist in microbial layers on the surface of the media and over which sewage or other pollutant liquids are trickled so that the colonies may feed in an aerobic environment upon the organic matter present.

The degree of treatment attained is largely dependent upon the area of microbial film to which the liquid is exposed and it therefore follows that the surface area of the media per unit volume should be as large as possible, consistent with adequate aeration and permeability.

This is best achieved with single-sized filter media. Percolating filters of optimum surface area and void ratio can be manufactured, usually of plastic, or alternatively, filter beds may be constructed from natural aggregates which may not possess optimum characteristics but have the advantage of low unit cost.

Biological percolating filters formed with natural aggregates typically consist of between 1.5 m and 3.0 m thickness of single-sized coarse aggregate over which the liquid to be treated is distributed. BS 1438:1971 recommends that the aggregate should be selected from the range 14 mm to 63 mm, the actual size specified depending on the operating conditions of the system. A layer of 100 mm size aggregate is often used to support the filter.

The aggregates may be naturally occurring or crushed and should preferably be angular and of rough surface texture so as to maximise the surface area. For example, a 25 mm single-size slag was found to have a specific surface area of 206 m^2/m^3 whereas the corresponding figure for crushed gravel was 176 m^2/m^3.

The aggregates must be strong enough to sustain their own weight and in particular must be durable and resistant to cyclic wetting over periods of many years. It is found in practice that most strong rocks such as granite or basalt are suitable for filter media, together with some artificial materials such as clinker and slag. Softer or potentially soluble aggregates, such as some types of limestone, have to be selected with care and are frequently found to give borderline results in the magnesium sulphate soundness test.

BS 1438:1971 sets out the particular requirements for aggregates for biological percolating media in respect of size, shape, flakiness, cleanness and durability. Durability is tested according to this standard by 20 cycles of the sodium (not magnesium) sulphate soundness test.

13.6 Filters for civil engineering structures

13.6.1 General. Protective filters for civil engineering structures usually consist of one or more layers of free-draining sand and/or gravel material placed on or within a less pervious soil to remove seepage water and prevent erosion of the soil or damage to overlying structures from uplift pressure. The soil which is to be protected by the filter is commonly referred to as the base or the base material. A given base material can be protected by a filter layer having a grading related to that of the base material, normally according to the Terzaghi criteria for filter design. Filters may consist either of a single layer, or of several layers each with a different grading in which case they are known as zoned filters. The filter or filter layers may be classed as uniform (i.e. having a narrow range of major particle sizes), or well-graded (i.e. having a broad range of particle sizes). In large structures such as embankment dams, it is more usual to design the filter layers to make the best use of the available natural materials, rather than to process large quantities of material to suit a pre-conceived design. Deficiencies in grading may be compensated by varying the thickness of the filter.

A full account of the criteria and methods used in the design of civil engineering filters is outside the scope of this work but selected references are given in the bibliography.

13.6.2 Specification of aggregates. The specification of filters, drains, etc. and their related aggregates is often rather loose and occasionally gives rise to serious difficulties. The following suggestions may prove useful to those formulating specifications.

The specification of aggregates for filters and drains should avoid the use of general terms such as pervious and free-draining, unless used in conjunction with specific requirements for grading, soundness, and permeability that assure the necessary physical properties. When grading requirements are part of a specification, the time and place of the sampling for testing should be clearly stated otherwise confusion arises, and enforcement of designers' intentions is difficult. Aggregates which break down and develop a greater proportion of fines during handling, placement, and compaction may meet a grading specification at the plant or when they arrive at the works but may fail when tested after compaction. The important criteria are the grading and permeability of the material as it will be in the finished work. Hence additional samples for testing should be taken after the filter materials have been placed and compacted.

With regard to the acceptance criteria which can be applied to a potential filter aggregate, very little is published. However the following guidelines might prove useful. The aggregates should be hard, durable, clean and should not contain deleterious materials of such a form or in sufficient quality or quantity to adversely affect the efficient operation or longevity of the filter. Durability should include resistance to frost, corrosion and solubility. Examples of deleterious materials are: clay, flaky and/or elongate particles, excessively porous and/or laminated materials, and chemically unsound or readily soluble materials.

A typical specification might thus combine some or all of the following particular requirements:

(i) Oven dried relative density not less than 2.5

(ii) Water absorption not greater than 3% by weight

(iii) Aggregate Impact Value not more than 30
or
Aggregate Crushing Value not more than 30
or
10% Fines value not less than 100 kN

(iv) Maximum flakiness and elongation indices not greater than 30

(vi) Aggregate Abrasion Value not greater than 20
or
Los Angeles Abrasion not greater than 40

(vii) Magnesium sulphate soundness value not more than 12

13.7 Concluding remarks

It is hoped that this short account will have made it clear that the properties required of filter aggregates are dominated by the highly diverse purposes for which they may be intended. Indeed in some cases, such as sewage treatment, their main role is not truly filtration at all, but to support microbial colonies active in a biological process. In other cases, the shape and size of the filter, its operating conditions (including the chemical environment), and both the character and quantity of available natural materials, can each affect design, making a general specification impossible. The over-riding needs are for durability and dimensional stability, which will usually have to be assessed from a wide range of physical and chemical properties. This calls for a degree of judgement and experience, on the part of the designer, which may be exceptional in the entire field of aggregate technology.

Selected bibliography

American Water Works Association. Water quality and treatment. 3rd ed. 1971. New York, McGraw-Hill. ISBN 07 001539-2.

British Standards Institution. 1971. Specification for media for biological percolating filters, BS 1439:1971 (amended Aug. 1972, Dec. 1980).

British Standards Institution. 1983. Specification for air-cooled blast furnace slag coarse aggregate for concrete, BS1047:1983.

Bruce, A. M. 1968. The significance of particle shape in relation to percolating filter media. *J. Brit. Granite and Whinstone Fedn.* **8,** (2), 1–15.

Degremont Company (editors). Water treatment handbook. 1973. New York, Taylor-Carlisle.

Department of Transport. 1976. Specification for road and bridge works 5th ed. HMSO, London.

Fair, G. M. *et al.* Water and wastewater engineering. Vol. 2: Water purification and wastewater treatment and disposal. 1968. New York, Wiley. ISBN 471251313.

Holden, W. S. (editor). Water treatment and examination. 1970. Edinburgh, Churchill Livingstone. ISBN 7000 1437 3.

Huisman, L. & Wood, W. E. 1974. Slow sand filtration. Geneva, World Health Organisation.

Ives, K. J. 1975. Specifications for granular filter media. *Effluent and Water Treatment J. June 1975.* **15** (6), 296, 298–299, 301–305.

Lamb, T. W. & Whitman, R. V. 1969. Soil Mechanics (see esp. pp 292–4), New York, John Wiley.

Twort, A. C., Hoather, R. C. & Law, F. M. Water supply. 2nd ed. 1974. London, Edward Arnold. ISBN 0713133 18.

US Corps of Engineers, Dept of Army, 1952. Seepage control, soil mechanics design. Washington DC.

US Department of the Interior. Earth Manual 2nd ed. 1974 (see esp. pp 305–9).

Glossary

To save space, many definitions in this glossary have been simplified and abbreviated as far as was judged possible for comprehension of the main text of the report. For more detailed definitions, readers are referred to existing technical dictionaries, such as the Penguin Dictionary of Civil Engineering (3rd edition, 1980) and the Penguin Dictionary of Geology (2nd edition, 1979), both published by Penguin Books Limited, Harmondsworth, Middlesex, England. Also to BS 6100:1984, Glossary of Building and Civil Engineering Terms, and Barker (1983), Dictionary of Concrete (see refs Chapter 8)

accessory minerals: minerals which are only minor constituents of a rock and are not essential to its broad classification, though they may be used to distinguish varieties of rock

accretion: the gradual addition of sediment by natural agencies, such as the spreading of silt and clay on a flood-plain by streams or the deposition of sand on a beach by waves

aeolian: deposits whose constituents were transported and dropped by the wind, such as dune sand and some volcanic debris

aggregate: particles of rock which, when brought together in a bound or unbound condition, form part or whole of a building or civil engineering structure

——, "all in": a well graded mixture of fine aggregate and coarse aggregate

——, coarse: aggregate which is mainly retained on a 5 mm BS 410 test sieve and containing only so much finer material as is permitted for the various sizes described in BS 882

——, fine: aggregate mainly passing a 5 mm BS 410 test sieve and retained on the 75 μm sieve, and containing only so much coarser material as is permitted for the various grading zones described in BS 882

——, natural: aggregate produced from naturally occurring sands and gravels

alluvium: a general term for unconsolidated detrital material such as clay, silt, sand and gravel, deposited by rivers and streams as sorted or semi-sorted sediment in the stream-bed or on the flood plain

alteration: chemical changes in the rock as a result of weathering or hydrothermal action (q.v.). Usually accompanied by loss of strength and/or soundness

amphibole: a group of rock-forming silicate minerals typically containing calcium, iron, magnesium and sometimes aluminium. Hydroxyl ion (OH) is always present. Hornblende (q.v.) is the best-known example

amphibolite: methamorphic rock composed primarily of amphibole with some plagioclase feldspar. The texture may be schistose or linear, i.e. parallel alignment of amphibole needles

amygdale: ellipsoidal to irregularly-shaped cavity in volcanic and some hypabyssal rocks formed during the evolution of gas. These cavities are generally infilled by minerals not present in the rock body e.g. zeolites, calcite, different forms of silica, chlorite

amygdaloidal: amygdale-bearing

analcite: a variety of zeolite (q.v.); a hydrous sodium aluminium silicate, nearly always a secondary mineral (q.v.) infilling veins or amygdales (q.v.)

andesite: a dark coloured, fine-grained extrusive rock, commonly containing scattered conspicuous crystals of pale-coloured feldspars; the extrusive equivalent of diorite

anisotropy: variation in physical or textural properties with direction

aquifer: a permeable geological formation which is capable of storing and yielding ground-water

arenaceous: sandy; said of a sediment consisting wholly or partly of sand-size fragments, or having a sandy texture

argillite: a generic term for rocks formed from indurated (q.v.) silt and clay, e.g. mudstone, siltstone, shale

argillaceous: clayey or silty; term applied to sediments consisting wholly or partly of clay or silt-sized particles

arkose: sandstone with feldspar content in excess of 25 per cent

asphalt: a natural or artificial mixture in which bitumen is associated with a substantial proportion of mineral matter

asphaltic cement: bitumen, a mixture of lake asphalt and bitumen, or lake asphalt and flux oils or pitch or bitumen, having cementing qualities suitable for the manufacture of asphalt pavements

augite: the most abundant of the pyroxene (q.v.) family of rock-forming minerals

back breakage: separation of fragments behind the quarry face

backfilling: the tipping of material (normally waste material) into a former mineral working in order to recreate a usable land surface

ball mill: a grinding machine consisting of a short, horizontal, rotating cylinder charged with steel balls and mineral fragments. Breakage is produced by the tumbling action of the balls during rotation

barite (or barytes): natural barium sulphate ($BaSO_4$); a dense (S.G. 4.5) mineral much used, for example, as an aggregate in radiation shielding

basalt: a dark or medium-dark, commonly extrusive (lava), locally intrusive (as in a dyke or sill), glassy to fine-grained igneous rock; the extrusive equivalent of gabbro. SiO_2 content 45–62%. Rich in ferromagnesian minerals (>60% by volume) and labradorite feldspar

base course: in a road pavement, the layer below the wearing course (q.v.) and above the road base; it serves to protect the less durable base materials from damage and provides a flat surface on which the wearing course is laid

bauxite: a mixture of hydrated aluminium oxides formed by the *in-situ* breakdown by tropical weathering, usually of basic igneous rocks. The ore of aluminium. Used as calcined bauxite (q.v.) in friction courses

bedding: the arrangement of distinct sedimentary rock layers or beds one upon another; the flat (planar) surface the separates successive beds is a bedding plane

benches, benching: horizontal levels to which successive quarry faces are taken

beneficiation: the processing of rocks and minerals to remove unwanted constituents

binder: any soil or aggregate cementing agent, e.g. water, clay, cement, lime, bitumen, synthetic resins

bioherm: a mound or dome shaped reef composed almost entirely of the remains of sedentary organisms, such as corals, algae, molluscus and gastropods and surrounded or enclosed by rock of a different kind

biotite: a dark-coloured iron-bearing member of the mica group of rock forming minerals. Biotite occurs as an original constituent of many igneous and metamorphic rocks

bitumen: a viscous liquid, or a solid consisting essentially of hydrocarbons and their derivatives, which is soluble in carbon disulphide; it is substantially non-volatile and softens gradually when heated. It is black or brown in colour and possesses waterproofing and adhesive properties. It is obtained by refinery processes from petroleum, and is also found as a natural deposit or as a component of naturally occurring asphalt, in which it is associated with mineral matter

bitumen emulsion: an emulsion in which, with the aid of suitable emulsifying agents, bitumen is dispersed in water, or in an aqueous solution

bitumen macadam: (see macadam)

bituminous: containing road tar, bitumen, pitch or mixtures thereof

Blake crusher: a jaw crusher with the moving jaw pivoted near the top to allow maximum leverage to be exerted on the coarsest material

bleeding: the release of water from a concrete or mortar mix after placing

blinding: the obstruction of a screen surface which occurs when the apertures become blocked by an accumulation of very fine material. Also a layer of mortar or concrete used to protect or regulate a surface prior to construction

boss: an igneous intrusion usually a few square kilometres in surface extent that is roughly circular in plan and is thought to extend to considerable depth

boulder clay: a glacial deposit consisting of sub-angular pebbles and boulders of all sizes embedded in stiff or hard reworked clay or rock flour. The term 'till' is preferable because it covers the wide range of lithologies included here and does not imply the presence of either boulders or true clay

bound aggregate: aggregates which are bound or coated with cementitious or bituminous binders, i.e. as in concrete or bituminous macadam

brachiopod: the lamp-shell; a solitary marine invertebrate with two bilaterally symmetrical, commonly calcareous, shells usually unequal in size and shape. A frequent fossil in limestones

braided deposit: sediment deposited from a tangled network of several small, branching and reuniting shallow stream channels (braided stream) resembling the strands of a complex braid. Such sediments are thought to indicate the inability of the stream to carry all of its sediment load

breccia: a rubble-rock: a coarse-grained clastic (q.v.) rock composed of large, angular broken rock-fragments held together in a finer-grained matrix. Breccia is similar to conglomerate (q.v.) except that most of the fragments have sharp edges and unworn corners; it can be of any origin, mode of accumulation or composition. Hence brecciated: a rock converted into, characterised by or resembling a breccia

bucket capacity: the volume able to be retained in an excavator bucket

building sands: sand with a grading suitable for use in mortars

calcined bauxite: bauxite (q.v.) which has been heated to produce corundum (Al_2O_3) as one of the main mineral phases, giving the material very good polish-resisting properties when used as a roadstone wearing course

Cambrian: the earliest period of the Palaeozoic era between about 590 and 505 million years ago, and the corresponding system of rocks. Named after Cambria, the Roman name for Wales

carbonate rocks: a generic term for rocks formed predominantly from the carbonates of calcium, magnesium, iron etc., occurring either singly or in combination. Limestone (calcium carbonate) is the most familiar example

Carboniferous: the geological period of time (360 to 286 million years ago) and the corresponding system of rocks, which in Britain contains the Carboniferous Limestone, an important source of crushed rock aggregate

cement: natural or synthetic material which binds rock particles together. In sedimentary rocks this may be silica, calcium carbonate, clay or iron oxide etc.

chalcedony:	cryptocrystalline (q.v.) variety of silica
chlorite:	family of greenish platy clay minerals frequently occurring as alteration products of ferromagnesian minerals and as a prograde constituent in low grade metamorphic rocks. Hydrous silicates of aluminium, ferrous iron and magnesium
classification:	in mineral processing, a method of sizing fine particles by exploiting differences in settling velocity in water, or occasionally in air. A "classifier" is a device which effects this type of separation
clastic:	a term relating to a rock or sediment composed principally of broken fragments of pre-existing rocks or minerals that have been transported individually from their place of origin
clay minerals:	a complex group of finely crystalline to amorphous hydrous silicates essentially of aluminium, formed chiefly by alteration or weathering of primary silicate minerals such as the feldspars, pyroxenes and amphiboles. The three most common clay minerals are kaolin, illite and montmorillonite
cleavage:	an ease-of-splitting direction in minerals, controlled by the atomic lattice. In deformed rocks such as slates etc. it is an ease-of-splitting direction induced by the deformation
coated chippings:	aggregate chippings which have been coated thinly with bituminous material for scattering over a wearing course or use in surface dressing
coated macadam:	a road material consisting of graded aggregate that has been coated with a tar or bitumen, or a mixture of the two, and in which the intimate interlocking of the aggregate particles is a major factor in the strength of the compacted roadbase or surfacing
cold asphalt:	a close-textured type of coated macadam wearing course material, consisting of aggregate wholly passing 6 mm BS sieve for the fine grade, and wholly or substantially passing 10 mm BS sieve for the coarse grade, coated with a binder solely or substantially of bitumen, the composition of the mixture being so adjusted that the material can be spread and compacted while cold or warm, and if required, after storage
compaction:	the process of causing soil or aggregate particles (bound or unbound) to pack more closely together, thereby causing an increase in the density
compression jointing:	fractures in a rock produced by compressive forces
cone crusher:	a crushing machine consisting essentially of two upward pointing, concentric, conical crushing surfaces. The crushing action is produced by the eccentric rotation of the inner conical surface against the outer
conglomerate:	a coarse-grained, clastic sedimentary rock composed predominantly of rounded fragments (generally larger than 5 mm in diameter) set in a fine-grained matrix of sand, silt or natural cementing material
contact metamorphism:	changes brought about in a rock by contact with intruded and extruded molten igneous material (magma). Metamorphic changes are effected by heat, superheated mineralised fluids and by deformation arising from the emplacement of the igneous mass
Cretaceous:	the last period of the Mesozoic era (between 144 and 65 million years ago) and the corresponding system of rocks
crinoid:	popularly referred to as a (fossil) sea-lily, this echinoderm consists of a five-rayed globular body enclosed by calcareous plates and supported by a stem or column comprising numerous circular disc-like calcareous elements. The skeletal plates and discs disaggregate and represent the major constituents of crinoidal limestones
cristobalite:	a high temperature polymorph of the mineral quartz (SiO_2), i.e. possessing a different atomic structure but identical chemical composition

crusher-run aggregate: unsized material produced by a simple crushing procedure without screening

cryptocrystalline: very finely crystalline material in which the crystals are so small as to be indistinguishable except under powerful magnification

cut-back bitumen: petroleum bitumen whose viscosity has been reduced by the addition of a suitable volatile dilutent

cycle time: the period taken to achieve a certain operation. In quarrying, particularly applied to mobile extractive equipment

cyclone: a classifying (q.v.) device which utilises a vortex to effect a size separation

decomposition: in geology, the predominantly chemical changes in a rock, usually accompanied by physical deterioration, which result from weathering or hydrothermal alteration (q.v.)

decoupling: in blasting, applied when the explosive charge is not in proper physical contact with the rock, thereby reducing the effect of the explosion

deliquescence (ent): the property possessed by materials which can absorb so much water from the atmosphere that they may eventually dissolve in it

dense tar surfacing: a hot process wearing course material consisting of aggregate, filler and road tar, in such gradings and proportions that when spread and compacted it provides a close textured impervious mixture

detonation impedance: the resistance of the rock to blasting, which has the effect of retarding the action of the blast

delayed action blasting: the firing of multiple charges in a timed sequence by the use of delay detonators. Advantages are improved fragmentation, reduced ground vibration, better control of rock pile, etc.

detrital: relating to or formed from detritus, which is a collective term for loose rock and mineral fragmental material, such as sand, silt and clay, derived from older rocks by mechanical means, mainly abrasion and disintegration by erosion and weathering

Devonian: a period of the Palaeozoic era, preceded by the Silurian and succeeded by the Carboniferous and spanning the time between 395 and 360 million years ago; named after the county of Devon, where rocks of this age were first studied

diapir: a dome or anticlinal fold above which the overlying rocks have been ruptured by upward squeezing and mobilisation of the plastic core material, which in sedimentary strata is commonly salt or shale. Igneous rocks may also show diapiric structure

diggability: a general expression of the excavation characteristics of the mineral or rock to be dug

diorite: a range of dark-coloured plutonic (q.v.) igneous rocks intermediate in composition (SiO_2 content 52–66%) and characterised by the common presence of the amphibole hornblende, plagioclase feldspar and sometimes a small amount of free quartz

discontinuity: as used in this report, the term denotes any interruption in the mechanical integrity of the rock, such as by a joint, fissure or cavity

disintegration: the breaking down of a rock into smaller particles by physical processes, without significant chemical alteration

Dodge crusher: a jaw crusher in which the moving jaw is pivoted near the base of the unit to give a fixed discharge area

dolerite: a dark coloured fine to medium-grained igneous rock of basic composition (that is, without free quartz), found in intrusions of moderate size, so allowing moderately rapid cooling of the magma

dolomite: a double carbonate mineral of calcium and magnesium, $CaMg(CO_3)_2$; also a

	rock containing a significant proportion (usually over 15 per cent) of magnesium carbonate
dolostone:	a carbonate rock consisting entirely of the mineral dolomite (q.v.)
double-toggle jaw crusher:	a jaw crusher in which the moving jaw is hung from a non-eccentric pivot and has a simple lateral motion
drainage:	in the special geotechnical sense, means the removal of water (usually by gravity) from a foundation, embankment or other parts of a structure where the development of excessive water pressures could endanger stability
drop balling:	rock breakage by dropping a heavy weight on a cable suspended from a crane or excavator jib
dropstones:	pebbles and boulders which have sunk from floating ice masses into the soft usually fine-grained sediments being deposited from a body of water; the weight and impact of the dropstones commonly give rise to deformation of the soft unconsolidated sediments
dry-bound macadam:	crushed aggregates laid in two separate sizes, coarse and fine, compacted dry by rolling and/or vibration to a dense layer
dry-screened sand:	building sand which has been passed dry through an appropriate sieve to remove over-sized deleterious materials
dyke:	a tabular igneous intrusion that cuts across the planar structure of the adjacent host rocks (q.v.): a discordant usually wall-like rock mass which may extend many miles across country, following usually vertical or near-vertical discontinuities in the host rocks
effective size:	the effective size is D_{10}. In other words, 10% of the particles are finer and 90% coarser than the effective size; often used in soil classifications
elastic limit:	the stress level above which a material remains permanently deformed when the stress is removed
Eocene:	an epoch of the Tertiary (q.v.) period between 54 and 38 million years ago
epidote:	a group of minerals comprising the basic silicates of calcium, aluminium, iron and manganese. They commonly occur in regionally metamorphosed rocks
epidotisation:	the process whereby epidote is formed as the product of hydrothermal alteration of plagioclase feldspar
fabric:	the sum total of all the features of texture and structure in a rock
facies:	the particular lithological and palaeontogical characteristics of a sedimentary rock, from which its origin and conditions of formation may be deduced
fault:	a surface, or closely spaced surfaces, of rock fracture along which there has been displacement, which may range from a few millimetres to many kilometres. A fault plane is a fault surface, normally more or less planar
feldspar:	the most important single group of rock-forming silicate minerals. They are aluminosilicates of potassium, sodium or calcium, depending for example on whether orthoclase (q.v.) or plagioclase (q.v.) feldspars are present
filler:	in asphaltic technology, is inert material (e.g. cement, limestone dust or fly ash) finer than 75 μm. Its function is to fill in voids and modify the viscosity and the temperature–viscosity relationship of the bitumen binder. Also a possible aid to binder/aggregate adhesion
fine cold asphalt:	a wearing course of bitumen and fine aggregate which is spread and compacted while cold or warm
fines:	strictly material finer than 60 μm, i.e. the silt and clay-sized fraction, but in connection with aggregates it usually refers to material finer than 75 μm
flint:	cryptocrystalline (q.v.) variety of silica

flotation: a mineral processing operation which involves conditioning the surfaces of mineral grains in water so that certain minerals can be collected into air bubbles and carried to the surface to effect a separation

flowsheet: a diagrammatic representation of the flow of material through a processing plant, showing the sequence of operations and their interrelationships

flow structure: the parallel alignment of particles (usually crystals) in the direction of movement of a fluid medium, such as lava

fluvial: relating to a river; a deposit produced by the action of a river. Geologists tend to use fluviatile for the product of river action, for example, fluviatile sand

fluvioglacial: or glaciofluvial, may be applied to sediment transported and deposited by running water discharged from an ice mass

foliation: the planar arrangement of textural or structural features in any type of rock, for example, the schistosity in a metamorphic rock

fossil weathering: ancient weathered rock preserved below more recent material

fragmentation: a term associated with hard rock quarrying to describe the degree of mechanical breakdown produced by blasting

free silica: in igneous rocks, silica (q.v.) which is present as quartz or one of its polymorphs, as opposed to silica present in the combined state as one of the many silicate minerals

friction course: a uniformly graded asphalt material, with a high voids content and high texture, giving good drainage and skid resistance properties, used primarily on airfields

gabbro: a coarse-grained basic plutonic rock containing plagioclase, a pyroxene and very commonly olivine. SiO_2 content 45–52%, dark minerals over 60% by volume

gabion: a stone-filled rectangular wire mesh box. Gabions may be laid to form self-standing structures or as mattresses to form flexible linings for coastal and river bank protection

gap-graded: absence or near-absence in the proportions of particular size fractions in a grading; also known as 'skip' or 'intermittent' grading

garnet: a group of complex silicates of aluminium, iron, manganese, chromium, calcium and magnesium. Usually found in metamorphic rocks but also in some granites (q.v.) and pegmatities (q.v.)

geophysics: the study of the physical characteristics of the Earth by means of instruments, and methods to determine subsurface conditions by analysis of such physical properties as specific gravity, electrical conductivity, magnetic susceptibility, radio-activity, seismic wave propagation, heat flow, etc

gibbsite: a hydrous aluminium oxide commonly occurring in deposits of bauxite (q.v.) and as an alteration product of aluminium silicates

glacial deposits, glacial debris or glacial drift: heterogeneous material transported by glaciers or icebergs and deposited directly on land or in the sea without sorting of the constituents

glacial till: unsorted, unstratified and generally unconsolidated drift deposited directly from a glacier. Includes "boulder clay" (q.v.)

gneiss: a foliated rock formed by regional metamorphism in which bands or lenticles of granular minerals alternate with bands or lenticles in which minerals of flaky or elongate prismatic habit predominate

grade: size sorting category in which all the particles fall within specified size limits. Also used in geology to denote degrees of alteration, e.g. by metamorphism or weathering, usually with reference to an agreed scale

grading: the proportions of different sizes present in an aggregate, established by sieve analysis; particle size distribution granite: generally, any completely crystalline quartz-bearing plutonic rock with light-coloured feldspars and micas as essential constituents (over 60% by volume). There may be a speckling of dark minerals. SiO_2 content over 66%

granodiorite: a coarse-grained plutonic (q.v.) rock intermediate in composition between granite (q.v.) and diorite (q.v.). Light coloured minerals over 60% by volume, SiO_2 content over 66%

granulite: a metamorphic rock characterised by granular texture. Originated under conditions of high grade regional metamorphism

gravel: in the British Standard particle size classification (BS 1377:1975 and BS 5930:1981) the term denotes granular material in the size range 2 mm to 60 mm

greywacke: a dark coloured sandstone comprising mineral and rock fragments, poorly-sorted and bound together with clay cement

gritstone: the BS812:1975 trade group which includes all clastic rocks with sand-sized particles, such as arkose, greywacke and sandstone. The term grit(stone) is generally used in geology for a sand(stone) with coarse angular particles

grizzly: a coarse screening device in which the screen surface consists essentially of massive parallel bars

ground moraine: rock debris generally fairly evenly though thinly spread beneath a glacier or ice sheet to give a gently rolling till surface when the ice has wasted away

gypsum: hydrous calcium sulphate: $CaSO_42H_2O$, associated with other evaporite minerals in extensive beds interstratified with limestone, shales and clays. Used, for example, as a soil additive, as a retarder in Portland cement and in making plaster-board

gypsum plaster: a plaster consisting of calcium sulphate, either partially hydrated or anhydrous, made by the controlled heating of gypsum

gyratory crusher: a crushing machine consisting of an upward pointing eccentrically rotating cone set inside an inverted cone

haematite or hematite: a natural oxide of iron (Fe_2O_3) with a high specific gravity (4.9 to 5.3), sometimes used as an aggregate for special-purpose dense concretes

hammer mill: an impact crusher consisting of a fast moving rotor with fixed hammers. If the hammers are hinged the crusher may be called a "swing hammer mill"

heavy-media separation: a mineral processing operation in which rock fragments can be separated on the basis of density differences by being allowed to float or sink in a heavy medium. The medium consists of finely divided particles of a high density solid, such as magnetite, dispersed in water

Holocene: see Recent

high energy sedimentary environment: turbulent action, such as that created by waves and currents, which prevents the settling and accumulation and abrasion of pebbles and sand grains

horizontal screen: a vibrating screen which utilises directed impulses to transport particles across its surface

hornblende: one of the most abundant rock-forming minerals of the amphibole (q.v.) family

hornfels: a fine-grained rock made of a mosaic of mutually interfering silicate minerals without preferred orientation and typically formed by contact metamorphism. Commonly yields strong aggregate when homogeneous and uniformly granular

host rock: the pre-existing rock in which later intrusions (q.v.) are emplaced

hydrosphere: a collective name for all the water on the earth's surface, in the atmosphere and underground

hydrothermal alteration: process of alteration to secondary minerals (q.v.) through the action of hot ground water, i.e. late stage igneous activity

hypabyssal:	a term applied to intrusive igneous rocks intermediate between plutonic (q.v.) and volcanic, generally medium grained
ice contact deposit:	normally stratified glacial drift deposited in contact with melting glacier ice; the deposit may exhibit an ice-contact slope marking the position of the ice mass at the time it was laid down
ice front:	the usually cliff-like limit of a mass of glacier ice where the supply of ice is balanced by wasting through thawing (on land) or erosion (by the sea)
igneous:	describes a rock or mineral that solidified from molten or partly molten material.
illite:	a broad term for one of the commonest groups of clay minerals, formed by the alteration of micas, alkali felspars etc. under alkaline conditions
indurated:	rock or soil hardened and compacted by the action of pressure, heat and cementation after deposition. Induration is a progressive, natural process
industrial screening:	the separation of particles by sieving (screening) on an industrial, as distinct from a laboratory, scale
initiation:	in blasting, the method used to fire the charge, or the instant at which the explosion begins
intermontaine:	located between mountains or mountain ranges
intrusion:	the process of emplacement of molten rock (magma) in pre-existing rock, also the igneous rock body so formed within the surrounding rock (the country rock). The term may also be applied to mobilised sediments, which may be injected along discontinuities or form plug-like masses under the influence of gravity
iron pyrites:	(see pyrite)
ironstone:	imprecise term, but usually denoting an impure iron carbonate occurring as nodules in some clays, e.g. London Clay
isopach, isopachyte:	a line drawn on a map through points of equal thickness of a specified rock unit
jaw crusher:	a crushing device consisting of one near-vertical fixed crushing face and an inclined mobile face which moves so as to alternately increase and decrease the gap between the two faces
jig, jigging:	a mineral processing machine/operation in which particles are separated on a density basis by water pulsation
joint:	a plane of discontinuity, fracture or parting in a rock, normally involving no displacement. Joints commonly occur as parallel sets, cutting the rock mass into joint blocks
Jurassic:	a period of the Mesozoic era between 213 and 144 million years ago, and the corresponding system of rocks. Named from the Jura mountains of France
kaolinite:	a common white or greyish white clay mineral, formed by hydrothermal alteration (and to some extent, the chemical weathering) of feldspars and other alumino-silicate minerals through the process known as kaolinisation. The process adversely affects the usefulness of rocks for concrete or road aggregate
kettlehole:	commonly basin or bowl-shaped depression in glacial drift (q.v.), without surface drainage, but often containing a lake or bog; formed by the melting of masses of stagnant (or 'dead') ice left behind by a retreating glacier within or upon the glacial drift. Kettles may be a few metres to a few tens of metres across
lacustrine:	relating to materials formed in or by lakes, for example lacustrine beach deposits
lake asphalt:	a highly viscous natural asphalt found in well defined surface deposits
laterite:	high weathered residual soil material rich in oxides or iron and aluminium, traditionally useful for brickmaking and exceptionally for low quality aggregate where hard iron-pan (ferricrete) layers have developed

leaching:	selective removal of soluble constituents from a soil or rock mass by the action of percolating water
lean mix concrete:	generally a concrete with a cement: aggregate ratio of 1:10 or greater, i.e. a low cement content, often incorporates an all-in aggregate (q.v.).
ledge rock:	*in situ* rock
lignite:	a brownish-black coal in which the alteration of vegetable matter has proceeded further than in peat but not so far as in bituminous coal
lime:	quick lime is calcium oxide made by heating limestone above 900°C. Slaked lime (calcium hydroxide) is the product of the reaction between quick lime and water
limestone:	a sedimentary rock composed mainly of calcium carbonate occurring as the mineral calcite or occasionally as aragonite in recent deposits
lithified:	made into rock: lithification is the process of consolidating a loose sediment into a solid rock
lithology:	the description of the physical nature of a rock: including its grain size, mineralogical composition, structure, colour
log-normal distribution:	graphical presentation of data on log paper which provides a clearer picture of normal distribution curves
log-washer:	a mineral processing device usually used for scrubbing or washing adhering clay from solid particles
low-grade metamorphism:	metamorphic grade is measured by the amount of change that the original (parent) rock has undergone. Thus the conversion of shale to slate would indicate low-grade metamorphism whilst its continued metamorphism to a schist would be regarded as high-grade
Lower Palaeozoic:	early part of the Palaeozoic (q.v.) era, ranging in time from 590 to 395 million years ago and comprising the Cambrian, Ordovician and Silurian (q.v.) Systems
macadam:	crushed stone mechanically locked by rolling and cemented together by application of stone screenings and water. Bituminous macadam is crushed material in which the fragments are bound together by bituminous materials
magma:	naturally occurring deep molten rock material from which igneous rocks are formed
magnetite:	an iron ore mineral, Fe_3O_4. It occurs as a primary constituent of most igneous rocks
marble:	metamorphic rock produced from limestone through recrystallisation
marine aggregate:	sand and gravel which is excavated by dredger from the sea bed and taken ashore for processing and distribution
massive:	in geology, without stratification, cleavage or schistosity; particularly applied to igneous rock-bodies but can be used to describe thick or obscured bedding in stratified rocks
mastic asphalt:	a type of asphalt composed of suitably graded mineral matter and asphaltic cement in such proportions as to form a coherent, voidless, impermeable mass, solid or semi-solid under normal temperature conditions, but sufficiently fluid when brought to a suitable temperature to be spread by means of a float
mesh:	the apertures in a screen surface, or the size of the apertures
Mesozoic:	the era following the Palaeozoic (q.v.), ranging in time from 230 to 70 million years ago. It comprises the Triassic, Jurassic and Cretaceous systems (q.v.)
meta:	prefix implying instability or change, e.g. as in metamorphism (q.v.)
metamorphism:	mineralogical and structural changes of rocks in their solid state in response to altered physical and chemical conditions
——, contact:	the process of localised thermal metamorphism brought about by the intrusion

	(and to a lesser extent, extrusion) of magma into rocks; changes are effected at or near the contact by heat and materials emanating from the magma and by some deformation arising from the emplacement of the igneous mass
——, dynamic:	changes in rock structure and mineralogy brought about commonly on a regional scale by crushing and shearing in the earth's crust; high temperatures may be involved, leading to extensive recrystallisation of rocks
——, low grade:	metamorphic grade is measured by the amount of change that the original (parent) rock has undergone. Thus the conversion of shale to slate (q.v.) would indicate low grade metamorphism whilst its continued metamorphism to schist would be regarded as high-grade
——, regional:	metamorphism affecting an extensive region usually brought about by deep burial of material in the earth's crust
——, thermal:	essentially the chemical reconstitution of rocks brought about by elevated temperature influenced, to some extent, by the confining pressure of deep burial in the earth's crust (though this is not essential)
mica:	the group of layered-lattice silicates (including moscovite and biotite) characterised by very strong cleavage
micaceous:	mica-bearing
microcline:	a member of the feldspar group of minerals ($KAlSi_3O_8$) found generally in acid igneous rocks, pegmatites and metamorphic rocks
microdiorite:	a range of hypabyssal igneous rocks of intermediate composition. Microdiorites are the hypabyssal equivalent of andesites and consist usually of intermediate plagioclase feldspar, hornblende or biotite (q.v.) and occasionally a small amount of quartz
microgranite:	a granite in which a majority of the crystalline matter requires microscopic examination to differentiate individual crystals
mineral:	(i) a naturally formed chemical element or compound and normally having a characteristic crystal form
mineral:	(ii) a mass of naturally occurring mineral material, (such as metallic ores and aggregate minerals) which is judged, against arbitrary physical, chemical and economic criteria, to have foreseeable use
mineral reserve:	that part of a mineral resource considered producible at a profit at the time of classification
mineral resource:	a concentration of naturally-occurring materials in such form that economic extraction is currently or potentially feasible
mineralogy:	the science concerned with the study of minerals, including their occurrence, composition, forms, properties and structure
Moh's scale of hardness:	a scale of scratch resistance graduated in terms of the relative hardness of ten common minerals which in order of increasing hardness are: 1 talc, 2 gypsum, 3 calcite, 4 fluorite, 5 apatite, 6 orthoclase, 7 quartz, 8 topaz, 9 corundum, 10 diamond
monominerallic:	rock composed essentially of one mineral species
montmorillonite:	member of clay mineral group characterised by marked swelling properties
moraine:	a mound or ridge of unsorted, unstratified glacial drift, commonly till (q.v.), deposited directly by glacier ice
mortar:	a mixture consisting essentially of cement, sand and water, and the hardened product of such a paste
mudstone:	an indurated mud (or mud-rock), commonly massive, blocky, non-fissile and comprising about equal proportions of clay and silt

muscovite: a light coloured member of the mica (q.v.) group of rock forming minerals

normal frequency distribution: bell-shaped frequency distribution curve symmetrical with respect to the mean value

olivine: common rock-forming mineral in basic, ultra-basic and low-silica igneous rocks; typically dark olive-green to brown magnesium iron silicate: $(MgFe_2)SiO_4$. Alters to serpentine

oolite: a sedimentary rock commonly a limestone (hence oolitic limestone), made chiefly of ooliths cemented together. Ooliths are spherical or ovoid accretionary bodies resembling the egg-roe of a fish, made of concentric layers typically deposited around a nucleus (such as a shell fragment or sand grain) in shallow moving water. Ooliths range in size from 0.25 to 2.00 mm, but are typically between 0.5 mm and 1.0 mm in diameter

opal: cryptocrystalline (q.v.) variety of silica

ophiolite: an assemblage of ultrabasic to basic rocks, frequently layered, with an upward succession from plutonic to hypabyssal and volcanic (q.v.). These represent sections of oceanic crust and the upper part of the underlying mantle incorporated into a sedimentary sequence during strong disturbances caused by plate underthrusting

Ordovician: a period of the Palaeozoic era between 505 and 435 million years ago, and the corresponding system of rocks. Named from the Ordovices, an ancient Celtic tribe of Central Wales

orthoclase: a potassium feldspar (q.v.) $KAlSi_3O_8$ occurring as an essential constituent of the more acid igneous rocks and also found in some metamorphic rocks

orthoquartzite: a very hard sandstone in which the constituent quartz grains have been cemented by secondary silica so that the rock breaks across individual grains rather than around them. Often loosely termed 'quartzite' (q.v.)

outcrop: the total area over which a particular rock unit or structure occurs at the ground surface or immediately below the superficial deposits (q.v.), whether visibly exposed or not

outwash: stratified deposits, mainly sand and gravel, washed from a glacier by melt-water streams and deposited as an outwash cone or outwash plain, the coarser material being deposited nearer to the ice. An outwash plain mainly of sand is called a sandur (plural: sandar)

Palaeozoic: the era ranging in time from 590 to 213 million years ago, It comprises the Cambrian, Ordovician, Silurian, Devonian, Carboniferous and Permian Systems (q.v.)

patterned ground: general term embracing more or less symmetrical arrangements of pebbles, stones and boulders which in plan assume polygonal, circular, reticulate or striped forms, through the action of intensive frost action, usually in superficial deposits in cold, typically periglacial, climates

pavement: the whole constructed thickness of a road whether of concrete, asphalt, macadam, stabilised soil, etc

pegging: the obstruction of a screen surface by single fragments which become wedged in the apertures

pegmatite: a very coarse grained igneous rock, conventionally refers to a rock of granite (q.v.) composition. Also used for other plutonic rocks whose names are used as a prefix, e.g. gabbro-pegmatite

periglacial: a term describing the influence of the cold temperature around an ice mass on the climate, topography, and natural processes, in which frost action is a primary factor

Permian: the final period of the Palaeozoic (q.v.) era between 286 and 248 million years

	ago, and the corresponding system of rocks. Named from the province of Perm in Russia
petrography:	the branch of geology dealing with the description and systematic classification of rocks especially by the microscopical study of thin sections
phyllite:	a cleaved metamorphic rock, coarser grained and less perfectly cleaved than slate (q.v.), but finer grained and better cleaved than schist. Formed by low temperature regional metamorphism (q.v.)
pitch:	the residue, liquid when hot and almost solid when cold, obtained from the distillation of tars, etc. When no source is specified, it is implied that it is obtained from coal tar
pitch-bitumen binder:	penetration grade binder (q.v.) containing pitch and predominance of bitumen suitable for use in rolled asphalt or dense bitumen macadam
plagioclase:	the group of soda-lime feldspars (q.v.). They are one of the most common rock forming mineral groups, having a continuous gradation in chemical composition between albite $NaAlSi_3O_8$ and anorthite $CaAl_2Ai_2O_8$
plaster shooting:	small-scale blasting of boulders, using a surface charge
plasticiser:	an additive to mortar which has the ability to entrain air in the mix and improve its working qualities
plasticity index:	the numerical difference between the liquid and plastic limits (q.v.), representing the range of moisture content at which the soil is plastic. Together with the liquid limit it gives an indication of the sensitivity of the soil to changes in moisture condition
Pleistocene:	the earliest epoch of the Quaternary Period and during which the Ice ages occurred. Spans the period between 2 million and 10 thousand years ago
plutonic:	applies to igneous rocks formed at great depth, which are characteristically medium or coarse-grained and of granite-like texture
pneumatic impact breaker:	a tool for breaking rock, concrete etc. consisting of a chisel driven by a pneumatic vibrator, either hand held or machine controlled
point load strength:	rock material strength determined by loading a rock specimen to failure between two conical points. The result is expressed as a point-load index (Is) which is the force at failure divided by the square of the length of the loaded axis
poorly graded:	in engineering terms this implies the absence or a low proportion of essential grade sizes. In geological terms it implies the presence of a proportion of all essential grade sizes
pop drilling:	the breaking up of large boulders by drilling holes for small explosive charges
porosity:	the ratio between the volume of voids in a material and its total volume
porphyry:	an igneous rock that contains conspicuous large crystals (phenocrysts) in a fine-grained groundmass. Hence porphyritic, the texture of a porphyry
pozzolan:	naturally occurring deposits which, when finely ground, combine chemically with hydrated limes and so can be used in mortars
pre-Cambrian:	that period of time (about 4000 million years) from the consolidation of the earth's crust to the beginning of the Palaeozoic era, and the corresponding system of rocks
primary blasting:	describes the initial blast, where more than one stage of blasting is needed to obtain the required fragmentation
psammite:	metamorphic rock derived from an impure sandstone
pulp:	a suspension of mineral particles in water; i.e. a slurry
pyrite:	iron sulphide (FeS_2), the most abundant of the sulphide minerals; it occurs most

commonly as cubes and nodules in sedimentary rocks and as veins associated with other minerals such as copper with which it may be mined. Undesirable in aggregates because it decomposes giving rise to iron-oxide in mass concrete, and may oxidise to ferrous sulphate, which causes sulphate attack

pyroclastic:	descriptive of broken rock material formed by volcanic explosion or by being thrown from a volcanic vent. A pyroclastic rock may be made of rock fragments of a wide size range and largely unsorted
pyroxene:	a group of rock-forming silicates, somewhat similar in general composition to amphibole (q.v.) but without hydroxyl (OH). Augite is the best-known example
quartering:	reduction of a sample to a desired volume by dividing a cone of material into four quarters and combining alternate quadrants. This is repeated until the desired sample quantity is obtained
quartz:	crystalline silica, SiO_2; an important durable rock-forming mineral (see also silica)
quartzite:	metamorphic rock consisting mainly of quartz and formed by recrystallisation of sandstone or chert by thermal or regional metamorphism. Also loosely used synonymously with 'orthoquartzite' (q.v.)
quartzo-feldspathic minerals:	minerals of the quartz and feldspar groups, generally imparting a light colour and relatively low specific gravity to the rock
Quaternary:	the latest era of geological time, from 2 million years ago to the present, largely represented in Britain by superficial deposits such as glacial drift (q.v.)
Recent:	the current epoch in the earth's history, comprising the 10 000 years or so since the end of the last glaciation. Also called Holocene
reduction-ratio:	a term used in crushing to indicate the ratio between the maximum particle sizes in the feed and the product
reef limestone:	a limestone consisting of the fossilised remains of reef-building organisms, such as corals, bryozoans, algae, sponges and other shells
refined lake asphalt:	lake asphalt from which unwanted materials, such as water and vegetable matter, have been removed
regional metamorphism:	see metamorphism
regression:	the retreat of the sea from a land area and the consequent evidence preserved in the strata; for example beach deposits may overlie offshore deposits in a geological sequence
remote sensing:	the measurement and recording (often in image form) of information about an object or group of objects that is not in physical contact with the recording device, for example airborne photography. Other techniques used are television cameras, ultra-violet and infra-red detectors, radio frequency and microwave receivers, radar systems and so on
residual soil:	a soil developed *in situ* from residual materials, originally consisting of fresh rock
rhyolite:	a fine grained to glassy acid volcanic rock, Mineralogically similar to granite. The more glassy members of the group are termed obsidian
riffling:	sample reduction using riffle box. This is a box comprising a number of internal chutes directing alternatively towards the two receiving trays. By this means each sample pass is halved and the material of one tray is rejected
ripper:	a tractor or bulldozer with a hooked tine or array of tines which can be forced into the ground hydraulically to break it up as the tractor moves forward
ripping:	the process of excavating rock with a ripper
rippability:	the facility with which a rock can be excavated by ripping
riprap:	a material used to protect slopes aginst the action of water, consisting of broken stone placed on the slope surface

road base:	the main structural element in a road pavement. Its function is to spread the concentrated loads from traffic over an area of subgrade large enough to sustain them
rock-forming minerals:	those minerals which are widespread in nature and which contribute significantly to the common rock groups, serving to define and classify the common rocks
rock pile:	the mass of broken rock resulting from a primary blast
rod mill:	a horizontal cylindrical mill charged with steel rods and mineral fragments (c.f. ball mill)
rolled asphalt:	a material used as a dense wearing course, base or roadbase material. It consists of a mixture of aggregate and asphaltic cement
sabkha:	a salt flat (usually coastal) which is inundated only occasionally
sand:	in the British Standard particle size classification (BS 1377:1975 and BS 5930:1981) sand is a granular material in the size range 0.06 mm to 2 mm. In the sense of 'concreting sand' however, the nominal upper size limit is 5 mm and there are constraints on the particle size distribution (see BS 882:1974)
sandstone:	a sedimentary rock made of abundant fragments of sand size set in a fine-grained matrix or cementing material. The sand particles are usually of quartz. The term sandstone may be used to describe any clastic (q.v.) rock containing individual grains visible to the unaided eye
sand waves:	large, linear, subaqueous sand dunes or sand ripples formed by currents on the bed of the sea or in a river; the smallest sand waves in sedimentary rocks are ripple marks, larger ones are megaripples. May be applied to wind-blown ripples—that is sand dunes
scalping:	the separation of the coarsest fragments from a particulate material by screening
schist:	metamorphic rock characterised by a parallel arrangement of the bulk of the constituent minerals. The common minerals which give rise to the layered structure are the micas (q.v.)
screen:	a particle sizing device like a sieve, consisting of a surface (usually flat) which is perforated by apertures of characteristic size and shape. Screening is a sizing operation effected by means of a screen
scrubber/scrubbing:	a mineral processing machine/operation in which adherent fine fragments or clay coatings are washed from the surface of larger particles
secondary minerals:	minerals formed by alteration or replacement of the original rock minerals e.g. serpentine after olivine (q.v.)
sedimentary rock:	a rock resulting from the consolidation of loose sediment that has accumulated in layers, or a chemical rock formed by precipitation or an organic rock consisting largely of the remains of plants and animals
sericite:	a secondary mica usually resulting from the alteration of other rock forming minerals. Chemically similar to muscovite (q.v.)
settling pond:	a reservoir of still water in which very fine material is allowed to settle
shaking table:	a device for separating mineral grains on the basis of differences in their density. It consists of a slightly inclined vibrated deck over which a stream of water moves at right angles to the direction of the vibration. Low density material is carried along the deck into an area where it can be recovered
shale:	fine-grained sedimentary rock composed of clay minerals and other finely divided material. These rocks are characterised by a well-marked bedding-plane fissility
shelling-out:	particles or fragments breaking off rock as a result of frost action or other physical or chemical changes
silica:	the chemically and physically resistant dioxide of silicon, SiO_2, which occurs

	naturally as quartz, flint, chert, opal or chalcedony and combines in silicates as an essential constituent of many rock-forming minerals
sill:	a tabular or sheet-like minor igneous intrusion that conforms with the bedding or other layered structure of the host (or country) rock
silt:	a deposit which has the average grain size between that of sand and clay (q.v.)
siltstone:	an indurated fine grained rock in which the amount of silt grade material exceeds that of clay; it tends to be flaggy, hard and durable
Silurian:	a period of the Palaeozoic era between 435 and 395 million years ago, and the corresponding system of rocks. Named from the Silures, an ancient Celtic tribe of the Welsh Borderland
single-toggle jaw crusher:	a jaw crusher in which one jaw is suspended from an eccentric pivot and moves both vertically and laterally
slate:	a low grade regionally metamorphosed argillaceous (q.v.) rock which has developed a well marked cleavage (q.v.) but has undergone little recrystallisation, so that the rock is still very fine grained
slime:	very fine material, often clay particles, in suspension which usually presents a processing and dewatering problem. "Desliming" is an operation which involves removing very fine material from a pulp or slurry
slurry:	a suspension of mineral particles in water
slurry seal:	a mixture of binder, fine aggregate and mineral filler with water added to produce a material of slurry consistency
soil creep:	slow downhill movement of soil and rock debris under the influence of gravity
solifluction:	slow, viscous, downslope flow of waterlogged soil and superficial materials, especially in cold climate conditions
spheroidal weathering:	the progressive splitting away, from a central core, of usually fine-grained rock of concentric or spherical shells ("onion-skins") of chemically weathered material. Also brought about by mechanical weathering through the freezing of water which has penetrated progressively into blocks, boulders or pebbles, or by differential expansion and contraction by alternating heating and cooling, usually on a daily cycle in response to insolation (i.e. heating by the sun)
spotting:	mechanical ability of extraction equipment to pick up individial rock fragments
stratified:	sedimentary material laid down in layers or beds; bedded
stratum:	a distinct layer or bed of sedimentary material. Commonly used in the plural form: strata
strip mining:	opencast excavation of minerals by the method of removal in defined sections or strips
stucco work:	plaster used for coating walls
subaerial:	said of natural processes taking place in the atmosphere usually near to the surface of the land, such as aeolian deposition (sand dunes) and erosion of all kinds
sub-base:	in a road pavement, a layer of usually granular material below the road base. Its functions include insulating the pavement from the subgrade (q.v.), increasing the pavement thickness to protect against frost and to provide an adequate surface for construction plant
subgrade:	rock or soil horizon immediately beneath a pavement (q.v.)
subsoil:	the weathered soil or rock immediately below the topsoil
superficial, or surficial (Am.), deposits:	deposits formed on or close to the present land surface by processes (e.g. glaciation) usually of Quaternary data. Their distribution and thickness are related essentially to the surface relief and not to the structure of the underlying bedrock

surface dressing:	one or two coats of single-sized aggregate rolled into the wearing-course to provide a non-skid, abrasion-resistant surface
tar:	a viscous liquid, black in colour, having adhesive properties, obtained by the destructive distillation of coal, wood, shale etc. Where no specific source is stated it is implied that the tar is obtained from coal
tar emulsion:	an emulsion in which, with the aid of suitable emulsifying agents, tar is dispersed in water, or in an aqueous solution
tarmacadam:	a road material consisting of stone coated in tar or a tar bitumen mixture. It has very little fine aggregate and a high proportion of voids
tension jointing:	fractures in a rock resulting from tensile stresses
Tertiary:	the era following the Mesozoic (q.v.), ranging in time from 65 to 2 million years ago, and the corresponding series of rocks
teschenite:	an alkali-rich variety of gabbro generally containing soda rich amphiboles or pyroxenes (q.v.), calcic plagioclase and analcite
texture depth:	a measure of the macrotexture of a paving surface; it determines the rate at which the skidding resistance of the surface changes with change in vehicle speed
thermal metamorphism:	see metamorphism, thermal
throw distance:	the distance to which blasted materials are thrown by the explosion
till:	unstratified, unsorted drift deposited directly by a glacier without reworking by water from the glacier; comprises a heterogeneous mixture of clay, sand, gravel and boulders
toe breakage:	separation of fragments from the base of the quarry face
transgression:	invasion of land areas by extension of the sea (brought about by a rise in sea level or subsidence of the land) and the evidence of this that is demonstrated by the strata. New marine deposits may spread over a former land surface in unconformable (q.v.) relationship to it, or deep-water environments may occupy areas formerly occupied by shallow-water conditions
Triassic:	the earliest period of the Mesozoic (q.v.) era between 248 and 213 million years ago, and the corresponding series of rocks. Named from the three fold division of the period which can be made at the type locality in Germany
tridymite:	high temperature polymorph of quartz
trommel:	a rotating, cylindrical, screening device
tuff:	a compacted, commonly stratified, pyroclastic deposit of volcanic dust, ash and up to half its bulk of sediments, mainly sand, silt and clay
unbound aggregate:	aggregate which is not bound or coated with cementitious or bituminous binders (i.e. various forms of wet or dry mix macadam as used in bases and sub-bases for road pavements)
unconformity:	an interruption in a geological sequence representing an interval of geological time during which no sediments or other rocks were formed. The underlying older rocks may have been deeply eroded, tilted, folded or even metamorphosed (q.v.) before sedimentation was resumed, resulting in non-parallelism of the structure above and below the unconformity
unconsolidated:	a broad term describing sediments which have not been hardened by deep burial and/or cementation and correspond to the engineering meaning of 'soil', i.e. clays, silts, sands and gravels
uniformity coefficient:	the ratio between the D_{60} size on a grading curve and the D_{10} size, the D_{60} size being that at which 60 per cent of the material is finer, the D_{10} size being that with 10 per cent of finer material. The ratio D_{60}/D_{10} gives a measure of the soil grading, with a well graded (q.v.) soil having a coefficient of 5 or more

unsoundness:	the inadequate resistance of a material to chemical attack or to repeated physical changes such as temperature, moisture content, stress etc.
vein quartz:	a coarsely crystalline variety of quartz characteristic of hydrothermal (q.v.) veins
vermiculite:	a layered silicate mineral of the chlorite (q.v.) family which on heating expands greatly to give a light cellular material much used for thermal insulation
vesicular, vesiculate:	describes the texture of a rock, especially a lava, containing abundant bubbles or vesicles formed by expanding gases when the lava was fluid; the vesicles may subsequently become filled with secondary minerals (suxh as calcite or quartz) to form amygdales
water-bound aggregate, or water-bound macadam:	usually a road base or sub base in which a layer of aggregate has sand watered in to fill the voids
wearing course:	in a road pavement, the uppermost layer. It contains the highest quality material in the pavement and its function is to provide the desired riding and non-skid properties to the road
weathering:	a comprehensive term for all the processes by which rock and soil are altered under the direct influence of the hydrosphere and atmosphere
well-graded:	in civil engineering, natural or artificial aggregate with a wide range of particle sizes distributed in a continuous sequence. In geology, in contrast, denotes a single sized grading, e.g. aeolian dune sand
wet-mix:	pre-mixed water-bound macadam (q.v.) consisting of crushed aggregate, usually 50 mm down to filler, with a carefully controlled amount of water, usually 2–5% by weight
wet-mix bituminous macadam:	a method of road surfacing in which wet aggregate is mixed with a small quantity of hydrated lime or Portland cement. This is then mixed with an acidified cut back bitumen binder to coat the particles. It has the advantage that expensive aggregate drying plant is not required
whinstone:	a colloquial term which includes any dark fine-grained igneous rock, e.g. andesite, basalt, dolerite
winning:	extraction of minerals by stripping of overburden and open excavation of underlying mineral matter
workability:	for practical purposes, the ease with which a concrete or mortar mix can be handled and placed in its final form
zeolite:	a group of hydrated silicates of calcium and aluminium, sometimes with sodium and potassium. They result in general from the alteration of feldspars and other aluminous minerals in igneous rocks
zircon:	zirconium silicate, $ZrSiO_4$. The mineral occurs as a primary constituent of igneous rocks, especially the more acid, e.g. granite

Index

Current Topics in Microbiology
196 and Immunology

The Role of Nitric Oxide in Physiology and Pathophysiology

Edited by H. Koprowski and H. Maeda

With 21 Figures and 7 Tables

Springer-Verlag
Berlin Heidelberg New York
London Paris Tokyo
Hong Kong Barcelona
Budapest

Hilary Koprowski
Department of Microbiology and Immunology
Head of the Center of Neurovirology
at Thomas Jefferson University
1020 Locust Street
Philadelphia, PA 10107
USA

Hiroshi Maeda
Department of Microbiology
Kumamoto University School of Medicine
Honjo 2-2-1
Kumamoto 860
Japan

Cover Illustration: This drawing illustrates the multi-faceted function of Nitric Oxide in the human organism and lists, at the same time, known inhibitors of Nitric Oxide synthase and a scavenger substance (PTIO) which binds directly to NO. Picture by courtesy of Neilson Carlin.

Cover design: Künkel und Lopka, Ilvesheim

ISSN 0070-217X
ISBN 3-540-58214-2 Springer-Verlag Berlin Heidelberg New York
ISBN 0-387-58214-2 Springer-Verlag New York Berlin Heidelberg

Production: PRODUserv Springer Produktions-Gesellschaft, Berlin
Typesetting: Thomson Press (India) Ltd, New Delhi
SPIN: 101 28 452 27/3020-5 4 3 2 1 0 – Printed on acid-free paper.

Preface

Fascination with nitric oxide dates back only a few years to when scientists discovered that this, the smallest known molecule with biological functions, has many properties. A wide range of activities attributed to nitric oxide, which is a product of the two enzyme constituent nitric oxide synthase and induced nitric oxide synthase, have been intensely studied by scientists throughout the world.

The present volume contains papers presented at a workshop on "The Role of Nitric Oxide in Physiology and Pathophysiology" at Thomas Jefferson University, Philadelphia, Pennsylvania, on October 25, 1993. The authors came from different parts of the world to participate in the workshop and presented up-to-date results of their research into nitric oxide. The editors would like to express their thanks for these contributions which will enrich the literature on the importance of nitric oxide in physiology and pathology.

HILARY KOPROWSKI
HIROSHI MAEDA

List of Contents

List of Contributors

(Their addresses can be found at the beginning of their respective chapters.)

Inducible Nitric Oxide Synthase: Regulation Subserves Function

C. Nathan

1 Inducible Isoform of Nitric Oxide Synthase Is a Distinct Gene Product

Nitric oxide synthase (iNOS) was purified (Stuehr et al. 1991) and cloned (Xie et al. 1992) from interferon-γ (IFNγ)- and lipopolysaccharide (LPS)-activated mouse macrophages. The primary structure, deduced from the cDNA and confirmed by HPLC-electrospray ionisation mass spectrometry, established that this inducible enzyme differs from the constitutive NOSs (cNOSs) cloned from cerebellar and endothelial libraries. Western blotting detected a highly similar enzyme in a wide variety of cell types after activation, including cardiac myocytes, where NO appeared to suppress rhythmicity (Roberts et al. 1992). Southern blotting and genomic cloning (Chartrain et al.1994) strongly suggest that there is a single iNOS gene. Thus, iNOS is a unique gene that can be expressed widely following activation.

Cornell University Medical College, Box 57, 1300 York Ave, New York, NY 10021, USA

2 Inducible Nitric Oxide Synthase Binds Calmodulin Without Dependence on Elevation of Calcium

Catalysis by cNOSs depends on binding of calmodulin (CaM), which in turn depends on an elevation of Ca^{2+} ($EC_{50} \geq 200$ n*M*) above the levels typically found in resting cells (≤ 100 n*M*). In contrast, pure iNOS functions without addition of exogenous CaM or Ca^{2+}, despite containing a canonical CaM-binding motif (amino acid residues 503–532). This paradox was explained by the ability of iNOS to bind CaM as a subunit at trace levels of Ca^{2+} (39 n*M*). CaM was identified in pure iNOS by reactivity with anti-CaM mAb, migration on SDS-PAGE, reverse phase-HPLC retention time, tryptic map, partial amino acid sequence, and electrospray ionization mass spectrometry (Cho et al. 1992).

3 Synergistic Transcriptional Induction of Inducible Nitric Oxide Synthase Is Mediated Through an Upstream Regulatory Region Rich in Transcription Factor Binding Site Consensus Sequences: Key Role of Nuclear Factor-κB

Instead of being regulated by agonist-induced Ca^{2+} transients, iNOS is controlled chiefly by transcription, as shown by the dependence of iNOS enzyme activity, iNOS mRNA synthesis, and iNOS antigen expression on exposure of macrophages to a synergistic combination of signals, such as IFNγ and LPS, and the sensitivity of each of these processes to nontoxic concentrations of actinomycin D. Induction of iNOS in mouse peritoneal macrophages by IFNγ and LPS required synthesis of an intermediary protein(s) and may involve tyrosine phosphorylation (Xie et al. 1993). From a mouse genomic library, we cloned a 1749 base pair fragment from the 5′ flanking region of the iNOS gene and used S1 nuclease mapping and primer extension to identify the mRNA transcription start site within it. The mRNA initiation site is preceded by a TATA box and at least 22 oligonucleotide elements homologous to consensus sequences for the binding of transcription factors involved in the inducibility of other genes by cytokines or bacterial products. These include ten copies of IFNγ response element (γ-IRE); three copies of γ-activated site (GAS); two copies each of nuclear factor-κB (NF-κB), IFN-α-stimulated response element (ISRE), activator protein-1 (AP-1), and tumor necrosis factor response element (TNFRE); and one X box. Plasmids in which all or the downstream one-half or one-third of this region of iNOS were linked to a reporter gene encoding chloramphenicol acetyltransferase were transfected into RAW264.7 macrophage-like cells. All these constructs conferred inducibility of the iNOS promoter by LPS, but only the construct containing all

1749 base pairs conferred synergistic inducibility by IFNγ plus LPS (XIE et al. 1993). Analysis of further constructs indicates that the downstream NF-κB site in the iNOS promoter, NF-kBd, is necessary to confer inducibility by LPS. A nuclear protein which binds to NF-kBd after LPS-induction was identified as an NF-kB heterodimer by competition, supershift, and UV cross-linking, and induction of iNOS mRNA was blocked by the NF-kB inhibitor, pyrollidine dithiocarbamate (XIE et al. 1994).

4 Antiviral Activity of Inducible Nitric Oxide Synthase: Potential Explanation for Widespread Inducibility of an Autotoxic Enzyme

It is a mystery why evolution has conferred on so many types of cells the capacity to respond to immune or inflammatory signals with the induction of an enzyme that, once tnanslated, can function for days to generate large amounts of a potentially toxic radical. We reasoned that NO may defend against a class of pathogens by which many cells types can be infected and may be a toxin against which it is difficult for pathogens to evolve to a resistant state. Consistent with this view, KARUPIAH et al. (1993) demonstrated that iNOS can account for a major portion of the antiviral activity of IFN-γ in mouse macrophages infected with ectromelia, herpes simplex or vaccinia viruses and is critical to the recovery of mice from ectromelia. The mechanism has not been defined, but one key target molecule may be ribonucleotide reductase (KWON et al. 1991).

5 Inducible Nitric Oxide Synthase Is Also Suppressible

The potential autotoxicity of iNOS for the host is exemplified by its critical role in experimental arthritis (MCCARTNEY-FRANCIS et al. 1993). Thus, it is not surprising that the widespread capacity to induce iNOS is balanced by an equally widespread ability of host cells to produce cytokines that can antagonize its induction. These include transforming growth factor-β (TGF-β), interleukin-4, and macrophage deactivation factor (DING et al. 1990; VODOVOTZ et al. 1993; BOGDAN et al. 1994). The potency of TGF-β in this regard may result from the multiplicity of levels at which it suppresses iNOS, including decreased stability and translation of iNOS mRNA and decreased stability of iNOS protein (VODOVOTZ et al. 1993).

In summary, the fact that iNOS is regulatable at transcriptional and multiple post transcriptional levels may reflect the important dual role of this enzyme in host defense and autotoxicity.

References

Bogdan C, Vodovotz Y, Paik J, Xie QW, Nathan C (1994) Mechanism of suppression of nitric oxide synthase expression by interleukin-4 in primary mouse macrophages. J Leuk Biol 55: 227–233

Chartrain NA, Geller DA, Koty PP, Sitrin NF, Nussler AK, Hoffman EP, Billiar TR, Hutchinson NI, Mudgett JS (1994) Molecular cloning, structure and chromsomal localization of the human inducible nitric oxide synthase gene. J Biol Chem 269: 6765–6772

Cho HJ, Xie QW, Calaycay J, Mumford RA, Swiderek KM, Lee TD, Nathan C (1992) Calmodulin is a subunit of nitric oxide synthase from macrophages. J Exp Med 176: 599–604

Ding A, Nathan CF, Graycar J, Derynck R, Stuehr DJ, Srimal S (1990) Macrophage deactivation factor and transforming groth factors-β-1, -2, and -3 inhibit induction of macrophage nitrogen oxide synthesis by interferon-γ. J. Immunol 145: 940–944

Karupiah G, Xie QW, Buller RML, Nathan C, Duarte C, MacMicking J (1993) Inhibition of viral replication by interferon-γ-induced nitric oxide synthase. Science 261: 1445–1448

Kwon NS, Stuehr J, Nathan CF (1991) Inhibition of tumor cell ribonucleotide reductase by macrophage-derived nitric oxide. J Exp Med 174: 761–768

McCartney-Francis N, Allen JB, Mizel DE, Albina JE, Xie QW, Nathan CF, Wahl SM (1993) Suppression of arthritis by an inhibitor of nitric oxide synthase. J Exp Med 178: 749–754

Roberts AB, Vodovotz Y, Roche NS, Sporn MB, Nathan CF (1992) Role of nitric oxide in antagonistic effects of transforming growth factor-β and interleukin-1β on the beating rate of cultured cardiac myocytes. Mol Endocrinol 6: 1921–1930

Stuehr DJ, Cho HJ, Kwon NS, Weise M, Nathan CF (1991) Purification and characterization of the cytokine-induced macrophage nitric oxide synthase: an FAD- and FMN- containing flavoprotein. Proc Natl Acad Sci USA 88: 7773–7777

Vodovotz Y, Bogdan C, Paik J, Xie QW, Nathan C (1993) Mechanisms of suppression of macrophage nitric oxide release by transforming growth factor-β. J Exp Med 178: 605–164

Xie QW, Cho H, Calacay J, Mumford RA, Swiderek KM, Lee TD, Ding A, Troso T, Nathan C (1992) Cloning and characterization of inducible nitric oxide synthase from mouse macrophages. Science 256: 225–228

Xie QW, Whisnant R, Nathan C (1993) Promoter of the mouse gene encoding calcium-independent nitric oxide synthase confers inducibility by interferon-γ and bacterial lipoposaccharide. J Exp Med 177: 1779–1784

Xie QW, Kashiwabara Y, Nathan C (1994) Role of transcription factor NF-κB/Rel in the induction of nitric oxide synthase. J Biol Chem 269: 4705–4708

Diffusible Messengers and Intercellular Signaling: Locally Distributed Synaptic Potentiation in the Hippocampus

D.V. Madison and E.M. Schuman

Diffusible messengers like nitric oxide (NO) have been proposed to act as intercellular signals in the production of long term potentiation (LTP; for review, see Schuman and Madison 1993). The postsynaptic generation of such a diffusible messenger leads to the prediction that synapses in the vicinity of the messenger-producing synapse will also be influenced. We have tested this prediction directly by recording simultaneously from two postsynaptic CA1 neurons and recording synaptic strength in response to Schaffer collateral stimulation before and after LTP induction (by pairing 1 Hz stimulation with postsynaptic depolarization) in one neuron (see also Bonhoefer et al. 1989). In experiments where we impaled two pyramidal neurons in close proximity (132.9 ± 39.6 μ*M* intersomatic distance, ISD), the paired cell exhibited robust LTP (186.0 ± 15.6 mean percentage of baseline) and the nearby neighboring neuron also exhibited a significant increase in synaptic strength (130.6 ± 9.5%; *n*=19). Postsynaptic injection of the NO synthase inhibitor L-Me-Arg into the paired cell blocked LTP production in that cell (95.8 +/– 6.9%) as well as in the nearby cell (102.2 ± 10.1%; ISD = 143.3 ± 43.6 μ*M*; n=18). This spreading of potentiation was spatially restricted as it was not observed when the neighboring cell was > 500 μ*M* away (paired cell LTP, 170.1 ± 11.8%; distant neighbor, 97.6 ± 0.7%; ISD = 674.0 ± 38.3 μ*M*; *n*=15). Since previous findings showed that LTP can be blocked by various postsynaptic manipulations, we were interested in whether these processes are also important in the neighboring cell. We found that a combination of dialysis, Ca^{2+} chelators, and membrane hyperpolarization in the nearby neighboring cell blocked the spreading of enhancement usually observed (paired 195.7 ± 43.1; neighbor, 105.6 ± 24.7; *n*=8). These results suggest that the neighboring cell may play an active role in the enhancement described, either as a target site for a diffusible signal, or perhaps for the generation of an additional messenger(s).

No matter what the exact molecular mechanisms underlying the intercellular communication of potentiation, these data suggest that the formation of synaptic changes previously thought to be restricted to synapses onto a single cell, can also result in synaptic changes at nearby synapses (see also Gally et al. 1980). As

---

Beckman Center, Department of Molecular and Cellular Physiology, School of Medicine, Stanford University Medical Center, Stanford, CA 94305–5426, USA

such, diffusible signals may act as retrograde synaptic signals or alternately may serve the function of amplifying synaptic transmission at synapses in close proximity to the site of messenger generation. These results may help to understand processes both in the developing and adult brain. During development, locally distributed potentiation may be important in the formation of functional arrays of neurons analogous to anatomically segregated anatomical structures such as cortial columns (MILLER et al. 1989). Thus, distribution of potentiation could serve to induce temporary functional domains in regions where no such anatomical specialization exists. To the extent that neurons in anatomical proximity frequently share common output targets, this type of enhancement may also serve the general purpose of locally amplifying synaptic signals which underlie common neuronal functions.

References

Schuman EM, Madison DV (1993) Nitric oxide as an intercellular signal in long-term potentiation. Semin Neurosci 5: 207–215

Bonhoefer T, Staiger V, Aertsen A (1989) Synaptic plasticity in rat hippocampal slice cultures; local "Hebbian" conjunction of pre- and postsynaptic stimulation leads to distributed synaptic enhancement. Proc Natl Acad Sci USA 86: 8113–8117

Gally JA, Montague PR, Reeke GN, Edelman GM (1990) The NO hypothesis; possible effects of a short-lived, rapidly diffusible signal in the development of function of the nervous system. Proc Natl Acad Sci USA 87: 3547–3551

Miller KD, Keller JB, Stryker MP (1989) Ocular dominance column development analysis and stimulation. Science 245: 605–615

Nitric Oxide Formation in the Vascular Wall: Regulation and Functional Implications

R. BUSSE, I. FLEMING, and V. B. SCHINI

1 Introduction

Numerous investigations have indicated that in both health and disease the short-lived radical nitric oxide (NO) is a key effector in the vascular system (Fig. 1). NO is generated by a five-electron oxidation of one of the terminal guanidino nitrogen atoms of L-arginine, catalyzed by NO synthases, and reaches the surrounding target cells by simple diffusion. The principal physiological source of NO in the vascular system is the endothelium, which constitutively expresses a NO synthase. The most important functions of endothelium-derived NO are the control of blood flow, and hence the supply of oxygen to organs, and the control of blood cell interaction with the vascular wall. Under certain pathophysiological conditions, endothelial cells are no longer the main source of NO. Indeed, in most types of vascular cells (e.g., vascular smooth muscle cells, macrophages, fibroblasts and endothelial cells) a NO producing pathway is induced following exposure to cytokines, such as interleukin-1β (IL-1β) and tumor necrosis factor-α (TNF-α), the levels of which are elevated in response to infection and injury. NO generated by this pathway in large amounts may account, at least in part, for the cytotoxic effect of macrophages and thus play a crucial role in host defense. The finding that NO can have such contrasting effects, i.e., to be protective and yet also cytotoxic, may be explained by the involvement of two NO-generating systems regulated by distinctly different mechanisms.

Zentrum der Physiologie, Klinikum der JWG-Universität, Theodor-Stern-Kai 7, 60590 Frankfurt/Main, Germany

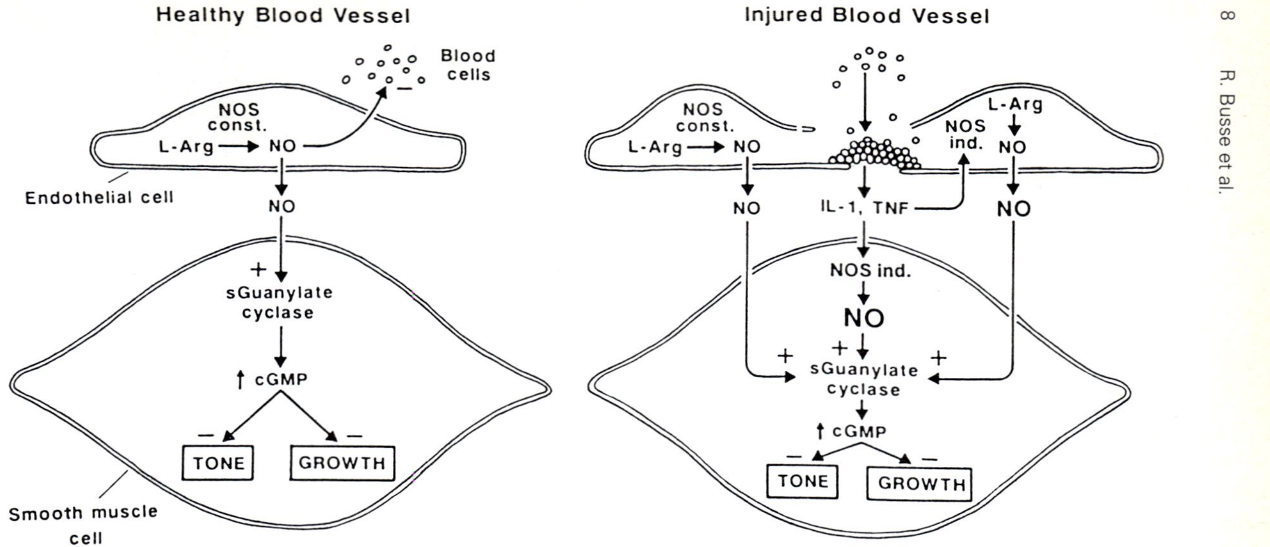

Fig. 1. In healthy blood vessels, nitric oxide (*NO*) is produced in the endothelial cells by the constitutive NO synthase (*NOS const.*). Destruction of the endothelium (e.g., by balloon catheterization) and exposure of the subendothelium rapidly results in the recruitment and activation of various blood cells. This inflammatory response is associated with the release of cytokines, such as interleukin-1 (*IL-1*) and tumor necrosis factor-α (*TNF-α*), which stimulate synthesis of the inducible NO synthase (*NOS ind.*) in the injured vessel wall

2 Nitric Oxide Biosynthetic Pathways in the Vascular Wall

2.1 Constitutive Nitric Oxide Synthase

The constitutive endothelial NO synthase displays binding sites for NADPH, flavin adenine dinucleotide, flavin mononucleotide and (6R)-5,6,7,8-tetrahydrobiopterin (BH_4) (LAMAS et al. 1992; SESSA et al. 1992; MARSDEN et al. 1992) and is strictly regulated by binding of the Ca^{2+}/calmodulin complex (BUSSE and MÜLSCH 1990a; FÖRSTERMANN et al. 1991). Calmodulin, in contrast with its role in other enzymes, has been shown to facilitate the reduction of the prosthetic heme group in NO synthase by transmitting electrons derived from NADPH (ABU-SOUD and STUEHR 1993).

NO synthase activity can be detected in the cytosol but activity is predominantly associated with the plasma membrane (HECKER et al. 1994; FÖRSTERMANN et al. 1991). The preferential association of the enzyme with membranes is likely to be a consequence of the myristoylation of the NH_2 terminal region of NO synthase (BUSCONI AND MICHEL 1993; SESSA et al. 1993). Localization of the enzyme in the plasma membrane may be of critical importance for the transduction mechanisms by which physical forces such as shear stress and pulsatile stretch elicit the formation of NO.

In situ, endothelial cells continuously produce low amounts of NO (GRIFFITH et al. 1984; MARTIN et al. 1986). This "basal" production can also be detected in cultured endothelial cells but decreases with increasing time in culture. It has recently been demonstrated that this phenomenon is paralleled by a decreased expression of the endothelial NO synthase (BUSSE et al. 1994). Expression of NO synthase mRNA and protein can, however, be up-regulated following the exposure of endothelial cell cultures to shear stress in vitro (BUSSE et al. 1994). Moreover a shear stress-responsive element in the promotor of the endothelial NO synthase gene has recently been identified, suggesting that a continuous physical stimulus such as shear stress transcriptionally regulates the expression of NO synthase in vivo (MARSDEN et al. 1992).

In addition to its basal release, both receptor-dependent and -independent mechanisms significantly enhance the release of NO from endothelial cells (for review see FURCHGOTT and VANHOUTTE 1989) (Fig. 2). Among the physiologically important receptor-dependent agonists are bradykinin, adenine nucleotides, 5-hydroxytryptamine (serotonin), substance P, and thrombin. These compounds activate specific receptors at the endothelial cell surface leading to a rapid increase in the intracellular concentration of free Ca^{2+} and, as a consequence of the Ca^{2+}/calmodulin dependency of the NO synthase, increase NO formation.

It is important to note, however, that the time course of the agonist-induced increase in NO synthase activity is significantly longer than that of the increase in intracellular Ca^{2+}. Thus, although a transient Ca^{2+} peak may be responsible for the initial increase in NO synthase activity, a second intracellular mechanism may be responsible for ensuring the maintained production of NO. Most of the

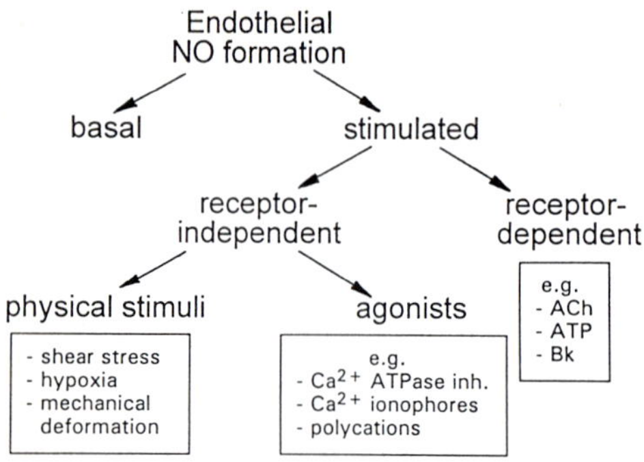

Fig. 2. The principal pathways by which nitric oxide (*NO*) can be formed in endothelial cells

physiologically relevant, receptor-dependent agonists activate the Na^+/H^+ exchanger leading to an increase in intracellular pH and it would appear that a sustained increase in NO formation can be observed following intracellular alkalinization. Indeed not only was the constitutive NO synthase found to be highly pH-sensitive in the physiological range but inhibition of the Na^+/H^+ exchanger significantly decreased bradykinin-stimulated NO formation (FLEMING et al. 1994).

2.2 Inducible Nitric Oxide Synthase(s)

An inducible NO synthase can be expressed in most types of vascular cells (see above) following their exposure to various inflammatory mediators such as cytokines (e.g., IL-1β, TNF-α) and bacterial lipopolysaccharides (for review see NATHAN 1992) (Table 1).

The inducible NO synthase is a predominantly cytosolic enzyme with essentially the same cofactor requirements as the constitutive enzyme. The major difference between this and the constitutive isoenzyme is that in the inducible NO synthase calmodulin is tightly bound in a Ca^{2+}-independent and noncovalent manner (CHO et al. 1992) and can be considered as an enzyme subunit, thus rendering its activity independent of changes in intracellular Ca^{2+}.

Table 1. Agents which influence expression of the inducible nitric oxide synthase

Inducers	Enhancers	Inhibitors
IL-1β	cAMP	TGFβ
TNF-α	EGF	$PDGF_{AB}$
IFN-γ	Basic FGF	$PDGF_{BB}$
LPS	Plasmin	IGF
		Thrombin

IL-1β, interleukin-1β; TGF-β, transforming growth factor-β; TNF-α, tumor necrosis factor-α; EGF, epidermal growth factor; PDGF, platelet-derived growth factor; IFN-γ, interferon-γ; FGF, fibroblast growth factor; IGF, insulin-like growth factor; LPS, bacterial lipopolysaccharide.

Once expressed, the enzyme is therefore always maximally activated. This observation may explain why the inducible L-arginine-NO pathway is associated with sustained production of large amounts of NO (IYENGAR et al. 1987)

The induction process requires a time lag of 4–6 h and involves activation of tyrosine kinases (MARCZIN et al. 1993; CORBETT et al. 1993), mRNA transcription and initiation of protein synthesis (for review see NATHAN 1992). In parallel with the initiation of inducible NO synthase synthesis, cytokines, by inducing the expression of guanosine triphosphate cyclohydrolase 1, also increase the de novo formation of the essential cofactor BH_4 (WERNER et al. 1989; FLEMING et al. 1991; GROSS and LEVI 1992). Inhibition of BH_4 synthesis prior to the exposure of vascular cells to cytokines completely prevents subsequent NO production, demonstrating that BH_4 synthesis is an absolute requirement for induction of NO synthesis.

Furthermore, NO synthase induction is facilitated by growth factors (such as basic fibroblast growth factor and epidermal growth factor, and the major fibrinolytic enzyme plasmin) which alone are without effect on inducible NO synthase expression (SCOTT-BURDEN et al. 1992; DURANTE et al. 1993). Increases in intracellular levels of cyclic adenosine 3',5'-monophosphate (cAMP) have a similar effect in that they enhance expression of the inducible NO synthase (KOIDE et al. 1993). The mechanisms underlying such an effect are unclear but may be due to activation of a cAMP regulatory element in the promotor of the inducible NO gene. It has, however, become apparent that the promotor of the gene encoding the murine inducible NO synthase does not contain a cAMP-responsive element (XIE et al. 1993), thus suggesting that cAMP exerts its effect by other, indirect, mechanisms. For example, cAMP has been reported to stabilize inducible NO synthase mRNA, prolonging its half life, rather than influencing its transcriptional activation (KUNZ et al. 1994).

The activation of many types of mammalian cells by a single cytokine and/or lipopolysaccharide is sufficient to cause the expression of the inducible NO synthase, but the combination of two or more stimuli leads to a synergistic enhancement of expression (KILBOURN and BELLONI 1990). Such findings suggest that the presence of threshold concentrations of several different cytokines at sites of vascular injury may result in the local generation of NO in amounts sufficient to elicit biological responses in neighboring cells. Although NO production by the inducible NO synthase is, at first glance, a process which is not subjected to any cellular regulation, it would appear that its expression can indeed be controlled by a variety of mechanisms. Platelets appear to play a significant role in the regulation of vascular NO formation, since substances released from aggregating platelets, such as platelet-derived growth factor, transforming growth factor-β and insulin-like growth factor, efficiently prevent expression of the inducible NO synthase in cultured vascular smooth muscle cells (SCHINI et al. 1992; DURANTE et al. 1994). In addition to these growth factors, the major coagulation protease thrombin prevents the induced formation of NO in vascular smooth muscle cells (SCHINI et al. 1993). Therefore, local activation of platelets and the coagulation cascade may prevent excessive formation of NO in vascular smooth muscle following injury, thus avoiding the cytotoxic effects of large amounts of NO.

3 Effector Pathways of Nitric Oxide

Nitric oxide is a multifunctional effector molecule in the vascular system which readily diffuses through lipid membranes and, in the presence of oxygen, is (within seconds) inactivated by conversion to nitrite and/or nitrate. Thus, the short half-life of NO constrains its biological action only to the generator cell itself and cells in its immediate vicinity. The major effector pathways mediating the action of NO in vascular cells have been summarized in Fig. 3.

The best characterized target for NO is the soluble guanylyl cyclase (Craven and DeRubertis 1978). NO binds to the iron present in the heme of the catalytic domain of the enzyme and by increasing its activity leads to an enhanced formation of cyclic guanosine 3',5'-monophosphate (Katsuki et al. 1977). This biochemical pathway accounts for the vasodilatory properties of NO in all types of blood vessels studied to date and also for its ability to prevent platelet activation (Gruetter et al. 1979; Mellion et al. 1981). These responses are eventually due to an attenuation of Ca^{2+} signaling (Busse et al. 1987; Karaki et al. 1988).

NO can also react with non-heme iron in target cells and such a reaction has been involved in host defense against infection. Cytotoxic activated macrophages, by releasing large amounts of NO, precipitate the loss of intracellular iron resulting in the inhibition of certain vital enzymes which require iron as an essential cofactor. Such enzymes include the NADH: ubiquinone oxidoreductase and succinate: ubiquinone oxidoreductase (both of which are involved in mitochondrial respiration), the critic acid cycle enzyme aconitase and ribonucleotide reductase (the rate limiting enzyme in DNA replication) (Drapier and Hibbs 1986; Hibbs et al. 1988).

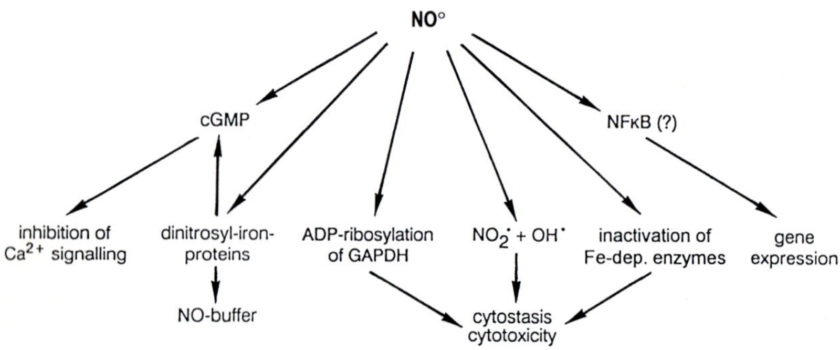

Fig. 3. The effects of nitric oxide (*NO*). *NFκB*, nuclear factor κB; Fe dep, non-heme iron-dependent enzymes

Other important cellular targets for NO are sulfhydryl (SH)-containing proteins. NO derivatives, such as the nitrosonium ion (NO^+), interact with these SH groups to produce biologically active S-nitrosoproteins (STAMLER et al. 1992b) (see following chapter by J.S. Stamler, pp 19–36). S-nitrosylation appears to be implicated in both the beneficial and non beneficial actions of NO. Among the thiol-containing proteins of potential physiological significance for S-nitrosylation are serum albumin and tissue-type plasminogen activator (STAMLER et al. 1992a,b). These S-nitrosylated proteins, like NO, are potent vasodilators and inhibitors of platelet aggregation, effects mediated via the activation of soluble guanylyl cyclase in the target cells. However, the half-lives of these compounds are significantly longer (>30 min) than that of NO (STAMLER et al. 1992b; KEANEY et al. 1993). These NO-containing compounds may therefore serve as adducts, which by stabilizing the highly reactive NO prolong its activity and/or facilitate its biological action thus extending its sphere of influence. In addition, NO, by S-nitrosylating the glycolytic enzyme glyceraldehyde-3-phosphate dehydrogenase decreases its activity and increases its auto-ADP-ribosylation (MOLINA-Y-VEDIA et al. 1992; DIMMELER et al. 1992). The impairment of this glycolytic enzyme by NO may, in association with the inhibition of iron-sulfur enzymes like aconitase and electron transport proteins of the respiratory chain (see above), contribute to the cytotoxic effects of NO.

Furthermore, NO may also modulate vascular functions by controlling the expression of genes encoding certain vasoactive proteins. Indeed, in cultured endothelial cells NO has been shown to down-regulate the expression of the chemoattractant MCP-1 (monocyte chemoattractant protein) (SCHRAY-UTZ et al. 1993), the potent vasoconstrictor endothelin-1, and the chemoattractant and vascular smooth muscle growth factor platelet derived growth factor (KOUREMBANAS et al. 1993). The demonstration that NO is able to modulate the expression of certain genes underlines the significance of endothelium-derived NO, both in short-term and long-term vascular homeostasis.

4 Role of Nitric Oxide in Vascular Physiology and Pathophysiology

In humans and experimental animals, the administration of structural analogues of L-arginine, which competitively inhibit NO synthase, results in a rapid and sustained increase in arterial blood pressure and a decrease in blood flow (REES et al. 1989a; AISAKA et al. 1989; VALLANCE et al. 1989). This pressor effect, which is associated with the constriction of several peripheral vascular beds, can be reversed by the infusion of L-arginine (REES et al. 1989a; AISAKA et al. 1989; GARDINER et al. 1990; GALLE et al. 1993) and is a consequence of the inhibition of basally released endothelium-derived NO. The augmentation of contractile responses in isolated preconstricted arteries following removal of the endothelium or incubation with NO synthase inhibitors suggests that vascular tone is a net effect of constrictor and dilator influences (Rees et al. 1989b; Mülsch and Busse 1990; Moore et al. 1990). Taken together, these observations indicate that NO derived from endothelial cells regulates the tone of the underlying vascular smooth muscle, and hence contributes to the control of regional blood flow. The most important physiological stimulus for the basal release of NO from the endothelium is the fluid shear stress imposed on the luminal surface of endothelial cells by the streaming blood. Indeed, in vitro investigations in the intact coronary vascular bed and in arterial segments under biossay conditions have shown that an increase in fluid shear stress triggers the release of endothelium-derived NO (Hecker et al. 1993; Lamontagne et al. 1992) (Fig. 4).

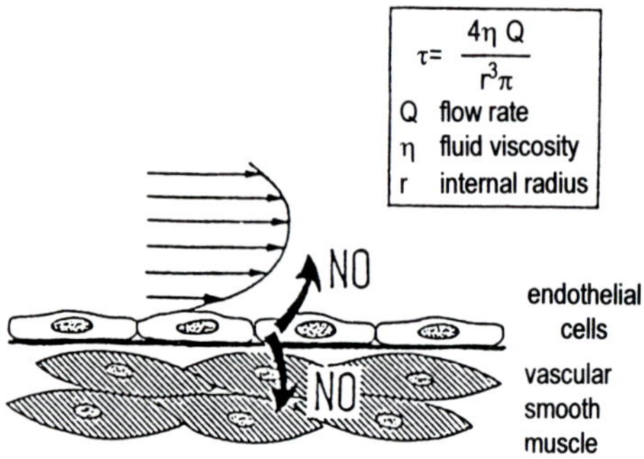

a

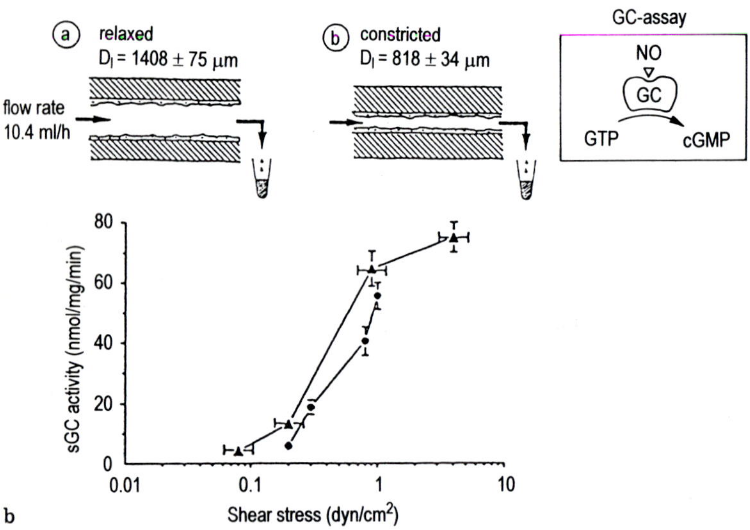

b

Fig. 4a, b. The relationship between fluid shear stress (τ) and the endothelial cell surface. Shear stress in rabbit femoral artery segments was increased either by **a** increasing flow through the vessel or **b** by decreasing vessel diameter. The graph shows the relationship between increases in shear elicited by vesoconstriction (triangles) or increased flow (circles) and nitric oxide (*NO*) release from rabbit femoral artery segments. NO production was determined by stimulation of soluble guanylyl cyclase (*sGC*) and expressed as net increase in enzyme activity over basal values. Shear was estimated under the simplifying assumption of a parabolic velocity profile

Shear stress-stimulated NO formation by endothelial cells, which is associated with an enhanced expression of the endothelial NO synthase, may contribute to the beneficial effects of physical exercise on the cardiovascular system (NISHIDA et al. 1992; SESSA et al. 1994). The physiologial significance of this enhanced formation of NO may be considered as a means of acutely increasing local blood flow, and thus supply of oxygen, in response to the greater metabolic requirements of tissues under working conditions (POHL et al. 1994). In addition to this "fundamentally vital" basal release of endothelium-derived NO, its formation is enhanced by physiologically relevant mediators such as hormones, neurotransmitters, and platelet-derived substances (see above). This additional release may play a significant role in the prevention of excessive vasoconstriction and abnormal coagulation following trauma to the blood vessel wall. NO may also contribute to the regulation of vascular growth and the remodeling of the blood vessel wall following trauma, since low levels of NO have been shown to effectively inhibit the proliferation of cultured vascular smooth muscle cells (GARG and HASSID 1989; SCOTT-BURDEN et al. 1992).

Since NO exerts a variety of effects on vascular function, it follows that a reduction as well as an increase in vascular NO formation has wide ranging consequences. For example, a decrease in NO production by a dysfunctional endothelium has been implicated in the pathogenesis of hypertension. The loss of the antiproliferative effect of NO may contribute to the vascular hypertrophy which characterizes a number of pathological states including atherosclerosis (for review see BUSSE and FLEMING 1993). An overproduction of NO, such as that observed during endotoxic shock, following the induction of the inducible NO synthase in vascular smooth muscle cells (BUSSE and MÜLSCH 1990b), may at first glance be desirable for defense against the invading pathogen. However, it eventually results in an unrelenting hypotension and hyporeacivity to vasoconstrictor agents (THIEMERMANN and VANE 1990; JULOU-SCHAEFFER et al. 1991).

In summary, it appears that the release of endothelium-derived NO, a potent antagonistic agent counteracting the development of local vasospasm, coagulation and aberrant proliferation of the underlying smooth muscle cells, plays an essential role in the short- and long-term homeostasis of the blood vessel wall. The induction of nonendothelial NO forming pathways in the vascular wall may be Janus-like, in that whereas an excessive and uncontrolled widespread production within the vasculature may be lethal, a more confined induction of NO synthase at sites of endothelial injury may be beneficial and compensate for the loss of endothelium-derived NO.

References

Abu-Soud HM, Stuehr DJ (1993) Nitric oxide synthases reveal a role for calmodulin in controlling electron transfer. Proc Natl Acad Sci USA 90: 10769–10772

Aisaka K, Gross SS, Griffith OW, Levi R (1989) N^G-methylarginine, an inhibitor of endothelium-derived nitric oxide synthesis, is a potent pressor agent in the guinea pig – does nitric oxide regulate blood pressure in vivo. Biochem Biophys Res Commun 160: 881–886
Busconi L, Michel T (1993) Endothelial nitric oxide synthase: *N*-terminal myristoylation determines subcellular localization. J Biol Chem 268: 8410–8413
Busse R, Fleming I (1993) The endothelial organ. Curr Opin Cardiol 8: 719–727
Busse R, Mülsch A (1990a) Calcium-dependent nitric oxide synthesis in endothelial cytosol is mediated by calmodulin. FEBS Lett 265: 133–136
Busse R, Mülsch A (1990b) Induction of nitric oxide synthase by cytokines in vascular smooth muscle cells. FEBS Lett 275: 87–90
Busse R, Lückhoff A, Bassenge E (1987) Endothelium-derived relaxant factor inhibits platelet activation. Naunyn-Schmiedebergs Arch Pharmacol 336: 566–571
Busse R, Hecker M, Fleming I (1994) Control of nitric oxide and prostacyclin synthesis in endothelial cells. Arzneimittel Forschung/Drug Res 44[Suppl]: 392–396
Cho HJ, Xie QW, Calaycay J, Mumford RA, Swiderek KM, Lee TD, Nathan C (1992) Calmodulin is a subunit of nitric oxide synthase from macrophages. J Exp Med 176: 599–604
Corbett JA, Sweetland MA, Lancaster JR, McDaniel ML (1993) A 1-hour pulse with IL-1β induces formation of nitric oxide and inhibits insulin secretion by rat islets of Langerhans: evidence for a tyrosine kinase signaling mechanism. FASEB J 7: 369–374
Craven PA, DeRubertis FR (1978) Restoration of the resposiveness of purified guanylyl cyclase to nitrosoguanidine, nitric oxide, and related activators ba heme and hemeproteins. J Biol Chem 253: 8433–8443
Dimmeler S, Lottspeich F, Brune B (1992) Nitric oxide causes ADP-ribosylation and inhibition of glyceraldehyde-3-phosphate dehydrogenase. J Biol Chem 267: 16771–16774
Drapier JC, Hibbs JB (1986) Murine cytotoxic activated macrophages inhibit aconitase in tumor cells. Inhibition involves the iron-sulfur prosthetic group and is reversible. J Clin Invest 78: 790–797
Durante W, Schini VB, Catovsky S, Kroll MH, Vanhoutte PM, Schafer AI (1993) Plasmin potentiates induction of nitric oxide synthesis by interleukin-1β in vascular smooth muscle cells. Am J Physiol 264: H617–H624
Durante W, Schini VB, Kroll MH, Catovsky S, Scott-Burden T, White JG, Vanhoutte PM, Schafer AI (1994) Platelets inhibit the induction of nitric oxide synthesis by interleukin-1β in vascular smooth muscle cells. Blood 83: 1831–1838
Fleming I, Gray GA, Schott C, Stoclet JC (1991) Inducible but not constitutive production of nitric oxide by vascular smooth muscle cells. Eur J Pharmacol 200: 375–376
Fleming I, Hecker M, Busse R (1994) Intracellular alkalinization induced by bradykinin sustains activation of the constitutive nitric oxide synthase in endothelial cells. Circ Res 74: 1220–1226
Förstermann U, Pollock JS, Schmidt HHHW, Heller M, Murad F (1991) Calmodulin-dependent endothelium-derived relaxing factor/nitric oxide synthase activity is present in the particulate and cytosolic fractions of bovine aortic endothelial cells. Proc Natl Acad Sci USA 88: 1788–1792
Furchgott Rf, Vanhoutte PM (1989) Endothelium-derived relaxing and contracting factors. FASEB J 3: 2007–2018
Galle J, Bauersach J, Bassenge E, Busse R (1993) Arterial size determines the enhancement of contractile responses after supression of endothelium-derived relaxing factor formation. Pflugers Arch 422: 564–569
Gardiner SM, Compton AM, Bennett T, Palmer RMJ, Moncada S (1990) Control of regional blood flow by endothelium-derived nitric oxide. Hypertension 15: 486–492
Garg UC, Hassid A (1989) Nitric oxide-generating vasodilators and 8-bromo-cyclic guanosine monophosphate inihibit mitogenesis and proliferation of cultured rat vascular smooth muscle cells. J Clin Invest 83: 1774–1777
Griffith TM, Edwards DH, Lewis MJ, Newby AC, Henderson AH (1984) The nature of endothelium-derived vascular relaxant factor. Nature 308: 645–647
Gross SS, Levi R (1992) Tetrahydrobiopterin synthesis. J Biol Chem 267: 25722–25729
Gruetter CA, Barry BK, McNamara BD, Gruetter DY, Kadowitz PJ, Ignarro LJ (1979) Relaxation of bovine coronary artery and activation of coronary arterial guanylate cyclase by nitric oxide, nitoprusside and a carcinogenic nitrosoamine. J Cycl Nucl Res 5: 211–224
Hecker M, Mülsch A, Bassenge E, Busse R (1993) Vasoconstriction and increased flow: two principle mechanisms of shear stress-dependent endothelial autacoid relase. Am J Physiol 265: H828–H833
Hecker M, Mülsch A, Bassenge E, Förstermann U, Busse R (1994) Subcellular localization and characterization of nitric oxide synthase(s) in endothelial cells - Physiologic implications. Biochem J 299: 247–252

Hibbs JB, Taintor RR, Vavrin Z, Rachlin EM (1988) Nitric oxide: a cytotoxic activated macrophage effector molecule. Biochem Biophys Res Commun 157: 87–94

Iyengar R, Stuehr DJ, Marletta MA (1987) Macrophage synthesis of nitrite, nitrate, and *N*-nitrosamines: precursors and role of the respiratory burst. Proc Natl Acad Sci USA 84: 6369–6373

Julou-Schaeffer G, Gray GA, Fleming I, Parratt JR, Stoclet JC (1991) Activation of the L-arginine pathway is involved in vascular hyporeactivity induced by endotoxin. J Cardiovasc Pharmacol 17: S207–S212

Karaki H, Sato K, Ozaki H, Murakami K (1988) Effects of sodium nitroprusside on cytosolic calcium level in vascular smooth muscle. Eur J Pharmacol 156: 259–266

Katsuki S, Arnold WP, Mittal CK, Murad F (1977) Stimulation of guanylate cyclase by sodium nitroprusside, nitroglycerin and nitric oxide in various tissue preparations and comparison to the effects of sodium azide and hydroxylamine. J Cycl Nucl Prot Phosphoryl Res 3: 23–35

Keaney JF, Simon DI, Stamler JS, Jaraki O, Scharfstein J, Vita JA, Loscalzo J (1993) NO forms an adduct with serum albumin that has endothelium-derived relaxing factor-like properties. J Clin Invest 91: 1582–1589

Kilbourn RG, Belloni P (1990) Endothelial cell production of nitrogen oxides in response to interferon-gamma in combination with tumor necrosis factor, interleukin-1, or endotoxin. J Natl Cancer Ins 82: 772–776

Koide M, Kawahara Y, Nakayama I, Tsuda T, Yokoyama M (1993) Cyclic AMP elevating agents induce an inducible type of nitric oxide synthase in cultured vascular smooth muscle cells. J Biol Chem 268: 24959–24966

Kourembanas S, McQuillan LP, Leung GK, Faller DV (1993) Nitric oxide regulates the expression of vasoconstrictors and growth factors by vascular endothelium under both normoxia and hypoxia. J Clin Invest 92: 99–104

Kunz D, Mühl H, Walker G, Pfeilschifter J (1994) Two distinct signalling pathways trigger the expression of inducible nitric oxide synthase in rat mesangial cells. Proc Natl Acad Sci USA 91: 5387–5391

Lamas S, Marsden PA, Li GK, Tempst P, Michel T (1992) Endothelial nitric oxide syntahse: molecular cloning and characterization of a distinct constitutive enzyme isoform. Proc Natl Acad Sci USA 89: 6348–6352

Lamontagne D, Pohl U, Busse R (1992) Mechanical deformation of vessel wall and shear stress determine the basal release of endothelium- derived relaxing factor in the intact rabbit coronary vascular bed. Circ Res 70: 123–130

Marczin N, Papapetropoulos A, Catravas JD (1993) Tyrosine kinase inhibitors suppress endotoxin- and IL-1β-induced NO synthesis in aortic smooth muscle cells. Am J Physiol 265: H1014–H1018

Marsden PA, Schappert KT, Chen HS, Flowers M, Sundell CI, Wilcox JN, Lamas S, Michel T (1992) Molecular cloning and characterization of human endothelial nitric oxide synthase. FEBS Lett 307: 287–293

Martin W, Furchgott RF, Villani GM, Jothianandan D (1986) Depression of contractile responses in rat aorta by spontaneously released endothelium-derived relaxing factor. J Pharmacol Exp Ther 237: 529–538

Mellion BT, Ignarro LJ, Ohlstein EH, Pontecorvo EG, Hyman AL, Kadowitz PJ (1981) Evidence for the inhibitory role of guanosine 3',5'-monophosphate in ADP-induced human platelet aggregation in the presence of nitric oxide and related vasodilators. Blood 57: 946–955

Molina-y-Vedia L, McDonald B, Reep B, Brüne B, DiSilvio M, Billiar TR, Lapetina EG (1992) Nitric oxide-induced S-nitrosylation of glyceraldehyde-3-phosphate dehydrogenase inhibits enzymatic activity and increases endogenous ADP-ribosylation. J Biol Chem 267: 24929–24932

Moore PK, Alswayeh OA, Chong NWS, Evans RA, Gibson A (1990) L-N^G-nitro arginine (L-NOARG), a novel L-arginine-reversible inhibitor of endothelium-dependent vasodilatation in vitro. Br J Pharmacol 99: 408–412

Mülsch A, Busse R (1990) N^G-nitro-L-arginine (N^5-[imino(nitroamino)methyl]-L-ornithine) impairs endothelium-dependent dilations by inhibiting cytosolic nitric oxide synthesis from L-arginine. Naunyn-Schmiedebergs Arch Pharmacol 341: 143–147

Nathan C (1992) Nitric oxide as a secretory product of mammalian cells. FASEB J 6: 3051–3064

Nishida K, Harrison DG, Navas JP, Fisher AA, Dockery Sp, Uematsu M, Nerem RM, Alexander RW, Murphy TJ (1992) Molecular cloning and characterization of the constitutive bovine aortic endothelial cell nitric oxide synthase. J Clin Invest 90: 2092–2096

Pohl U, Lamontagne D, Bassenge E, Busse R (1994) Attenuation of coronary autoregulation in the isolated rabbit heart by endothelium derived nitric oxide. Cardiovasc Res 28: 414–419

Rees DD, Palmer RM, Moncada S (1989a) Role of endothelium-derived nitric oxide in the regulation of blood pressure. Proc Natl Acad Sci USA 86: 3375–3378

Rees DD, Palmer RMJ, Hodson HF, Moncada S (1989b) A specific inhibitor of nitric oxide formation from L-arginine attenuates endothelium-dependent relaxation. Br J Pharmacol 96: 418–424

Schini VB, Durante W, Elizondo E, Scott-Burden T, Junquero DC, Schafer AI, Vanhoutte PM (1992) The induction of nitric oxide synthase activity is inhibited by TGF-β1, $PDGF_{AB}$ and $PDGF_{BB}$ in vascular smooth muscle cells. Eur J Pharmacol 216: 379–383

Schini VB, Catovsky S, Durante W, Scott-Burden T, Schafer AI, Vanhoutte PM (1993) Thrombin inhibits induction of nitric oxide synthase in vascular smooth muscle cells. Am J Physiol 264: H611–H616

Schray-Utz B, Zeiher AM, Busse R (1993) The expression of monocyte chemoattractant protein (MCP-1) mRNA in human endothelial cells is modulated by nitric oxide. FASEB J 7: A130

Scott-Burden T, Schini VB, Elizondo E, Junquero DC, Vanhoutte PM (1992) Platelet-derived growth factor suppresses and fibroblast growth factor enhances cytokine-induced production of nitric oxide by cultured smooth muscle cells; effects on cell proliferation. Circ Res 71: 1088–1100

Sessa WC, Harrison JK, Barber CM, Zeng D, Durieux ME, D' Angelo DD, Lynch KE, Peach MJ (1992) Molecular cloning and expression of cDNA encoding endothelial cell nitric oxide synthase. J Biol Chem 267: 15274–15276

Sessa WC, Barber CM, Lynch KR (1993) Mutation of N-myristoylation site converts endothelial cell nitric oxide synthase from a membrane to a cytosolic protein. Circ Res 72: 921–924

Sessa WC, Pritchard K, Seyedi N, Wang J, Hintze TH (1994) Chronic exercise in dogs increases coronary vascular nitric oxide production and endothelial cell nitric oxide synthase gene expression. Circ Res 74: 249–353

Stamler JS, Jaraki O, Osborne J, Simon DI, Keaney J, Vita J, Singel D, Valeri CR, Loscalzo J (1992a) Nitric oxide circulates in mammalian plasma primarily as an S-nitroso adduct of serum albumin. Proc Natl Acad Sci USA 89: 7674–7677

Stamler JS, Simon DI, Osborne JA, Mullins ME, Jaraki O, Michel T, Singel DJ, Loscalzo J (1992b) S-nitrosylation of proteins with nitric oxide—synthesis and characterization of biologically active compounds. Proc Natl Acad Sci USA 89: 444–448

Thiemermann C, Vane JR (1990) Inhibition of nitric oxide synthesis reduces the hypotension induced by bacterial lipopolysaccharides in the rat in vivo. Eur J Pharmacol 182: 591–595

Vallance P, Collier J, Moncada S (1989) Effects of endothelium-derived nitric oxide on peripheral arteriolar tone in man. Lancet 2: 997–1000

Werner ER, Werner-Felmayer G, Fuchs D, Hausen A, Reibnegger G, Wachter H (1989) Parallel induction of tetrahydrobiopterin biosynthesis and indoleamine 2,3-dioxygenase activity in human cells and cell lines by interferon-gamma. Biochem J 262: 861–868

Xie Qw, Whisnant R, Nathan C (1993) Promotor of the mouse gene encoding calcium-dependent nitric oxide synthase confers inducibility by interferon gamma and bacterial lipopolysaccharide. J Exp Med 177: 1779–1784

S-Nitrosothiols and the Bioregulatory Actions of Nitrogen Oxides Through Reactions with Thiol Groups

J.S. STAMLER

1 History and Overview

Chemists have long been aware of the red color that develops upon treatment of thiols with nitrous acid. Shortly after the turn of the last century, TASKER and JONES (1909) reported on the synthesis of benzene thionitrite, which exhibits a red color. The authors further noted that the compound was highly unstable and rapidly decomposed to (biphenyl) disulfide and nitric oxide (NO˙) gas. Thermal and photolytic decomposition of thionitrites was later shown to involve homolytic fission, as inferred from these early experiments (LECHER and SIEFKEN 1926; RAO et al. 1967; JOSEPHY et al. 1984). TASKER and JONES (1909) also described the thionitrite (or *S*-nitrosothiol; RS-NO) formed from ethane-thiol treatment with nitrosyl chloride (NOCl). This compound was shown to be significantly more stable than the corresponding benzene thiol derivative, but also disappeared with evolution of nitric oxide. Thus, the well documented importance of the electron withdrawing effect of the thiyl (RS) group in hastening the homolytic decomposition of RS-NO had been appreciated well over 50 years ago. In 1969, MIRNA and HOFMAN provided additional insight into the physical properties of biological RS-NOs. These studies demonstrated the trend for greater stability of thionitrites at low pH. At the same time, differences in the stability of thionitrites derived from cysteine and

Divisions of Respiratory Medicine and Cardiovascular Medicine, Duke University Medical Center, Bell Building, Room 350, Durham, NC 27710, USA

glutathione were noted. While *S*-nitroso-cysteine rapidly decomposes through homolytic fission, the *S*-nitroso adduct of glutathione remains stable over a wide (physiological) pH range (MIRNA and HOFMANN 1969). Shortly thereafter, FIELD and colleagues, isolated the highly stable thionitrite derivative of *N*-acetylpenicillamine (FIELD et al. 1978). More importantly, this work also demonstrated that disappearance of RS-NO can follow heterolytic pathways, specifically, reactions in which RS-NO formally transfers NO^+ (or NO^-). Additional reactions, persumed to be heterolytic in mechanism, were subsequently reported by MASSEY and colleagues (1978) and OAE and coworkers (1978) and supported the growing use of thionitrites in organic synthesis as effective nitrosating agents. The notable stability of protein thionitrites has been appreciated most recently, and heterolytic fission of the S-N bond appears to predominate in many biological systems (STAMLER et al. 1992a,b,c; LIPTON et al. 1993; STAMLER 1994).

In a series of papers in the early 1980s, IGNARRO and coworkers demonstrated the smooth muscle relaxant and antiplatelet properties of RS-NO (IGNARRO and GRUETTER 1980; GRUETTER et al. 1981; IGNARRO et al. 1981). These newer biochemical functions were attributed to activation of the enzyme guanylate cyclase, which catalyzes the transformation of guanosine triphosphate (GTP) to cyclic guanosine monophosphate (cGMP) (IGNARRO and GRUETTER 1980; GRUETTER et al. 1981). The later discovery that cyclic GMP (cGMP) acted as a signal transduction mechanism for the L-arginine: $NO^\cdot$ pathway in mammalian cells, in which $NO^\cdot$ is conceived to activate guanylate cyclase by interaction with its heme center, strengthened the resolve that nitric oxide ($NO^\cdot$) itself is the bioactive moiety in RS-NO (IGNARRO 1989). This line of reasoning, taken together with an early focus on the biochemistry of *S*-nitroso-cysteine, which readily undergoes homolysis to yield $NO^\cdot$ in vitro (MIRNA and HOFMANN 1969; KANNER 1979; MYERS et al. 1990), led to the common view that all RS-NOs are inherently unstable molecules which decompose to liberate nitric oxide. It is clear from the above cited literature, however, that many of these recent views do not encompass the larger chemistry of thionitrites. Moreover, as will become more evident, the (rapid) homolytic liberation of $NO^\cdot$ by RS-NO is, in part, an artifact of in vitro systems containing contaminant metals, which mediate this pathway. In this perspective I will attempt to integrate the broader chemistry of RS-NO in a review of their biological properties in order to explain their diverse biological actions.

2 Physical Properties and Chemistry of *S*-Nitrosothiols

S-nitroso derivatives of primary aminothiols are usually red (IGNARRO et al. 1980; OAE and SHINHAMA 1983; STAMLER and LOSCALZO 1992). Tertiary thionitrites are green colored (FIELD et al. 1978; OAE and SHINHAMA 1983). Solutions of protein RS-NOs are orange or rose colored (STAMLER et al. 1992a). The stability of thionitrites varies as a function of the thiyl group (RS). With an increase in electron withdrawing effect of R, or the proximity of a primary amino group, the stability of RS-NO tends to decrease (TASKER and JONES 1909; MORRIS and HANSEN 1981; J.S. Stamler, personal observation). This may be a consequence of a change in thiol pK incurred by the adjacent group (in which heterolytic fission is governed by the basicity of RS^-), or the role the amino group may play in catalyzing the decomposition of RS-NO by way of a thionitrite to nitosamine rearrangement. Recent studies suggest that contaminant metals, and in particular copper ions, facilitate thionitrite decomposition (RAMDEV et al. 1993; MCNAINLY and WILLIAMS 1993); this affect may involve the intermediate formation of Cu-complexes (MCNAINLY and WILLAMS 1993), which are almost certain to involve copper amine coordination. The finding that moderate acidity (pH 1–5) markedly increases the stability of RS-NO may be explained, in part, by the pH limiting affects on metal complexation. However, RS-NOs exhibit very complex pH-dependent stability profiles in which acid catalyzed hydrolysis of RS-NO is balanced by product (NO^+) mediated *S*-nitrosation (Eq. 1), and alkaline conditions lend to heterolytic fission (Eq. 2)

$$\text{RS-NO} + H^+ \rightleftarrows \text{RSH} + NO^+, \qquad (1)$$

$$\text{RS-NO} + OH^- \rightarrow NO_2^- + RS^- + H^+. \qquad (2)$$

The importance of metals in decomposition reactions may also rationalize the much higher stability of homocysteine thionitrite than of its cysteine homologue (STAMLER et al. 1993; RAMDEV et al. 1993), as the sulfur is distanced from the terminal amine by one extra methylene, perhaps limiting Cu^{2+} coordination. Further, modification of the amino group of cysteine by *N*-acetylation, dramatically stabilizes the derivative RS-NO (WILLIAMS 1988); in this case, however, internal stabilization may also derive from ring isomerization.

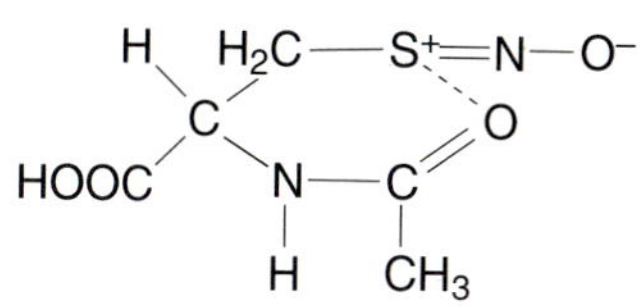

Bulky protecting groups also stabilize thionitrites. Examples include the thionitrite derivatives of *N*-acetyl-penicillamine and *t*-butyl-thionitrite (FIELD et al. 1978; OAE and SHINHAMA 1983). Further, protein RS-NOs are often more stable than the *S*-nitroso derivatives of amino acids and small peptides (STAMLER et al. 1992a). In the latter case, the local microenvironment of the protein thiol may stabilize the thionitrite by supporting an acidic millieu, through intramolecular interactions and/or by limiting the access of reactive nucleophiles. Although it may be easy to envision unstable protein nitrosothiols, those tested to date, i.e. derivatives of albumin (STAMLER et al. 1992a), tissue plasminogen activator (STAMLER et al. 1992a,d,e), cathepsin B (STAMLER et al. 1992a,e), and soybean extract (MASSEY et al. 1978) form comparatively stable *S*-nitroso derivatives at physiologic pH. Disproportionation of RS-NO is accelerated by O_2 and slowed by excess thiol (RSH); not only is the latter true for highly stable thionitrites such as *S*-nitroso-glutathione but also for unstable compounds such as *S*-nitroso-cysteine (RAMDEV et al. 1993). Thiols may stabilize RS-NO by scavenging reactive oxygen species and/or complexing metals, to limit their respective roles in thionitrite disproportionation. Indeed, the critical importance of (contaminant) metals in the mechanism of homolytic fission of RS-NO (RAMDEV et al. 1993; MCNAINLY and WILLIAMS 1993), taken with the relative resistance of thionitrites to hydrolysis (STAMLER et al. 1992c), indicates that the apparent "instability" of these species in vitro is largely artifactual: truly physiological systems contain millimolar concentrations of thiol and free metals are sparse.

RS-NOs exhibit ultraviolet absorption peaks at approximately 200 nm ($e \approx 10\,000\ M^{-1}\ cm^{-1}$), 335 nm ($e \approx 1000\ M^{-1}\ cm^{-1}$), approximately 500 nm ($e \approx 5\ M^{-1}\ cm^{-1}$), and 550 nm ($e \approx 5\ M^{-1}\ cm^{-1}$), (OAE and SHINHAMA 1983; BYLER et al. 1983; STAMLER et al. 1993). The extinction coefficients of protein RS-NO at $\approx$ 330 nm and 550 nm are on the order to three fold greater than RS-NOs (STAMLER et al. 1992a,d). The absorption maxima at $\approx$ 500 and 550 nm are unique to *S*-nitrosothiols while alkyl nitrites and nitrosoamines absorb in the lower ranges as well. Thionitrites possess infrared NO stretching frequencies in the range of 1490–1700 cm^{-1} and bending frequencies in the ranges of 610–660 cm^{-1} (OAE et al. 1978; OAE and SHINHAMA 1983; MASON 1969). The corresponding alkyl nitrites exhibit NO stretching vibrations at smaller wavelengths, which may be explained by the greater electronegativity of oxygen vs sulfur. ^{15}N-NMR spectra show shifts for RS-NO in the 750–790 ppm range as compared against an internal ^{15}N natural abundance spectrum of $NaNO_2$ at 587 ppm (BONNETT et al. 1975). Protein RS-NO exhibit ^{15}N-NMR shifts in the same range (STAMLER et al. 1992a,d).

The synthesis of thionitrites most commonly involves S-nitrosation chemistry (i.e., involving reaction or transfer of NO^+, the nitrosonium ion) (ALDRED and WILLIAMS 1982; STAMLER et al. 1992c)

$$RSH + XNO \rightleftarrows RSNO + HX.$$

The synthesis has been best characterized in acidic aqueous solution ($pH \leq 1$) with nitrous acid (HNO_2) (STAMLER et al. 1992a,d; WILLIAMS 1988; STAMLER and LOSCALZO 1992); however, treatment of thiols with N_2O_4 in inert solvents such as $CHCl_3$ or CCl_4 is a very convenient alternative method (OAE et al. 1977). Nitrosation has also been carried out with NOCl, alkyl nitrites (RONO) and dinitrogen trioxide (N_2O_3) (OAE and SHINHAMA 1983). Nitrosation of thiols with nitric oxide gas requires the presence of oxygen in vitro (PRYOR et al. 1982). This reaction is presumed to occur by way of reactive intermediates such as N_2O_3 and N_2O_4 (PRYOR et al. 1982; STAMLER et al. 1992c; WINK et al. 1993) and likely serves as one pathway for RS-NO formation in vivo. Direct S-nitrosation of thiols by metal (Fe or Cu) nitrosyl compounds has also been reported and is consistent with an eletrophilic attack by NO^+, donated by the metal nitrosyl compound (STAMLER et al. 1992c). Metal nitrosyl form effectively under neutral physiological conditions in the presence of nitric oxide, and may account, in part, for the presence of thionitrites in biological systems (STAMLER et al. 1992c). Nitrosoamines also fall into the catagory of biological nitrosonium (NO^+) equivalents (STAMLER et al. 1992c) and directly effect S-nitrosation, as exemplified in the well studied case of thiourea (WILLIAMS 1988

$$R_2NNO + (NH_2)_2CS \xrightarrow{H^+} (NH_2)_2CS^+NO + R_2NH.$$

Thionitrites may also form in a radical reaction between nitric oxide and thiyl radical (RS·) (WILLIAMS 1988). There is one report of RS-NO formation in the reaction involving nitroxyl (NO^-) and disulfide (RINDEN et al. 1989); however, the mechanistic details have not been worked out. There is also evidence for thiol nitrosation mediated by peroxynitrite ($OONO^-$) (WU et al. 1994; MOHR et al. 1994), which is produced in the diffusion-limited reaction of $NO^{\cdot}$ and superoxide. This pathway may help rationalize the high concentrations of RS-NO produced with inflammation (GASTON et al. 1993).

The kinetics of thionitrite formation are reasonably well characterized in the reaction between RSH and nitrite in acid solutions (BYLER et al. 1983; ALDRED and WILLIAMS 1982; AL-KAABI et al. 1982). This reaction exhibits a first order dependence on thiol, nitrous acid, and H^+. The forward reaction is interpreted as one involving S-nitrosation by NO^+ (or H_2ONO^+) at pH less than 1, and N_2O_3 at pH of greater than 2 (ALDRED and WILLIAMS 1982; RIDD 1978; WILLIAMS 1988). The rate expression for this reaction is as follows:

$$\text{Rate} = k_2\,[HNO_2]\,[H^+]\,[RSH].$$

Much less is known about the mechanistic details of nitrosation reactions in biological systems. The transnitrosation reactions afforded by thionitrites, including those with secondary amines, alkylamides and alcohols, have not been studied mechanistically; however, it is likely that the nitrosation occurs directly and not by way of alternative nitroso intermediates (ALDRED and WILLIAMS 1982).

Thionitrites can disproportionate both homolytically and heterolytically. For example, both low molecular weight and protein thiol may participate in transnitrosations (reaction or transfer of NO^+) with other thionitrites (SIMON et al. 1993; BUTLER and ASKEW 1993; PARK 1988; STAMLER et al. 1992c). Moreover, some hydrolysis of thionitrites is to be expected under the acidic conditions found in lysosomal compartments of the gastrointestinal tract. That such reactions could lead to formation of carcinogenic nitrosoamines has been a source of some concern. Alternatively, very high excess of thiol (or vicinal thiols on proteins) may promote the liberation of nitroxyl anion (NO^-) (STAMLER et al. 1992c; BYLER et al. 1983; PARK 1988). The heterolytic mechanism involved can be rationalized by the existence of a thionitrite resonance form in which the NO group accomodates the negative charge; by "activating" the derivatized sulfur atom, this pathway may also serve to potentiate disulfide formation

$$RS\text{-}N{=}O \overset{H^+}{\rightleftarrows} [RS^+ = N\text{–}O^-\,(H^+)] \xrightarrow{RS^-} RSSR + HNO.$$

The long forgotten report of hydroxylamine isolation from RS-NO (Tasker and Jones 1909) provides further support for this mechanism, as it is a major product of the reaction of singlet NO^- and thiol (Turk and Hollocher 1992)

$$HNO + 2RSH \rightarrow RSSR + NH_2OH.$$

What are the biological implications of this isomerism ($RS\text{-}N{=}O \rightleftarrows RS^+{=}N\text{–}O^-$)? In light of many vicinal SH groups essential for protein activity, the formation of RS-NO suggests a mechanism by which "nitric oxide" may regulate protein function by potentiating disulfide formation. Moreover, the propensity of RS-NO for transnitrosation reactions and disulfide formation, taken together with the strong thiol reducing environment in cells, suggests that heterolytic cleavage mechanisms may contribute significantly to RS-NO metabolism in biological systems (Stamler et al.1992a; Simon et al. 1993; Butler and Askew 1993; Stamler 1994). Thionitrites are also susceptible to oxidation, the latter process leading to formation of thionitrates ($RS\text{-}NO_2$) (Oae and Shinhama 1983). Nevertheless, the ease by which thionitrites are reduced suggests that reductive pathways are more likely in vivo.

3 Biological Actions of *S*-Nitrosothiols

The inhibition of bacterial growth is an essential function of nitrite during the food curing process. Early studies on the mechanism of bacterial inhibition show that an activation of nitrite is required and that thiol supports such activation (Castellani and Niven 1955; Johnston and Loynes 1971). Independent studies of Mirna and Hoffman indicate that thionitrites form in nitrite-cured meats and that such reactions are very dependent on pH (Mirna and Hoffman 1969). These observations led to the hypothesis that RS-NO may account for the bacteriostatic action of nitrite (Kanner 1979). In a series of important studies Incze and colleagues (1974) and Morris and Hansen established inhibitory effects of RS-NO on *Streptococcus faecium*, *Salmonella* strains, *Clostridium sporogenes* and *Bacillus cereus*, also learning much about the mechanism of action (Morris et al. 1984; Morris and Hansen 1981). In particular, the inhibitory actions of RS-NO on spore outgrowth are correlated with the electron withdrawing effects of the thiyl group (Morris and Hansen 1981) and were shown to be independent of RS-NO size, shape, charge, hydrophobicity and membrane permeability. In light of the above discussion on the chemistry of these compounds, these data suggest a mechanism of action independent of spontaneous homolytic cleavage of the RS-NO to yield $NO^{\cdot}$. In fact, reports by Morris and Hansen firmly establish that the antimicrobial action of RS-NO results from covalent modification of bacterial membrane sulfhydryl groups (Morris et al. 1984; Morris and Hansen 1981), and direct inhibition of sulfhydryl-dependent enzymes, such as glyceraldehyde phosphate dehydrogenase, may further contribute to their bacteriostatic activity (O'Leary and Solberg 1976). Moreover, cell impermeable RS-NO potently inhibits growth of the malarial parasite, *Plasmodium falciparum*, whereas a saturated solution of $NO^{\cdot}$ does not (ROCKETT et al. 1991). These data support the interpretation of NO^{+} transfer to critical membrane and cellular thiols leading to formation of protein RS-NO (not excluding subsequent evolution to disulfide), thereby disrupting vital microbial functions. Approximately 20 years later, Fung and coworkers would report that the vasodilatory effects of RS-NO could not be correlated with size, shape, charge, hydrophobicity or lipophilicity and were entirely independent of the rate of NO generation (KOWALUK and FUNG 1990); we have since reported the same for airway smooth muscle relaxation (GASTON et al. 1994b). Moreover, we and others have now shown that RS-NOs react with the cell membrane, both with enzymes that are poorly defined (KOWALUK and FUNG 1990; RADOMSKI et al. 1993) and in chemical reactions with surface thiol groups (SIMON et al. 1993; BUTLER and ASKEW 1993; LIPTON et al. 1993; STAMLER 1994). Notwithstanding these cogent observations, there remains a consensus of opinion that the actions of RS-NO occur solely through "spontaneous" liberation of nitric oxide, the effects of which are in turn mediated through activation of cytosolic guanylate cyclase.

Ignarro and coworkers are to be credited with the first demonstrations that RS-NO are likely intermediates in the metabolism of organic nitrites, nitrates and perhaps nitroprusside (IGNARRO and GRUETTER 1980; GRUETTER et al. 1981; IGNARRO et al. 1980, 1981; IGNARRO 1989). Notwithstanding current controversy on the metabolism of organic nitrates, there are several lines of reasoning supporting the role of RS-NO: (a) incubation of nitroglycerin, organic nitrites, and nitroprusside with thiol all lead to formation of thionitrites under physiological conditions (IGNARRO et al. 1981; FEELISCH 1991); (b) thionitrites markedly activate guanylate cyclase and elevate cyclic GMP in tissue (IGNARRO 1989; STAMLER et al. 1992c); (c) thiols potentiate vasodilatory (IGNARRO et al. 1981) and antiplatelet (LOSCALZO 1985; STAMLER et al. 1988; STAMLER and LOSCALZO 1991) effects of organic nitrites and nitroprusside and these effect can be mimicked by infusion of the corresponding RS-NO in animals (SCHARFSTEIN et al. 1993a; KEANEY et al. 1993); (d) RS-NOs are potent relaxants of vascular smooth muscle in vitro and decrease systemic arterial pressure when infused in animal models; further, they exhibit hemodynamic profiles that closely resemble those of nitroglycerin and nitroprusside (IGNARRO et al. 1981; KEANEY et al. 1993), and finally, (e) activation of purified guanylate cyclase by nitroglycerin, nitroprusside, and nitrite shows a critical dependence on the presence of thiol, under conditions promoting the formation of RS-NO (FEELISCH and NOACK 1987; IGNARRO 1989; STAMLER and LOSCALZO 1991). These findings certainly suggest that RS-NOs, at the very least, can act as intermediates in the activation of guanylate cyclase. A more direct role is certainly not excluded.

Interestingly, very few studies have actually examined the reaction between organic nitrates and sulfhydryl compounds. In early work, IGNARRO reported on the formation of *S*-nitroso-cysteine, when nitroglycerin was incubated with excess cysteine for prolonged periods (IGNARRO and GRUETTER 1980; IGNARRO et al. 1981). The mechanism involved was not studied, but formation of RS-NO occurs only with cysteine and *N*-acetylcysteine and not with other thiols (IGNARRO and GRUETTER 1980; FEELISCH and NOACK 1987). The most plausible first step in this reaction involves nucleophilic attacks by RS^- (thiolate) on the positively charged *N*-atom of the nitrate leading to formation of a thionitrate ($RS\text{-}NO_2$) (YEATES et al. 1985; LIPTON et al. 1993). This intermediate would then be subject to further reduction yielding a thionitrite. An interesting possibility might involve isomerization of the thionitrate to form a sulfonylnitrite (RSONO) through migration of the oxygen atom from nitrogen to sulfur. This species would then be susceptible to further nucleophilic attack by thiol yielding RS-NO; nitration of RSH by $OONO^-$ may lead to RS-NO by the same pathway (WU et al. 1994; MOHR et al. 1994). Notwithstanding these possibilities, other mechanisms of reduction may occur in cells or in plasma to explain the thiol-mediated catalysis of RS-NO formation reported in biological systems (FUNG et al. 1988; GASTON et al. 1993). The mechanism of RS-NO formation from nitroprusside and organic nitrites (IGNARRO and GRUETTER 1980; FEELISCH and NOACK 1987) is easier to reconcile and likely involves direct nucleophilic attacks on the positively charged nitrogen of the NO group present in both compounds.

It is the general consensus of opinion that nitroso-vasodilators mediate their vascular smooth muscle relaxant effects by activation of cytosolic guanylate cyclase (IGNARRO 1989). The mechanism of activation of guanylate cyclase has been rationalized through formation of the common active species nitric oxide, which interacts with the heme center of the enzyme in the pathway of activation (IGNARRO 1989). In actuality, the molecular details of this mechanism are scanty and poorly understood. In particular, the assumption that RS-NO undergoes homolytic cleavage yielding $NO^\cdot$ to activate the enzyme (IGNARRO and GRUETTER 1980) has been invalidated in certain experiments (KOWALUK and FUNG 1990; WILLIAMS 1988). Moreover, there is an extensive literature on the importance of enzyme thiol groups for its activity. The oxidation of protein thiols can lead directly to activation of guanylate cyclase, and, in concert with the reaction of the NO group at the heme center, leads to maximal enzyme activation (STAMLER et al. 1992c) NIROOMAND et al. 1989; KAMISAKI et al. 1986). Further, studies with nitroprusside reveal that activation of guanylate cyclase with nitroprusside promotes mixed disulfide formation involving thiol groups at the enzyme active site (NIROOMAND et al. 1989; KAMISAKI et al. 1986). Taken together, these observations suggest that S-nitrosation of guanylate cyclase may serve a role in activating the enzyme by facilitating the formation of disulfide. While this hypothesis remains to be tested, it is difficult to imagine—under most assay conditions in which thiol is included at millimolar concentrations—that NO group transfer to functional groups distinct from the heme iron would not be taking place.

The pharmacological effects of RS-NOs have been described in other tissues. *S*-nitroso-glutathione, *S*-nitroso-cysteine, *S*-nitroso-homocysteine, *S*-nitroso-penicillamine and *S*-nitroso-albumin are all potent relaxants of both guinea pig and human airway smooth muscle (JANSEN et al. 1991; GASTON et al. 1994b). As reported in vascular tissue, potency in no way correlates with lipophilicity, size or rate of nitric oxide generation (GASTON et al. 1994b). The role of guanylate cyclase in the mechanism of airway relaxation is less clear than in vascular smooth muscle (GASTON et al. 1994a,b). RS-NOs activate airway guanylate cyclase; however, the lack of effect of guanylate cyclase inhibitors in preventing RS-NO relaxation and temporal discrepancies between the rise in cyclic GMP and relaxation responses have resulted in much controversy in this area. RS-NOs have also been shown to relax smooth muscle in the gastrointestinal tract (SLIVKA et al. 1994) and corpus cavernosum of the penis (Saenz de Tejada, personal communication). In the last regard, there is evidence implicating RS-NO as the bovine retractor penis inhibitory factor (KERR et al. 1992).

4 *S*-Nitrosothiols in Control of Protein Function

In the early studies described above, the reaction of NO with protein thiol groups was envisioned as an adverse consequence of food curing with nitrite (MASSEY et al. 1978; O'LEARY and SOLBERG 1976). Renewed interest in the reaction of NO_x with the SH groups of proteins has been stimulated by the discovery of endogenous nitrogen oxide production in mammalian cells. We have recently focused on the possibility that posttranslational modification of proteins by NO, through reaction with SH groups, may regulate their function (STAMLER et al. 1992a,c,d,e; LIPTON et al. 1993). In a series of studies we demonstrated that proteins of diverse nature and function can be S-nitrosylated, that the derivatized proteins are quite stable at neutral pH, and that this modification confers on proteins smooth muscle (vascular and airway) relaxant and antiplatelet bioactivity (STAMLER et al. 1992a,d,e; SIMON et al. 1993; GASTON et al. 1993, 1994b). *S*-nitroso-albumin was subsequently identified in plasma as a reservoir of NO and is envisioned to serve a buffer-like function regulating levels of free nitric oxide for the determination of vasomotor tone (STAMLER et al. 1992b). When given by intravenous administration, *S*-nitroso-albumin markedly reduces mean arterial pressure and inhibits ex vivo platelet aggregation, supporting its potential biological role in modulating vasomotor tone (KEANEY et al. 1993). By comparison with nitroglycerin and nitroprusside, *S*-nitroso-albumin exhibits prolonged vasodilatory effetcs consistent with the notable stability of *S*-nitrosoproteins in general (KEANEY et al. 1993). More recently, *S*-nitrosoproteins were identified in human airway lining fluid in disease states associated with impairment of alveolar-capillary membrane integrity (GASTON et al. 1993). This finding is consistent with the presence of albumin and cysteine-rich mucus proteins in the airways. Although their biological role remains to be ascertained, RS-NO possesses bronchodilator activity and may limit the putative toxicity of $NO^{\cdot}$ in the O_2-rich environment of the lung (GASTON et al. 1993).

The idea that protein function may be regulated by NO has early support from several groups of investigators. As detailed above, MORRIS and HANSEN (1981) showed that the bacteriostatic effects of nitrogen oxides resulted from S-nitrosation of protein membrane thiols. In our earliest experiments we directly studied the effects of S-nitrosylation on the activity of two enzymes: tissue-type plasminogen activator (tPA) an endothelium derived fibrinolytic agent with a single free thiol in a domain distant from the active site, and cathepsin B, a sulfhydryl protease in which the thiol is critical for enzyme activity (STAMLER et al. 1992a,d,e). In the case of tPA (STAMLER et al. 1992d), S-nitrosylation enhanced enzyme activity, albeit slightly, while conferring vasodilatory and antiplatelet activity on the native protein. Most importantly, the fibrin specificity of tPA is preserved, providing a means by which NO can be selectively delivered to a fibrin clot, at which site a deficit of NO would be predicted because of endothelial damage. In contrast, the enzymatic activity of cathepsin B is inhibited by S-nitrosylation of active site thiol (STAMLER et al. 1992e). Similar NO-dependent inhibitory effects have been observed with other enzyme. These include aldolase (O'LEARY and SOLBERG 1976), alcohol dehydrogenase (PARK 1988), γ-glutamylcysteinyl synthetase (HAN et al. 1994) and glyceraldehyde-3-phosphate dehydrogenase (GAPDH) (O'LEARY and SOLBERG 1976; MOLINA Y VEDIA et al. 1992; MOHR et al. 1994), the latter event also associated with an ADP-ribosylation-like modification (MOLINA Y VEDIA et al. 1992; STAMLER 1994) in vivo. Studies on the mechanism of GAPDH modification reveal that S-nitrosylation of an active site thiol is requisite for subsequent modification by nicotine adenine dinucleotide phosphate (NAD^+) (MOHR et al. 1994). Thus, the formation of protein RS-NO serves a catalytic function. Interestingly, the S-NO intermediate of GAPDH formed during labelling with NAD is devoid of enzymatic activity (MOLINA Y VEDIA et al. 1992), analogous to cathepsin B. That only a fraction of S-nitrosylated protein undergoes covalent modification, taken together with the reversible nature of the S-NO modification, suggests that enzyme activity is regulated by S-nitrosylation (MOLINA Y VEDIA et al. 1992; MCDONALD and MOSS 1993; STAMLER et al. 1992a,c); nonenzymatic NAD-dependent modification of cysteine residues may represent a biological marker for intracellular protein S-nitrosylation. Intriguingly, other proteins such as actin and albumin undergo similar covalent modification of active site thiol, on treatment with NO (CLANCY et al. 1993; DIMMLER and BRUNE 1992).

These ideas have been recently extended to a cell system in which the thiol groups under study comprise the "redox modulatory" site of the N-methyl-D-asparte (NMDA)-type glutamate receptor and determine its ligand-gated activity (LIPTON et al. 1993). Conditions supporting S-nitrosylation of receptor thiol were shown to decrease the calcium influx induced by NMDA binding and the consequent toxicity to surrounding cells. It is not clear from these studies whether S-nitrosylation is sufficient to modulate the activity in the system or whether subsequent oxidation of protein RS-NO to disulfide is required. However, the finding that the reversal of inhibition of NMDA occurs in a two-phase process (LEI et al. 1992) suggests that both direct modification by NO (rapidly reversible) and the promotion of disulfide formation (long-lasting inhibition) occur to modulate the response. Analogous activation of a Ca^+ dependent potassium channel on smooth muscle that facilitates relaxation was recently reported (BOLOTINA 1994). That S-nitrosylation of critical thiols on the receptor produced an effect discordant from alkylation suggests that the molecular mechanism of activation involves disulfide formation. Specifically, only the NO group can accommodate a negative charge, and thereby promote the activation-associated oxidation that leads to a conformational switch. The presence of critical thiol groups on other proteins, enzymes—including guanylate cyclase—and surface receptors suggests a common mechanism by which NO can modulate protein function that also shares analogy with the more classical cellular signal transduction pathways. Thus, for example, S-nitrosylation of proteins may act as a rapid reversible signal transduction mechanism, akin, perhaps, to phosphorylation. If true, NO group transfer reactions may be enzymatically controlled in addition to the chemically controlled processes described here.

5 *S*-Nitrosothiols and NO˙ Mediated Toxicity

There is a great deal of concern over the potential toxicity of nitric oxide. Several mechanisms have been described by which NO may exert its cytotoxic effects. Nitric oxide may participate in nitrosative reactions leading to deamination of DNA and mutagenesis (WINK et al. 1991), and reactions of nitric oxide with superoxide lead to the formation of the highly toxic oxidant peroxynitrite (BECKMAN et al. 1990). In addition, the direct reaction of nitric oxide with iron sulfur centers in proteins may lead to the inhibition of their activity (HIBBS 1991). These concerns are supported by data showing that nitric oxide exposure can cause severe lung injury (WORLD HEALTH ORGANIZATION 1977; GASTON et al. 1994a), promote the formation of tumors (WORLD HEALTH ORGANIZATION 1977), and enhance neuronal destruction in stroke and other neurodegenerative disorders (SNYDER 1993). The reaction of nitrogen oxides (NO_x) with thiols may provide insights into divergent protective responses. Certainly, one of the main functions of biological thiols is to detoxify oxygen-derived free radicals and peroxides. By extension, thiols may play an important protective role in disarming nitric oxide. Levels of glutathione rise in lung tissue in response to oxidant stress, such as exposure to high levels of NO contained in cigarette smoke, and endogenous RS-NO levels increase in the lung in response to immune activation and inflammation (GASTON et al. 1993, 1994a). Moreover, the relative stability of these endogenous RS-NO (on the order of hours) may serve to limit the reactions of NO˙ with oxygen and superoxide that are associated with nitric oxide mediated toxicity (GASTON et al. 1993, 1994a; STAMLER et al. 1992c). The added finding that RS-NO possesses bronchodilator activity suggests a means by which the NO group is packaged in biological systems to preserve its bioactivity (GASTON et al. 1993, 1994b). Recent work indicates that NO can inhibit glutathione synthesis by inactivation of γ-glutamylcysteinyl synthetase (HAN et al. 1994), which catalyses the initial and rate-limiting step. Thus, the potential for regulation of cellular functions influenced by redox status notwithstanding, excessive NO may facilitate oxidative injury by depleting the thiol pool.

In the central nervous system, similar protective effects of RS-NO have been observed. Nitric oxide mediated toxicity has been attributed to the formation of peroxynitrite by way of reaction with superoxide (LIPTON et al. 1993). In contrast, formation of RS-NO provides a mechanism for limiting the reaction of NO· with $O_2^{-\cdot}$. As discussed above, S-nitrosation reactions may have additional protective effects by down-regulating NMDA receptor activity, which is a prerequisite for nitric oxide synthesis (LIPTON et al. 1993; SNYDER 1993). These findings suggest that, with excessive generation of nitric oxide by neurons, the molecule may feedback to inhibit its own production (MANZONI et al. 1992).

Paradoxically, biological thiols may exert toxic effects as well. Homocysteine, for example, will readily reduce oxygen, resulting in formation of superoxide and hydrogen peroxide (STAMLER et al. 1993; STARKEBAUM and HARLAN 1986); metal ions such as Cu^{2+} and Fe^{2+} may serve as catalysts in these reactions, possibly forming thiol-metal complexes (JOCELYN 1972)

$$Fe(II)\,(SR)_2 + O_2 \rightarrow Fe(III)\,(SR)_2 + O_2^{-\cdot},$$

$$O_2^{-\cdot} + RS^- \rightarrow O_2^{=} + RS^{\cdot},$$

$$Fe(III)\,(SR)_2 + RS^- \rightarrow Fe(II)\,(SR)_2 + RS^{\cdot},$$

$$2\,RS^{\cdot} \rightarrow RSSR$$

Reduced oxygen species are highly toxic to endothelial cells and are believed to play an important role in the vascular related morbidity seen in patients with hyperhomocysteinemia (STAMLER et al. 1993; STARKEBAUM and HARLAN 1986), an independent risk factor for atherosclerosis. Under normal physiological circumstances the interaction between endothelium-derived relaxing factor (EDRF) and homocysteine results in the formation of *S*-nitroso-homocysteine, a potent vasodilator and antiplatelet agent (STAMLER et al. 1993). With the thiol group "blocked" by NO, homocysteine is also prevented from generating hydrogen peroxide (STAMLER et al. 1993). However, with elevated levels of homocysteine incurred through an acquired or inborn alteration of metabolism, the cytotoxic properties of the molecule may embarrass EDRF production, setting a cycle in motion in which the antithrombotic, cytoprotective mechanism of S-nitrosation is increasingly compromised. Thus, S-nitrosation may represent a protective cell regulatory mechanism which simultaneously confers upon biological thiols EDRF-like bioactivity.

6 *S*-Nitrosothiols In Vivo

S-nitrosothiols have been identified in biological systems by several groups of investigators. Fung and coworkers identified low molecular weight RS-NOs in the plasma of human subjects treated with nitroglycerin (FUNG et al. 1988). These experiments also showed that metabolism of nitroglycerin was dependent, in part, on the thiol group of albumin, suggesting that protein RS-NOs may also form (FUNG et al. 1988; CHONG and FUNG 1990). We have extended these findings by showing that *S*-nitroso-proteins are present in the plasma of normal subjects (STAMLER et al. 1992b) and in the exudate of airway lining fluid in selected disease states (GASTON et al. 1993). In normal airways, *S*-nitroso-glutathione has been identified in concentrations sufficient to influence basal airway tone (GASTON et al. 1993). *S*-nitroso-glutathione also forms intracellularly in neutrophils on exposure to NO gas and with cell activation (CLANCY et al. 1994). Interestingly, Ribeiro and colleagues identified nitrosovasodilator activity in the salivary gland of *Rhodnius prolixus*, of which approximately 10%–20% was attributed to *S*-nitrosothiol (RIBEIRO et al. 1993). Finally, it has been estimated that a significant percentage of NO_x in the gastrointestinal tract exists in the form of RS-NO (MIRNA and HOFMANN 1969; KANNER 1979; MASSEY et al. 1978).

The hemodynamic effects of RS-NOs have received recent attention. In early reports, Ignarro's group showed vasodilator effects in the cat (IGNARRO et al. 1981). *S*-nitroso-*N*-acetylpenicillamine, *S*-nitroso-cysteine, *S*-nitroso-mercapto-ethylamine and *S*-nitroso-3-mercaptopropionic acid each decreased systemic arterial pressure in a dose-related fashion (IGNARRO et al. 1981). The onset of action was immediate in each case and the reported duration of action on the order of approximately 1–3 min. *S*-nitroso-cysteine was the most potent of the RS-NOs with a dose-response curve also lying to the left of that of nitroglycerin and nitroprusside. The potency of *S*-nitroso-cysteine has been confirmed by KEANEY and colleagues (1993). *S*-nitroso-cysteine caused marked changes in blood pressure when given by intravenous bolus infusion and very dramatic changes in coronary flow, indicative of its vasodilator effects on microvessels. In the same study, the authors investigated the effects of *S*-nitroso-albumin, which also exhibited vasodilator and antiplatelet properties reminiscent of ERDF. However, the kinetic profile and hemodynamic response was unique. *S*-nitroso-albumin had a greater duration of action than either nitroglycerin or *S*-nitroso-cysteine when given by intravenous infusion and a much longer duration of action on coronary blood flow when infused directly into the coronary artery. It appears that the actions of *S*-nitroso-albumin are mediated by way of NO groups transfer to more reactive, low-molecular weight thiol pools in plasma (STAMLER et al. 1992b; SCHARFSTEIN et al. 1993b), which in turn, may target the NO group to the vascular (or platelet) surface (FUNG et al. 1988; SIMON et al 1993; RADOMSKI et al. 1993). SCHAFER and coworkers (1991) studied the hemodynamic effects of *S*-nitroso-captopril in the anesthetized dog. Interestingly, the nitrovasodilator effect was ten- to 30-fold less potent than nitroglycerin when administered by bolus, but significantly more effective than the former when given by continuous infusion. The duration of action of *S*-nitroso-captopril was also significantly greater than that of nitroglycerin, which for unclear reasons, significantly exceeded the half-life of the drug. The attenuated pressure response to angiotensin in these animals indicates that this agent also manifests angiotensin converting enzyme inhibitor activity, which contributes to its mechanism of action. Recently, Vallance and coworkers infused *S*-nitroso-glutathione into the forearm of humans (Vallance, personal communication). The antiplatelet effects of the drug was significantly more impressive than its vasodilator response implicating different pathways of metabolism in the platelet and vascular wall.

The effects of RS-NOs on human gastrointestinal motility (SLIVKA et al. 1993) and airway tone (GASTON et al. 1993, 1994a, 1994b) have also been recently studied. When applied topically to the duodenum and colon, *S*-nitroso-*N*-acetylcysteine causes profound inhibition of intestinal motility (SLIVKA et al. 1994). Application directly to the sphincter of Oddi lowers sphincter of Oddi pressure and inhibits the intrinsic frequency of contraction. At the concentration of 100 μM applied in 10 ml of saline, the onset of action occurs in minutes and the duration of action was estimated on the order of 10–20 min. Studies examining the bronchodilator action of *S*-nitroso-*N*-acetylcysteine are currently underway. As in the gastrointestinal tract, topical administration (nebulized in saline) was not associated with systemic side effects. Early results in subjects with mild asthma should become available in the near future. RS-NOs have also been used on ureteral smooth muscle and corpus cavernosum, suggesting their potential uses in the treatment of ureteral spasm and erectile dysfunction (Saenz de Tejeda, personal communication).

7 Summary

The reactivity of selected RS-NOs has led to the misconception that these compounds are uniformly unstable under physiological conditions. Moreover, current evidence supports the notion that biological responses elicited by RS-NOs may result from either liberation of nitric oxide or from NO group transfer chemistry involving either NO^+ or NO^-. Some evidence suggests that such reactions may be enzymatically controlled. The data supporting the potential biological relevance of RS-NOs include : (1) evidence that these compounds form under physiological conditions; (2) their identification in insects, lower mammals, and several human biological systems; and (3) findings that RS-NOs possess a wide range of biological activities, including antimicrobial effects, vasodilation, platelet inhibition, bronchodilation and inhibition of intestinal motility, while being relatively resistant to reactions with O_2 and $O_2^{-\cdot}$ associated with $NO^\cdot$ toxicity. It is further noteworthy that biological activity of RS-NO is often not related to the propensity to liberate $NO^\cdot$, and these adducts are generally more potent and selective in their action than $NO^\cdot$ itself (STAMLER et al. 1989; COOKE et al. 1990; ROCKETT et al. 1991; JANSEN et al. 1991; LIPTON et al. 1993).

The data presented here support the idea that RS-NO may be involved in stabilizing nitric oxide-like bioactivity, in transporting and targeting the NO group to specific (thioregulatory) effector sites, in mitigating the cytotoxic effects of nitric oxide that result from reaction with oxygen species, and may serve to regulate protein function in a posttranslational modification akin, perhaps, to phosphorylation. The recently demonstrated NO group transfer reactions to plasma membrane proteins containing reactive sulfhydryls (Lipton et al. 1993; Stamler 1994) also raises the possibility of signal transduction initiated through more traditional "agonist-receptor" mediated pathways.

Acknowledgements. I am indebted to Professor David J. Singel for his critical review of this manuscript and for related discussions. JSS is a Pew Scholar in the Biomedical Sciences and the recipient of a Clinical Investigator. Award from the National Institutes of Health (HL 02582-011)

References

Al-Kaabi SS, Williams DLH, Bonnet R, Ooi S (1982) A kinetic investigation of the thionitrite from (±)-2-acetylamino-2-carboxy-1,1-dimethylethanethiol as a possible nitrosating agent. J Chem Soc Perkins II: 227–230

Aldred SE, Williams LH (1982) Kinetics and mechanisms of the nitrosation of alcohols, carbohydrates, and a thiol. J Chem Soc Perkin Trans II: 777–782

Beckman JS, Beckman TW, Chen J, Marshell PA, Freeman BA (1990) Apparent hydroxyl radical production by peroxynitrite: implications for endothelial injury from nitric oxide and superoxide. Proc natl Acad Sci USA 87: 1620–1624

Bolotina VM, Najibi S, palacio JJ, Pagano PJ, Cohen RA (1994) Nitric oxide directly activates calcium-dependent potassium channels in vascular smooth muscle. Nature 368: 850–853

Bonnett R, Holleyhead R, Johnson BI, Randall EW (1975) Reactions of acidified nitrite solutions with peptide derivatives: evidence for nitrosamine and thionitrite formation from ^{15}N N.M.R. studies. J Chem Soc Perkin I: 2261–2264

Butler AR, Askew SC (1993) The vascular action of S-nitroglutathione: evidence for NO transfer. Endothelium 1: 144A

Byler DM, Gosser DK, Susi H (1983) Spectroscopic estimation of the extent of S-nitrosothiol formation by nitrite action on sulfhydryl groups. J Agric Food Chem 31: 523–527

Castellani AG, Niven CF (1985) Factors affecting the bacteriostatic action of sodium nitrite. Appl Microbiol 3: 154–159

Clancy RM, Piziak-Leszcynska J, Abramson SB (1993) Nitric oxide stimulates ADP-ribosylation of action in human neutrophils. Biochem Biophys Res Commun 191: 847–852

Clancy RM, Yegudin J, Levartovsky D, Piziak-Leszcynska J, Abramson SB (1994) Nitric oxide reacts with intracellular glutathione and activates the hexose monophosphate shunt in human neutrophils: evidence for S-nitrosoglutathione as a bioactive intermediary. Pro Natl Acad Sci USA 91: 3680–3684

Chong S, Fung HI (1990) Thiol-mediated catalysis of nitroglycerin degradation by serum proteins. Drug Metab Dispos 18: 61–67

Chong S, Fung H (1991) Biochemical and pharmacological interactions between nitroglycerin and thiols. Biochem Pharmacol 42: 1433–1439

Cook JP, Stamler JS, Andon N, Davies PF, McKinley G, Loscalzo J (1990) Flow Stimulation endothelial cells to release a nitrosovasodilator that is potentiated by reduced thiol. Am J Physiol 28: H804–H812

Dimmler S, Brune B (1992) Characterization of a nitric oxide catalysed ADP ribosylation of glyceraldehyde-3-phosphate dehydrogenase. Eur J Biochem 1202: 305–310

Feelisch M (1991) The biochemical pathways of nitric oxide formation from nitrosovasodilators; Appropriate choice of exogenous NO donors and aspects of preparation and handling of aqueous NO solutions. J Cardiovasc Pharmacol 17 Suppl 3: S25–S33

Feelisch M, Noack E (1987) Nitric oxide formation from nitrovasodilators occurs independently of hemoglobin or non-heme iron. Eur J Pharmacol 142: 465–469

Field L. Dilts RV, Ravichandran R, Lenhert PG, Carnahan GE (1978) An unusually stable thionitrite from N-acetyl-D, L-penicillamine; X-ray crystal and molecular structure of 2-(acetylamino)-2-carboxy-1, 1-dimethylethyl thionitrite. J Chem Soc Chem Comm: 249–250

Fung HL, Chong S, Kowaluk E, Hough K, Kakemi M (1988) Mechanisms for the pharmacologic interaction of organic nitrates with thiols. Existence of an extracellular pathway for the reversal of nitrate vascular tolerance by *N*-acetylcysteine. J Pharmacol Exp Ther 245: 524–530

Gaston B, Reilly J, Drazen JM, Fackler J, Ramdev P, Arnelle D, Mullins M, Sugarbaker DJ, Chee C, Singel DJ. Loscalzo J, Stamler JS (1993) Endogenous nitrogen oxides and bronchodilator S-nitrosothiols in human airways. Proc Natl Acad Sci USA 90: 10957–10961

Gaston B, Drazen JM, Loscalzo J, Stamler JS (1994a) The biology of nitrogen oxides in the airways. State-of-the-Art. Am Rev Respir Dis 149: 538–551

Gaston B, Drazen JM, Jansen A, Sugarbaker DJ, Loscalzo J, Richards W, Stamler JS (1994b) Relaxation of human bronchial smooth muscle by S-nitrosothiols in vitro. J Pharmacol Exp Ther 268: 978–984

Gruetter CA, Gruetter DY, Lyon JE, Kadowitz PJ, Ignarro LJ (1981) Relationship between cyclic guanosine 3':5'-monophosphate formation and relaxation of coronary arterial smooth muscle by glycerol trinitrate, nitroprusside, nitrite and nitric oxide: effects of methylene blue and methemoglobin. J Pharmacol Exp Ther 219: 181–186

Han J, Stamler JS, Griffith O (1994) Inhibition of γ-glutamylcysteine synthetase by nitirc oxide donors. FASEB J 8: A1288

Hibbs JB (1991) Overview of cytotoxic mechanisms and defense of the intracellular environment against microbes. The biology of Nitric oxide II. Portland, Chapel Hills, pp 201–206

Incze K, Parkes J, Mihalyi V, Zukal E (1974) Antibacterial effect of cysteine-nitrosothiol and possible precursors thereof. Appl Microbiol 27: 202–205

Ignarro LJ (1989) Biological actions and properties of endothelium-derived nitric oxide formed and released from artery and vein. Circ Res 65:1–21

Ignarro L, Gruetter CA (1980) Requirement of thiols for activation of coronary arterial guanylate cyclase by glycerol trinitrate and sodium nitrite. Biochim Biophys Acta 631: 221–231

Ignarro LJ, Edwards JC, Gruetter DY, Barry BK, Gruetter CA (1980) Possible involvement of S-nitrosothiols in the activation of guanylate cyclase by nitroso compounds. FEBS Lett 110: 275–278

Iganarro LJ, Lippton H, Edwards JC, Baricos WH, Hyman AL, Kadowitz PH, Gruetter CA (1981) Mechanism of vascular smooth muscle relaxation by organic nitrates, nitrites, nitroprusside and nitric oxide: Evidence for the involvement of S-nitrosothiols as active intermediates. J Pharmacol Exp Ther 218: 739–749

Jansen A, Drazen J, Osborne JA, Brown R, Loscalzo J, Stamler JS (1992) The relaxant properties in guinea pig airways of S-nitrosothiols. J Pharmacol Exp Ther 261: 154–160

Jocelyn PC (1972) In: Biochemistry of the SH group. Academic, London

Johnson MA, Loynes R (1971) Inhibition of *Clostridium botulinum* by sodium nitrite in a bacteriologic medium and in meat. Can Inst Food Technol J 4: 179–184

Josephy PD, Rehorek D, Janzen EG (1984) Electron spin resonance spin trapping of thiyl radicals from the decomposition of thionitrites.Tetrahedron Lett 25: 1685–1688

Kamisaki Y, Waldman SA, Murad F (1986) The involvement of catalytic site thiol groups in the activation of soluble guanylate cyclase by sodium nitroprusside. Arch Biochem Biophys 251:709–714

Kanner J (1979) S-nitrosocysteine (RSNO), and effective antioxidant in cured meat. J Am Oil Chem Soc 56: 74–76

Keaney JF, Simon DI, Stamler JS, Jaraki O, Scharfstein J, Vita JA, Loscalzo J (1993) NO forms an adduct with serum albumin that has endothelium-derived relaxing factro-like properties. J Clin Invest 91: 1582–1589

Kerr SW, Buchanan LV, Bunting S, Mathews WR (1992) Evidence that S-nitrosothiols are responsible for the smooth muscle relaxing activity of the bovine retractor penis inhibitory factor. J Pharmacol Exp Ther 263:285–263

Kowaluk EA, Fung HL (1990) Spontaneous liberation of nitric oxide cannot account for in vitro vascular relaxation by S-nitrosothiols. J Pharmacol Exp Ther 255: 1256–1264

Lecher H, Siefken W (1926) Nitrosyl-derivate des zweiwertigen Schwefels, I: Das Nitrosylethylmercaptid. Ber Dtsch Chem Ges 59B: 1314–1326

Lei SZ, Pan ZH, Aggarwal SK, Chen HSV, Hartman J, Sucher NJ, Lipton SA (1992) Effect of nitric oxide production on the redox modulatory site of the NMDA receptor-channel complex. Neuron 8: 1087–1099

Lipton SA, Choi YB, Pan ZH, Lei SZ, Vincent Chen HS, Sucher NJ, Loscalzo J, Singel DJ, Stamler JS (1993) A redox-based mechanism for the neuroprotective and neurodestructive effects of nitric oxide and related nitroso-compounds. Nature 364: 626–632

Loscalzo J (1985) N-acetylcysteine potentiates inhibition of platelet aggregation by nitroglycerin. J Clin Invest 76: 703–708

Manzoni O, Prezeau L, Marin P, Deshager S, Bockaert J, Fagni L (1992) Nitric oxide-induced blockade of NMDA receptors. Neuron 8: 653–662

Mason J (1969) Trithioromethyl thionitrite. J Chem Soc A 1587–1592

Massey RC, Crews C, Davies R, McWeeny DJ (1978) A study of the competitive nitrosations of pyrrolidine, ascorbic acid, cysteine and *p*-Cresol in a protein-based model system. J Sci Food Agric 29: 815–821

McNainly J, Williams DLH (1993) Fate of nitric oxide from the decomposition of S-nitrosothiols. Endothelium 1: 141A

Mirna A, Hofmann K (1969) Uber den verbleib von Nitrit in fleischwaren. 1. Umsetzung von Nitrit mit sulhydryl verbindungen. Fleischwirtschaft 10: 1361–1364

Molina Y Vedia L, Mcdonald B, Reep B, Brune B, DiSilvio M, Billiar TR, Lapentina EG (1992) Nitric-oxide-induced S-nitrosylation of glyceraldehyde-3-phosphate dehydrogenase inhibits enzymatic activity and increases endogenous ADP ribosylation. J Biol Chem. 267: 24929–24932

Mohr S, Stamler JS, Brune B (1994) Mechanism of covalent modification of glyceraldehyde-3-phosphate dehydrogenase at its active site thiol by nitric oxide, peroxynitrite and related nitrosating agents. FEBS Letters 348: 223–227

Morris SL, Hansen JN (1981) Inhibition of *Bacillus cereus* spore outgrowth by covalent modifications of a sulfhydryl group by nitrosothiol and iodoacetate. J Bacteriol 148: 465–471

Morris SL, Walsh RC, Hansen JN (1984) Identification and characterization of some bacterial membrane sulfhydryl groups which are targets of bacteriostatic and antibiotic action. J Biol Chem 259: 13590–13594

Myers PR, Minor RL, Guerra R, Bates JN, Harrison DG (1990) Vasorelaxant properties of the endothelium-derived relaxing factor more closely resemble S-nitrosocysteine than nitric oxide. Nature 345: 161–163

Niroomand F, Rossle R, Mulsch A, Bohme A (1989) Under anaerobic conditions, soluble guanylate cyclase is specifically stimulated by glutathione. Biochem Biophys Res Commun 161: 75–80.

Oae S, Shinhama K (1983) Organic thionitrites and related substances. In: Organic preparations and procedures. Organic Prep Proced 15: 165–198

Oae S, Fukushima D, Kim YH (1977) Novel method of activating thiols by their conversion into thionitries with dinitrogen tetroxide. J Chem Soc Chem Comm: 407–408

Oae S, Kim YH, Fukushima D, Shinhama K (1978) New syntheses of thionitrites and their chemical reactivities. J Chem Soc Perkin 1: 913–917

O'Leary V, Solberg M (1976) Effect of sodium nitrite on inhibition of intracellular thiol groups and on the activity of certain glycolytic enzymes in *Clostridium porringers*. Appl Environ Microbiol 31: 208–212

Park JW (1988) Reaction of S-nitrosoglutathione with sulfhydryl groups in protein. Biochem Biophys Res Commun 152: 916–920

Pryor WA, Church DF, Govinden CK, Crank G (1982) Oxidation of thiols by nitric oxide and nitrogen dioxide: synthetic utility and toxicological implications. J Org Chem 47: 156–159.

Radomski MW, Rees DD, Durta A, Moncada S (1993) S-nitroso-glutathione inhibits platelet activation in vitro and in vivo. Br J Pharmacol, 107: 745–749

Ramdev P, Loscalzo J, Feelisch M, Stamler JS (1993) Biochemical properties and bioactivity of a physiologic NO reservoir. Circulation 88: 1–522

Rao PM, Copeck JA, Knight AR (1967) Reactions of thiyl radicals II. The photolysis of methyl disulfide vapor. Can J Chem 45: 1369–1374

Ribeiro JM, Hazzard JMH, Nussenzveig RH, Champagne DE, Walker FA (1993) Reversible binding of nitric oxide by a salivary heme protein from a bloodsucking insect. Science 260: 539–541

Ridd J (1978) Diffusion control and pre-association in nitrosation, nitration and halogenation. Adv Phys Organ Chem 16: 1–49

Rinden E, Maricq MM, Grabowski JJ (1989) Gas-phase ion-molecule reactions of the nitric oxide anion. J Am Chem Soc II: 1203–1210

Rockett KA, Auburn MM, Lowden WB, Clark IA (1991) Killing of *Plasmodium falciparum* in vivo by nitric oxide derivatives. Infect Immun 59: 3280–3283

Schafer JE, Lee F, Thomson S, Han BJ, Cooke JR, Loscalzo J (1991) The hemodynamic effects of S-nitrosocaptopril in anesthetized dogs. J Pharmacol Exp Ther 256: 704–709

Scharfstein JS, Keaney J, Stamler JS, Vita J, Loscalzo (1993a) Low molecular weight thiols transfer nitric oxide from an endogenous plasma reservoir to vascular smooth muscle. Clin Res 41: 232A

Scharfstein JS, Slivka A, Stamler JS, Loscalzo J (1993b) In vivo transfer of nitric oxide from a plasma reservoir to cysteine. Circulation 88: 1–523

Simon DI, Stamler JS, Jaraki O, Keaney J, Osborne JA, Francis SA, Singel DJ, Loscalzo J (1993) Antiplatelet properties of protein S-nitrosothiols derived from nitric oxide and endothelium-derived relaxing factor. Arterioscler Thromb 13: 791–799

Slivka A, Loscalzo J, Chuttani R, Kobzik L, Bredt D, Carr-Locke DL, Stamler JS (1994) Inhibition of sphincter of oddi function by the nitric oxide carrier, *S*-nitroso-*N*-actylcysteine in rabbits and humans. J Clin Invest (in press)

Snyder SH (1993) Janus faces of nitric oxide. Nature 364: 577

Stamler JS, Loscalzo J (1991) The antithrombotic effects of organic nitrates. Trends Cardiovasc Med 1: 346–353

Stamler JS (1994) Redox Signaling: Nitrosylation and related target interactions of nitric oxide. Cell 78: 931–936

Stamler JS, Loscalzo J (1992) Capillary electrophoretic detection of thiols and their S-nitrosated derivatives. Anal Chem 64: 779–785

Stamler JS, Cunningham M, Loscalzo J (1988) Reduced thiols and the effect of nitroglycerin on platelet function. Am J Cardiol 62: 377–380

Stamler JS, Mendelsohn M, Amarante P, Davies PF, Cooke JP, Loscalzo J (1989) N-acetylcysteine potentiates platelet inhibition by endothelium derived relaxing factor. Circ Res 65: 789–795

Stamler JS, Simon DI, Osborne JA, Mullins ME, Jaraki O, Michel T, Singel DJ, Loscalzo J (1992a) S-nitrosylation of proteins with nitric oxide: synthesis and characterization of biologically active compounds. Proc Natl Acad Sci USA 89: 444–448

Stamler JS, Jaraki O, Osborne J, Simon DI, Keaney J, Vita J, Singel D, Valeria RC (1992b) Nitric oxide circulates in mammalian plasma primarily as an S-nitroso adduct of serum albumin. Proc Natl Acad Sci USA 89: 7674–7677

Stamler JS, Singel D, Loscalzo J (1992c) Biochemistry of nitric oxide and its redox activated forms. Science 258: 1898–1902

Stamler JS, Simon DI, Jaraki O, Osborne JA, Francis J, Mullins M, Singel D, Loscalzo (1992d) S-nitrosylation of tissue-type plasminogen activator confers vasodilatory and antiplatelet properties on the enzyme. Proc Natl Acad Sci USA 89: 8087–8091

Stamler JS, Simon DI, Osborne JA, Mullins M, Jaraki O, Michel T, Singel D, Loscalzo J (1992e) Exposure of sulfhydryl containing proteins to nitric oxide and endothelium-derived relaxing factor confers novel bioactivity and modulates their intrinsic functional properties. In: Moncada S, Marletta MA, Higgs A, Hibbs JB (eds) Biology of nitric oxide I. Portland Chapel Hill, pp 20–23

Stamler JS, Osborne JA, Jaraki O, Rabbani LE, Mullins M, Singel D, Loscalzo J (1993) Adverse effects of homocysteine are modulated by endothelium-derived relaxing factor and related oxides of nitrogen. J Clin Invest 1: 308–318

Starkebaum G, Harlan JM (1986) Endothelial cell injury due to copper-catalyzed hydrogen peroxide generation from homocysteine. J Clin Invest 77: 1370–1376

Tasker HS, Jones HQ (1909) The action of mercaptans on acid chlorides, part II. The acid chlorides of phosphorous, sulfur and nitrogen. J Chem Soc 95: 1910

Turk T, Hollocher TC (1992) Oxidation of dithiothreitol during turnover of nitric oxide reductase: evidence for generation of nitroxyl with the enzyme from paracoccus denitrificans. Biochem Biophys Res Commun 183: 983–988

Williams DHL (1988) Nitrosation. Cambridge University Press, Cambridge

Wink DA, Kasprzak KS, Maragos CM, Elespuru RK, Misra M, Dunams TM, Cebula TA, Koch WH, Andrews AW, Allen S, Keefer LK (1991) DNA deaminating ability and genotoxicity of nitric oxide and its progenitors. Science 254: 1001–1003

Wink DA, Darbyshire JF, Nims RW, Saavedra JE, Ford PC (1993) Reactions of bioregulatory agent nitric oxide in oxygenated aqueous media: determination of the kinetics of oxidation and nitrosation by intermediates generated in the NO/O_2 reaction. Chem Res Toxicol 6: 23–27

World Health Organization (1977) Environmental health criteria 4 oxides of nitrogen. World Health Organization, Geneva

Wu M, Kaminski PM, Fayngersh RP, Groszek LL, Pritchard KA, Hintze TH, Stemerman MB, Wolin MS (1994) Involvement of nitric oxide and nitrosothiols in relaxation of pulmonary arteries to peroxynitrite. Am J Physiol 266: H2108–H2113

Yeates RA, Laufen H, Leitold M (1985) The reaction between organic nitrates and sulfhydryl compounds: a possible model system for the activation of organic nitrates. Mol Pharmacol 28: 555–559

A New Nitric Oxide Scavenger, Imidazolineoxyl *N*-Oxide Derivative, and Its Effects in Pathophysiology and Microbiology

H. Maeda[1], T. Akaike [1], M. Yoshida [2], K. Sato[1], and Y. Noguchi [1]

1 Introduction

The increased appreciation of nitric oxide (NO) as an important molecule in biology and medicine has been accompanied by an unprecedented interest in inhibitors of NO synthesis and NO scavengers. L-Arginine analogues (Fig. 1) have been used widely as competitive inhibitors of NO synthase (NOS), thus inhibiting the function of NO indirectly. Some of these analogues have even been used in human trials, although there are potential problems with in vivo use. L-Arginine analogues may interfere with the urea cycle and protein synthesis, and long-term suppression of both the constitutive (cNOS) and the induced forms (iNOS) of the enzyme might result in impairment of the neural and circulatory systems. Although L-arginine analogues can suppress the biological effect of endothelium-derived relaxing factor (EDRF) stimulators, such as acetylcholine, proof that NO is an EDRF awaits the demonstration of chemical entrapment of NO at the site of action with simultaneous disappearance of its vasorelaxing activity. Meanwhile NO gas or dilute NO solutions have been shown to relax vascular tone, and the identity of EDRF with NO appears essentially unambiguous.

[1] Department of Microbiology, Kumamoto University School of Medicine, Honjo 2-2-1, Kumamoto 860, Japan

[2] Department of Urology, Kumamoto University School of Medicine, Honjo 2-2-1, Kumamoto 860, Japan

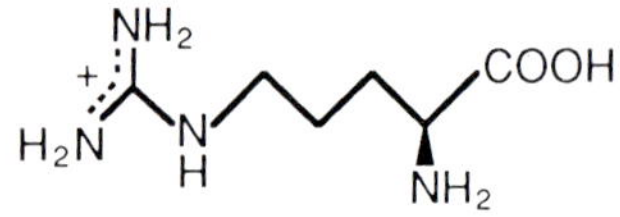

L-Arginine

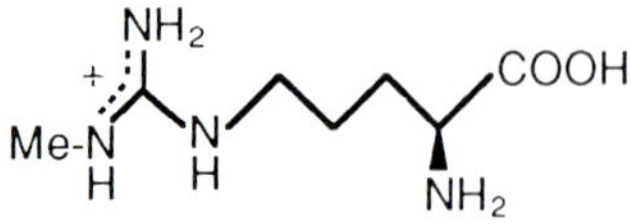

N^{ω}-Monomethyl-L-arginine (L-NMMA)

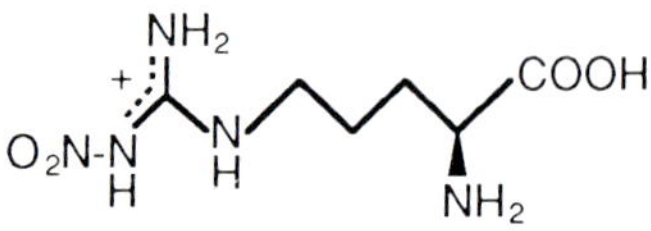

N^{ω}-Nitro-L-arginine (L-NNA)

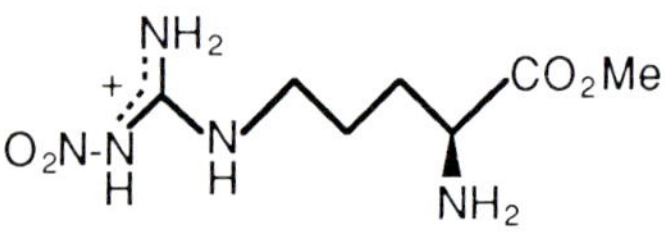

N^{ω}-Nitro-L-arginine methylester (L-NAME)

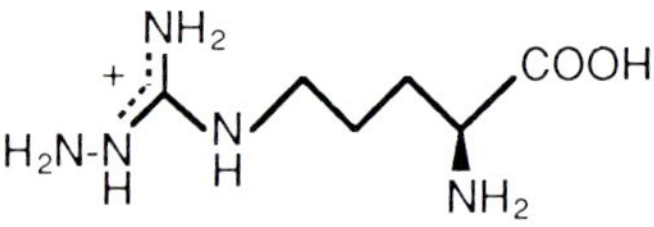

N^{ω}-Amino-L-arginine (L-NAA)

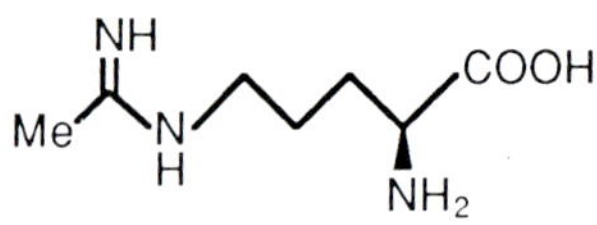

L-N^{5}-(1-Iminoethyl)-ornithine (L-NIO)

Fig. 1. Structure of L-arginine and various L-arginine analogues, that inhibit nitric oxide synthase

We have recently described an imidazolineoxyl *N*-oxide (PTIO) compound that selectively reacts with NO by converting it to NO_2, while simultaneously and completely abrogating the biological effect of NO. This prompted a detailed investigation of the chemical mechanism by which NO interacts with PTIO, focusing on unique electron spin resonance (ESR) signals, and further functional analysis of NO in vitro and in vivo. We investigated two pathological conditions, endotoxic shock and vascular permeability enhancement in solid tumors, in which NO seems to be involved. Both of these pathological situations are clearly inhibited by PTIO, suggesting the direct causal role of NO and the potential usefulness as a therapeutic strategy based on the administration of PTIO. We have also studied the controversial cytotoxic effect of NO using the bacteria *Staphyloccus aureus* and *Cryptococcus neoformans*.

The biological function of NO has been analyzed primarily through its inhibition of NOS, which has been achieved using glucocorticoids to inhibit NOS induction, (MONCADA et al. 1992), trifluoperazine to inhibit calmodulin-dependent function (for neuronal and endothelial cNOS) (BREDT and SNYDER 1990), and diphenylene iodonium, di-2-thienyliodonium, or iodoniumdiphenyl to inhibit NADPH, an essential electron donor for NOS (STUEHR et al. 1991). However since NADPH is required in many biochemical reactions, such inhibitors have only limited value in an in vivo setting (GATLEY and SHERRATT 1976). Compounds such as 2,4-diamino-6-hydroxyprimidine and *N*-acetylserotonin, which inhibit biosynthesis of tetrahydrobiopterin, a cofactor of NOS, might selectively inhibit iNOS activity in vivo (GROSS and LEVI 1992; SAKAI et al. 1993), but all of these reagents have potential drawbacks.

Other compounds such as SH- or iron-containing compounds, including glutathione (STAMLER et al. 1992), albumin or hemoglobin (KANNER et al. 1992), might compete with the target molecules of NO or serve as an NO carrier in biological phenomena in which NO is crucially involved. This possibility is discussed in the chapter by Stamler. Here, we focus on PTIO derivatives and functional analyses of NO using PTIOs in vitro, ex vivo, and in vivo.

2 Selective Reaction of Imidazolineoxyl *N*-Oxide Derivatives with Nitric Oxide

Figure 2 shows the chemical structure of PTIO and the proposed reaction equation, in which R can be replaced with various groups. Synthesis of these PTIO derivatives has been described (AKAIKE et al. 1993). The derivatives 2-phenyl-4,4,5,5-tetramethylimidazoline-1-oxy-3-oxide (PTIO), carboxy-PTIO and carboxymethoxy-PTIO are the most well studied. Figure 3 shows changes in the electron spin resonance (ESR) spectra obtained from PTIO alone before reaction with NO, the purified reaction product, i.e. 2-phenyl-4,4,5,5-tetramethyl-imidazoline-1-oxyl (PTI), and the reaction mixture of PTIO and NO. Computer simulation resolves the latter spectrum into two components similar to the spectra in (a) and (b). Thin-layer chromatography revealed a single spot of PTI after complete reaction of PTIO and NO, with a different mobility from PTIO. The Rf value of the PTI product was 0.48, with a yellow color, whereas Rf of the original PTIO was ~0.30, with a blue color as determined in a *n*-hexane/chloroform/methanol system (10:5:1, v/v). The amount of PTI formed was equal to that of PTIO consumed, and NO reacted with PTIO stoichiometrically under anaerobic

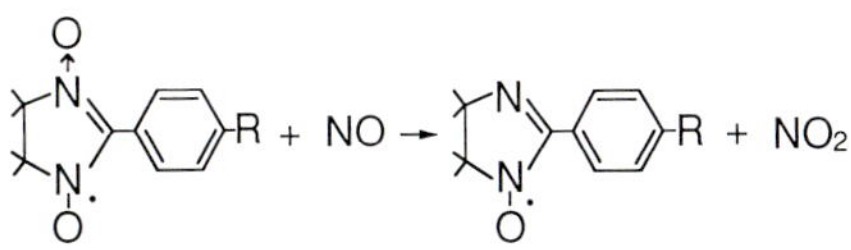

Fig. 2. Structures of imidazolineoxyl *N*-oxide (*PTIO*) derivatives (left) and reaction scheme of PTIOs with nitric oxide (*NO*) (*R*,H:2-phenyl-4,4,5,5-tetramethyl-imidazoline-1-oxyl-3-oxide (PTIO); *R*,COOH: carboxy-PTIO; *R*,OCH_2CHOOH, carboxymethoxy (CM)-PTIO)

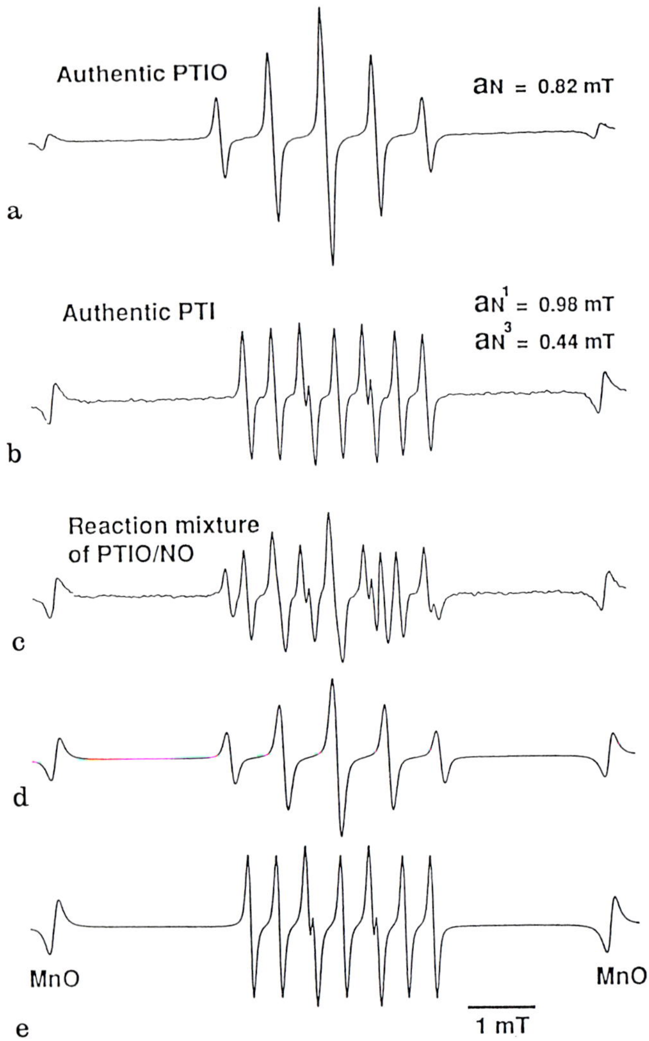

Fig. 3. Electron spin resonance spectra of imidazolineoxyl *N*-oxide (*PTIO*) reaction product (*PTI*) and reaction mixture of PTIO and nitric oxide (*NO*). **a**, PTIO; **b**, PTI; **c**, spectrum of the reaction mixture of PTIO and NO; **d** and **e**, computer simulation of **c**. (From AKAIKE et al. 1993, with permission)

condition. The results of mass spectroscopy and NMR studies of PTIO and PTI were consistent with the reaction scheme (Fig. 2). The reaction can be quantified either by ESR signal or by spectroscopy of PTIO (λ_{max} 358 nm, ε_{mol} 6130; λ_{max} 562 nm, ε_{mol} 760) and PTI (λ_{max} 432 nm, ε_{mol} 585) (AKAIKE et al. 1993; and our unpublished data). The minimum sensitivity of ESR spectroscopy for NO detection is ~ 0.1 μM.

3 Generation of Nitric Oxide from Rabbit Aorta by Stimulation with Vascular Relaxing Agents and Inhibition of Vascular Relaxation with Imidazolineoxyl *N*-Oxide: Ex Vivo Study

We investigated whether PTIOs react with NO to inhibit its biological function as an EDRF in aortic smooth muscle from rabbits. Stimulation by vascular relaxing agents such as acetylcholine or ATP is known to mediate generation of EDRF, i.e., NO, during this vasorelaxation. Female New Zealand white rabbits weighing 2.5-3 kg, were anesthetized with sodium pentobarbiturate and exsanguinated. The thoracic aorta was removed from the chest and, after removal of excess fat and connective tissue, cut into 5 mm wide rings. The rings were mounted vertically in a 20 ml organ bath filled with Krebs solution, and isometric tension development was recorded with an ink-writing recorder. The medium was maintained at 37 °C and aerated with a mixture of 95% O_2 and 5% CO_2. Tissues were precontracted with 0.15 mM phenylephrine, and acetylcholine- or ATP-induced relaxation was measured, as were the effects of a series of PTIOs, PTI, N^{ω}-nitro-L-arginine (L-NNA), and N^{ω}-monomethyl-L-arginine (L-NMMA) on the induced relaxation. The vascular tone of aortic rings was studied in the presence of 3.0 m*M* indomethacin after phenylephrine-induced contraction of the smooth muscle.

Evidence of NO generation from the vascular rings was obtained using ESR spectroscopy to quantify PTI formed from the reaction with PTIO during stimulation with acetylcholine in the incubation medium. Generation of NO was inhibited in the presence of the NOS inhibitor, L-NMMA (Fig. 4) (see AKAIKE et al. 1993 for details). All three PTIOs used inhibited this NO-related vasorelaxation induced by acetylcholine in a dose-dependent manner (Fig. 5). Similarly, vasorelaxation induced with ATP was also markedly inhibited by all PTIOs (data not shown). In contrast, virtually no inhibitory action was observed against

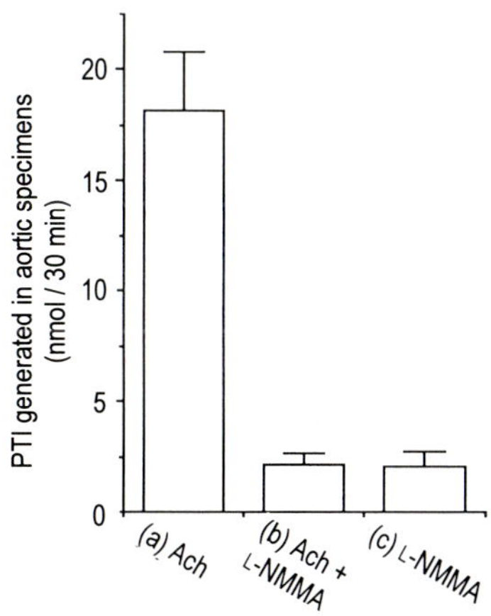

Fig. 4. Conversion of imidazolineoxyl *N*-oxide (*PTIO*) to reaction product (*PTI*) during vasorelaxation induced by acetylcholine (*Ach*). Aortic specimens, 5cm long, from rabbits were incubated with 0.5n*M* PTIO and Cu, Zn superoxide dismutase (SOD) (500 units/ml) in the presence or absence of 10 n*M* Ach or 1 mM (*L-NMMA*) in Krebs solution (1.5 ml) for 30 min at 37°C. Error bars indicate mean ±SEM. (From AKAIKE et al. 1993, with permission)

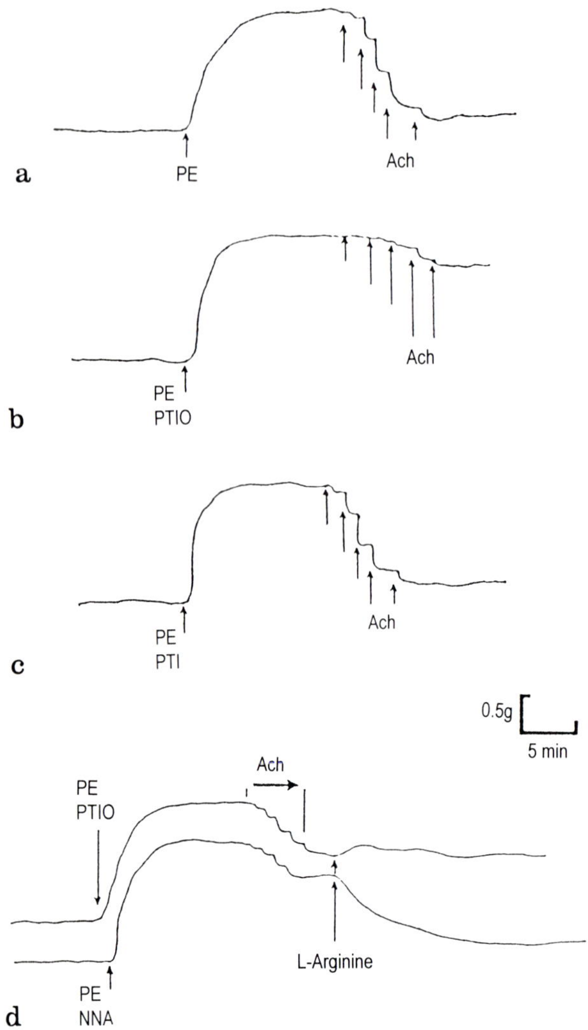

Fig. 5a–d. Effects of imidazolineoxyl *N*-oxide (*PTIO*) on acetylcholine (*Ach*) induced vasorelaxation of rabbit thoracic aorta. Tissues were precontracted with 0.15 n*M* phenylephrine after which Ach was added serially at concentrations of 0.001, 0.03, 0.1, 0.3 and 1.0 m*M* with or without PTIO and *PTI*. **a** Control without PTIO; **b** with 300 μ*M* PTIO; **c** with 300 μ*M* PTI; **d** effect of L-arginine (5 m*M*) on inhibition by 100 μ*M* PTIO (*upper tracing*) or 100 μ*M* N^{ω}-nitro-L-arginine (*NNA*) (*lower tracing*) of Ach-induced vasorelaxation. (From AKAIKE et al. 1993, with permission)

vasorelaxation with PTI, the reaction product of PTIO and NO. The inhibition of the smooth muscle relaxation by PTIOs was not abolished by the addition of L-arginine, the precursor of NO biosynthesis, whereas L-arginine significantly reversed the inhibition of vasorelaxation by L-NNA (Fig. 6). These results indicate that the inhibition of vascular relaxation by PTIOs was not due to direct inhibition of NOS. More specifically, L-arginine analogues such as L-NNA and L-NMMA

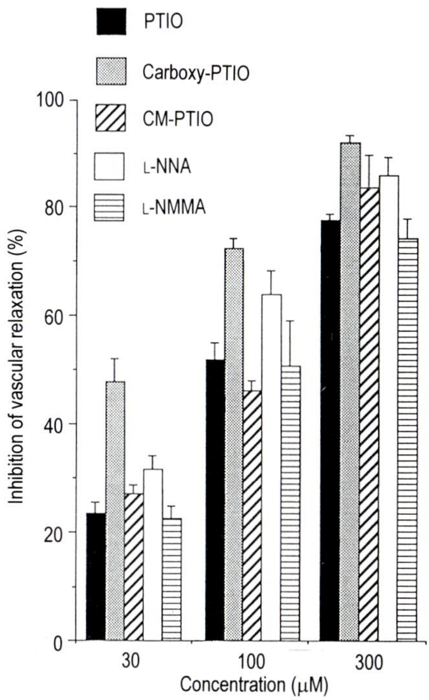

Fig. 6. Effects of concentration of various imidazoline-*N*-oxide (*PTIO*) derivatives on vascular relaxation of rabbit thoracic aorta. Bioassay conditions were the same as in Fig.5. Data are shown as mean ± SE (n = 3-4). (From AKAIKE et al. 1993, with permission)

are competitive inhibitors of NOS and thus the presence of L-arginine can reverse these inhibitory effects (KILBOURN and GRIFFITH 1992). In contrast, PTIO reacts directly with EDRF (NO) generated by cNOS in the endothelium. EDRF-dependent relaxation could be reinduced when acetylcholine was again administered to aortic tissues pretreated with imidazoline compounds for up to 4 hours after stimulation by acetylcholine or ATP, followed by washing out of these reagents. This observation suggests that the inhibitory effect of imidazolineoxyl *N*-oxide radicals on vasorelaxation did not result from cytotoxicity against endothelial cells of the aorta. The inhibitory potential of PTIOs is comparable to that of L-NMMA and L-NNA (Fig. 6).

4 Therapeutic Effect of Carboxy PTIO on Lipopolysaccharide-Induced Hypotension and Endotoxic Shock, In Vivo

We tested the effect of a water-soluble derivative of PTIO, i.e. carboxy-PTIO, on endotoxic shock rats (Wistar male rats weighing 220–250 g) (YOSHIDA et al. 1994). The lower abdominal aorta was cannulated after abdominal incision to measure mean arterial blood pressure (MABP) and heart rate of the rats, and blood samples were obtained intermittently for ESR analyses to measure the level of carboxy-PTIO and carboxy-PTI. Rats were given bolus intravenous injections of lipopolysaccharide (LPS), *E. coli* 026:B6 (Difco Laboratories, Detroit, MI). Blood pressure and heart rate were measured by a pressure transducer with a recorder unit. Soon after injection of LPS, at a dose of 10 mg/kg, MABP began to decline (Figs. 7, 8). After 3 h, it was lower than 60 mm Hg, and rats started to die. In contrast, when carboxy-PTIO was slowly infused via the left jugular vein, beginning at 90 min at 6 ml/h (1.7 mg/min per kg) and continuing for 60 min the decrease in MABP was inhibited at a carboxy-PTIO dose of more than

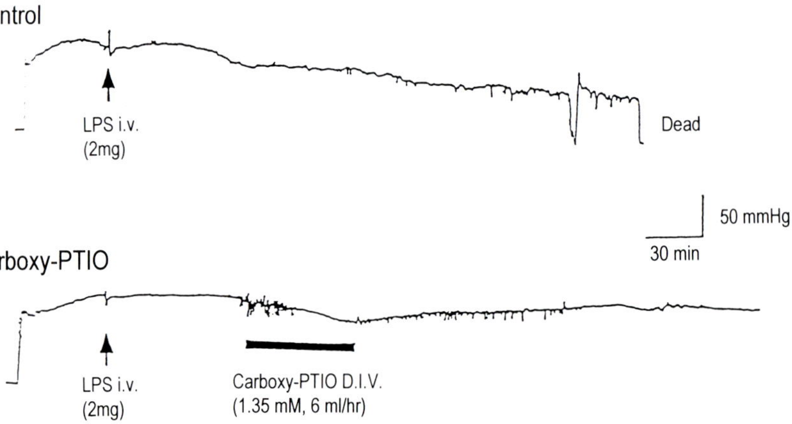

Fig. 7. Effects of carboxy-immidazolineoxyl *N*-oxide (*PTIO*) on lipopolysaccharide (*LPS*)-induced hypotension in rats

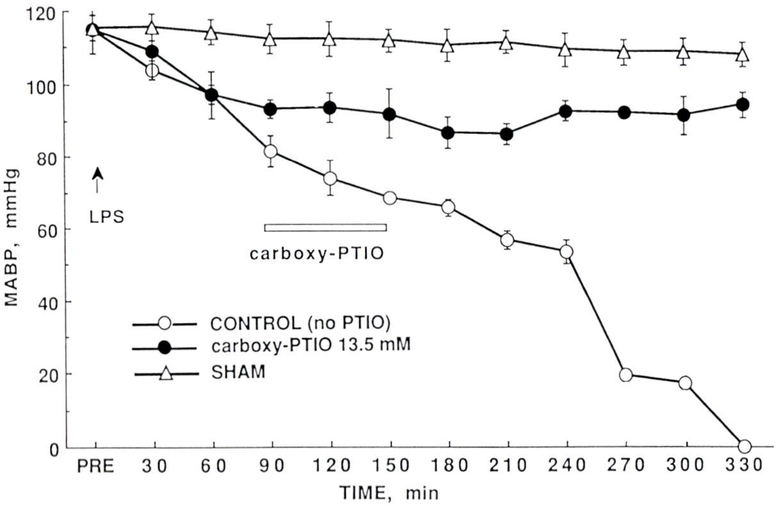

Fig. 8. Therapeutic effect of carboxy-imidazolineoxyl *N*-oxide (*PTIO*) in rats with lipopolysaccharide induced endotoxic shock. MABP, mean arterial blood pressure

0.056 mg/kg per min. At a dose of 0.17 mg/kg per min or higher, carboxy-PTIO infusion completely rescued the endotoxic shock rats.

ESR spectroscopy indicated decreased blood levels of carboxy-PTIO and increased levels of carboxy-PTI in the endotoxic shock rats (not shown). When the infusion was stopped, plasma levels of carboxy-PTIO declined gradually, and both carboxy-PTIO and -PTI were rapidly excreted in the urine (not shown). Total urinary recovery of carboxy-PTI and -PTIO was about 81% in 3 h (not shown).

These results indicate that NO is responsible for the hypotension induced by endotoxin and that carboxy-PTIO has potential therapeutic value against endotoxic shock in animals.

5 Vascular Permeability of Solid Tumors Mediated by Nitric Oxide and Inhibition by Imidazolineoxyl *N*-Oxide

Tumor tissues have unique vascular characteristics, including extensive angiogenesis (FOLKMAN 1990), which usually results in hypervasculture, irregular morphology and defective architecture, and enhanced permeability (MATSUMURA and MAEDA 1986; SUZUKI et al. 1987; MAEDA and MATSUMURA 1989; SKINNER et al. 1990; MAEDA 1991; MAEDA et al. 1992). This enchanced permeability is relevant in the selective delivery of macromolecules and lipids to tumor tissues, since such tissue exhibits much slower return of these substances into the lymphatics (IWAI et al. 1984; MATSUMURA and MAEDA 1986; MAEDA 1991; MAEDA et al. 1992). We have termed this phenomenon the "enhanced permeability and retention" (EPR) effect of solid tumors (MATSUMURA and MAEDA 1986; MAEDA and MATSUMURA 1989; MAEDA 1991; MAEDA et al. 1992).

The enhanced vascular permeability of solid tumors, which facilitates tumor growth and perhaps metastasis, is now known to be mediated by a number of factors, including tumor vascular permeability factor (SENGER et al. 1983), bradykinin and hydroxyprolyl^3 bradykinin (MAEDA et al. 1988; MATSUMURA et al. 1988, 1991), tumor necrosis factor and interleukin-2 (ETTINGHAM et al. 1988), and others. However, there appears to be no report on the effect of NO on vascular permeability of tumor tissues. Here we present evidence that vascular permeability is indeed mediated by NO (MAEDA et al. 1994).

Figure 9 shows the levels of Evans blue dye relased from sarcona 180 (S-180) solid tumors in ddY mice in the presence or absence of PTIO. PTIO in an oil formulation (0.5 ml, 43 mM) was administered orally and plasma concentrations were determined by ESR spectroscopy. Since plasma levels can be maintained for about 2.5 h, PTIO was administered every 2 h, and Evans blue (0.2% in physiological saline, 0.2 ml) was injected intravenously into tumor-bearing mice. Tumors were more than 3 mm but less than 7 mm in diameter. The results clearly show significant suppression of extravasation by PTIO in two different size groups (Fig. 9).

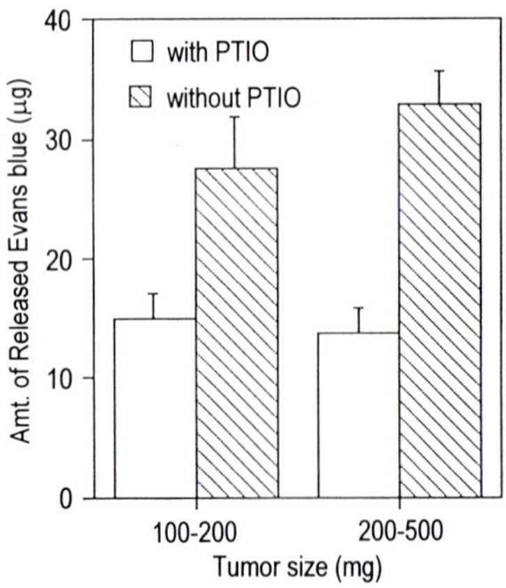

Fig. 9. Inhibitory effect of imidazolineoxyl *N*-oxide (*PTIO*) on vascular permeability in solid tumors. *Bar* indicate SD (n = 6-8)

We also tested the extravasating effect of NO in guinea pig skin, in which 0.05 ml of an oil formulation of NO diluted to different concentrations was injected intradermally in animals that had already received Evans blue intravenously. Dye extraction was performed with formamide at 60 °C for 48 h, followed by spectroscopic quantification (620 n*M*). The results showed significant extravasation by NO, although the potency of NO is 1/100–1/1000 of bradykinin (not shown). Extravasation caused by bradykinin was effectively suppressed by PTIO given intraperitoneally (MAEDA et al. 1994).

Data suggest that bradykinin stimulates the release of EDRF (NO) (PALMER et al. 1987; KELM et al. 1988; MYERS et al. 1992; TADJKARIMI et al. 1992). Our data show that NO plays a crucial role in extravasation in tumor tissue, and bradykinin may be an initiating trigger of this event, as discussed previously (MATSUMURA et al. 1988, 1991; MAEDA et al. 1988; MAEDA and MATSUMURA 1989). Further experiments will clarify whether the level of iNOS and for cNOS is elevated in tumor tissue.

6 Microbiocidal Activity of Nitric Oxide Using Imidazolineoxyl *N*-Oxide

It has been suggested that NO plays an important role in the killing of invading bacteria by macrophages and polymorphonuclear cells (PMR) since their NO generation parallels that of oxygen radicals and cell killing (HIBBS et al. 1987; DRAPIER 1991; STUEHR and NATHAN 1991). However, it remains controversial whether NO itself can kill bacteria. We therefore, investigated the effect of NO on *Cryptococcus neoformans* and *Staphylococcus aureus* 209P using PTIO (YOSHIDA et al. 1993).

Staphylococcus aureus was routinely cultured in tryptosoy broth overnight and washed twice with phosphate-buffered saline (PBS; pH 7.3). Cultures were

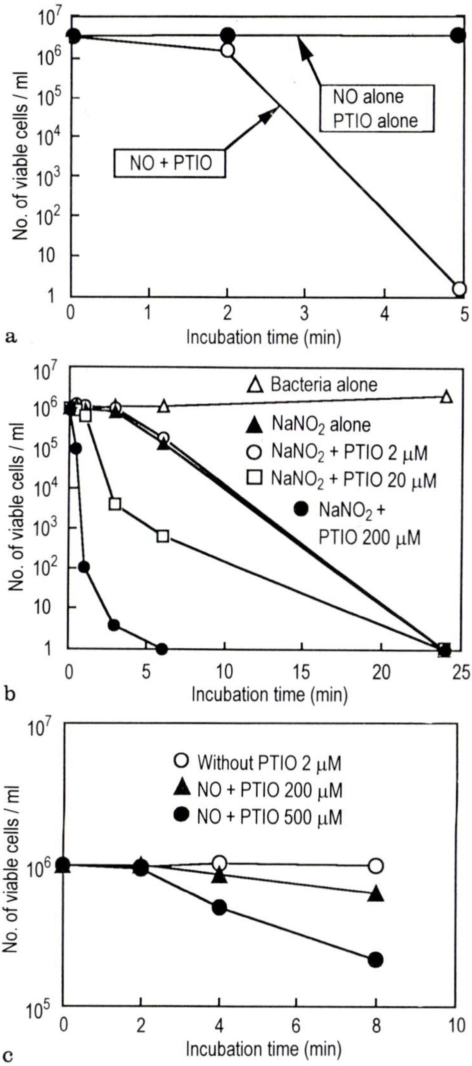

Fig. 10a–c. Effect of imidazolineoxyl *N*-oxide (*PTIO*) on nitric oxide (*NO*)-induced micro-biocidal action against **a** *Staphylococcus aureus* and *Cryptococcus neoformans*; **b,c** Bactericidal action of NO and PTIO. *S. aureus* (3 x 10^6 cells/ml) was reacted with NO (150 n*M*) in 0.01 *M* PBS (pH 7.3) in the presence or absence of 857 m*M* PTIO at room temperature. **b** Bactericidal effect of NO generated in a system forming nitrous acid. *C. neoformans* (serotype D) was added to a reaction mixture containing 1 m*M* $NaNO_2$ in 80 m*M* succinate buffer (pH 4.0) in the presence or absence of PTIO, followed by incubation for 24 h at room temperature. Number of viable fungal cells was quantitated by colony-forming assay on Sabouraud agar plates. Data represents the mean of triplicate plates. **c** Effects of authentic NO on *C. neoformans* serotype D in solutions at pH 4.0. NO (gas)-saturated solution was added to the suspended *C. neoformans* cells in 200 m*M* succinate buffer (pH 4.0) and the number of viable cells was measured as in **a**. No cell killing was seen at pH 7.0 and the A serotype showed a trend similar to that of the serotype D (YOSHIDA et al. 1993, with permission)

suspended in PBS and incubated with an aliquot of NO-saturated solution in the presence or absence of PTIO for different time periods (Fig. 10a). Cell suspensions (1 x 10^6 cells/ml) of *C. neoformans* were prepared in 0.15 *M* NaCl with varying concentrations of $NaNO_2$ and the suspensions were brought to pH 4.0 by the addition of succinic acid (Fig. 10b). At this pH, NO is known to be released from nitrite. *C. neoformans* was exposed to an NO solution (Fig. 10c) and the effect of PTIO was examined in the same manner as for *S. aureus*. In these assays PTIO alone did not show any cytotoxic activity. The number of viable cells in each reaction system was determined by colony-forming assay.

These results showed that NO alone did not exhibit significant microbiocidal

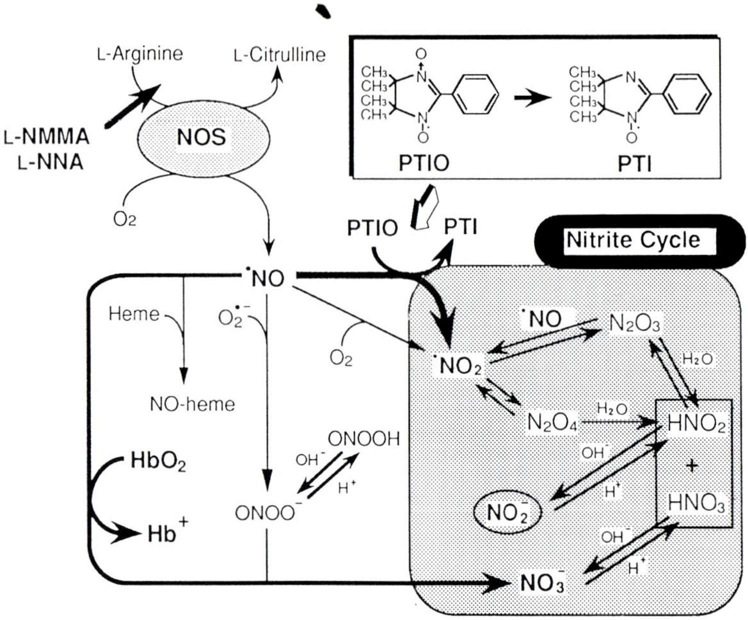

Fig. 11. Interaction of nitric oxide (*NO*) and imidazolineoxyl *N*-oxide (*PTIO*), and various NO intermediates in the nitrite cycle. (From YOSHIDA et al. 1993)

activity against either *S. aureus* or *C. neoformans*. Instead PTIO together with NO showed substantial cytotoxicity. Based on the known chemistry of NO and PTIO (Fig. 11), we speculate that one or more NO metabolites, such as NO_2 (N_2O_4) and N_2O_3, may be toxic but not NO per se. Another derivative, peroxynitrite [OONO], which is formed via the reaction of NO with O_2^- might be also a candidate responsible for NO-dependent microbiocidal action in vivo.

7 Concluding Remarks

The PTIO derivatives are a new class of NO scavenger. The chemistry of radical-radical reactions was fully demonstrated and the product analyses, including mass spectroscopy and NMR studies (not shown here), confirmed this notion (AKAIKE et al. 1993).

All arginine derivatives may have some drawbacks, especially in vivo. Heme derivatives are also known to react rapidly with NO, but they may also liberate NO and thus may not necessarily be a class of scavenger (RIBIEIRO et al. 1993). In this respect thiol-NO adducts may have the same problems as heme-NO adducts (STAMLER et al. 1992)

Identity between EDRF and NO was fully established in our study in vitro, ex vivo and in vivo systems using PTIO, and these results are consistent with those obtained with the L-arginine analogue L-NMMA, i.e., PTIO can prevent endotoxin-induced hypotension and shock and can rescue rats from lethal dose of endotoxin (YOSHIDA et al. 1994).

As was mentioned briefly, we found no real evidence that NO is microbiocidal; rather the reaction product(s) of NO and PTIO, e.g. NO_2 or N_2O_4, seems to exhibit more microbiocidal activity (YOSHIDA et al. 1993)

Using a solid tumor model in mice, we demonstrated that the vascular permeability enhancing activity of solid tumors appears to be mediated by NO, because PTIO formulated for slow release significantly suppressed the vascular permeability of Evans blue dye. In conjuction with this finding, PTIO also suppressed the vascular permeability enhancement induced by bradykinin in normal guinea pig skin (MAEDA et al. 1994). Our results indicate that the action of bradykinin on blood vessels is, at least in part, mediated by NO. PTIO will be a very useful reagent for the analysis of NO-mediated activities both in vitro and in vivo.

References

Akaike T, Yoshida M, Miyamoto Y, Sato K, Kohno M, Sasamoto K, Miyazaki K, Ueda S, Maeda H (1993) Antagonistic action of imidazolineoxyl N-oxides against endothelium-derived relaxing factor/. NO through a radical reaction. Biochemistry 32: 827–832

Bredt DS, Snyder SH (1990) Isolation of nitric oxide synthetase, a calmodulin-requiring enzyme. Proc Natl Acad Sci USA 87: 682–685

Drapier JC (1991) L-Arginine-derived nitric oxide and the cell mediated immune response. Res Immunol 142: 551–602

Ettinghausen SE, Puri RJ, Rosenberg SA (1988) Increased vascular permeability in organs mediated by the systematic administration of lymphokine-activated killer cells and recombinant interleukin-2 in mice. J Natl Cancer Inst, 80: 177–187

Folkman J (1990) What is the evidence that tumors are angiogenesis dependent? J Natl Cancer Inst 82:4–6

Gatley SJ, Sherratt HSA (1976) The effects of diphenyleneiodonium on mitochondrial reactions: Relation of binding of diphenylene [^{125}I] iodonium to mitochondria to the extent of inhibition of oxygen uptake. Biochem J 158: 307–315

Gross SS, Levi R (1992) Tetrahydrobiopterin synthesis: an absolute requirement for cytokine-induced nitric oxide generation by vascular smooth muscle. J Biol Chem 267: 25722–25729

Hibbs JB Jr, Taintor RR, Vartin Z (1987) Macrophage cytotoxicity: role for L-arginine deiminase and iminonitrogen oxidation to nitrite. Science 235: 473–476

Iwai K, Maeda H, Konno T (1984) Use of oily contrast medium for selective drug targeting to tumor: enhanced therapeutic effect and X-ray image. Cancer Res 44: 2114–2121

Kanner J, Harel S, Granit R (1992) Nitric oxide, an inhibitor of lipid oxidation by lipoxygenase, cyclooxygenase and hemoglobin. Lipids 27: 46–49

Kelm M, Feelisch M, Spahr R, Pipper M, Noack E, Schrader J (1988) Quantitative and kinetic characterization of nitric oxide and EDRF released from cultured endothelial cells. Biochem Biophys Res Commun 154: 236–244

Kilbourn RG, Griffith OW (1992) Overproduction of nitric oxide in cytokine-mediated and septic shock. J Natl Cancer Inst 84: 827–831

Maeda H, Matsumura Y (1989) Tumoritropic and lymphotropic principles of macromolecular drugs. Crit Rev Ther Drug Carrier Syst 6: 193–210

Maeda H (1991) SMANCS and polymer-conjugated macromolecular drugs: advantages in cancer chemotherapy. Adv Drug Deliv Rev 6: 181–202

Maeda H, Matsumura Y, Kato H (1988) Purification and identification of [hydroxyprolyl^3]-bradykinin in ascitic fluid from a patient with gastric cancer. J Biol Chem 263: 16051–16056

Maeda H, Seymour LW, Miyamoto Y (1992) Conjugates of anticancer agents and polymers: Advantages of macromolecular therapeutics in vivo. Bioconj Chem 3: 351–362

Maeda H, Noguchi Y, Sato K, Akaike T (1994) Enhanced vascular permeability in solid tumor is mediated by nitric oxide and inhibited by both new nitric oxide scavenger and nitric oxide synthase inhibitor. Jpn J Cancer Res 85: 331–334

Matsumura Y, Maeda H (1986) A new concept of macromolecular therapeutics in cancer chemotherapy: mechanism of tumoritropic accumulation of proteins and anti tumor agent SMANCS. Cancer Res 46: 6387–6392

Matsumura Y, Kimura M, Yamamoto T, Maeda H (1988) Involvement of kinin-generating cascade in enhanced vascular permeability in tumor tissue. Jpn J Cancer Res 79: 1327–1334

Matsumura Y, Maruo K, Kimura M, Yamamoto T, Konno T, Maeda H (1991) Kinin-generating cascade in advanced cancer patients and in vitro study. Jpn J Cancer Res 82: 732–741

Moncada S, Palmer RMJ, Higgs EA (1992) Nitric oxide: physiology, pathophysiology, and pharmacology. Pharmacol Rev 43: 109–142

Myers PR, Guerra R Jr, Harrison DG (1992) Release of multiple endothelium-derived relaxing factors from porcine coronary arteries. J Cardiovasc Pharmacol. 20: 392–400

Palmer RMJ, Ferrige AG, Moncada S (1987) Nitric oxide release accounts for the biological activity of endothelium-derived relaxing factor. Nature 327: 524–526

Ribieiro JMC, Hazzard JMH, Nussenzveig RH, Champagene DE, Walker FA (1993) Reversible binding of nitric oxide by a salivary heme protein from a bloodsucking insect. Science 260: 539–541

Sakai N, Kaufman S, Milstien S (1993) Tetrahydrobiopterin is required for cytokine-induced nitric oxide production in a murine macrophage cell line (RAW 264). Mol Pharmacol 43: 6–10

Senger DR, Galli SJ, Dvorak AM, Perruzzi CA, Harvey VS, Dvorak HF (1983) Tumor cells secrete a vascular permeability factor that promotes accumulation of ascitic fluid. Science 219: 983–985

Skinner SA, Tutton PJM, O'Brein E (1990) Microvasculature architecture of experimental colon tumor in rats. Cancer Res 50: 2411–2417

Stamler JS, Jaraki O, Osborne J, Simon DI, Keaney J, Vita J, Singel D, Valeri CR, Loscalzo J (1992) Nitric oxide circulates in mammalian plasma primarily as an *S*-nitroso adduct of serum albumin. Proc Natl Acad Sci USA 89: 7674–7677

Stuehr DJ, Nathan CF (1991) Nitric oxide: a macrophage product responsible for cytostasis and respiratory inhibition in tumor target cells. J Exp Med 169: 1543–1555

Stuehr DJ, Fasehun OA, Kwon NS, Gross SS, Gonzales JA, Levi R, Nathan CF (1991) Inhibition of macrophage and endothelial cell nitric oxide synthase by diphenyleneiodonium and its analogs. FASEB J 5: 98–103

Suzuki M, Takahashi T, Sato T (1987) Medial regression and its functional significance in tumor supplying host arteries. Cancer 59: 444–450

Tadjkarimi S, O'Neil GS, Luu TN, Allen SP, Schyns CJ, Chester AH, Yacoub MH (1992) Comparison of cyclic GMP in human internal mammary artery and saphenous vein: implications for coronary artery bypass graft patency; Cardiovasc Res 26: 297–300

Yoshida K, Akaike T, Doi T, Sato K, Ijiri S, Suga M, Ando M, Maeda H (1993) Pronounces enhancement of NO-dependent antimicrobial action by an NO-oxidizing agent, imidazolineoxyl N-oxide. Infect Immun 61: 3552–3555

Yoshida M, Akaike T, Wada Y, Sato K, Ikeda K, Ueda S, Maeda H (1994) Therapeutic effects of imidazolineoxyl *N*-oxide against endotoxin shock through its direct nitric oxide-scavenging activity. Biochem Biophys Res Commun 202: 930–932

The Role of Nitric Oxide in the Pathogenesis of Virus-Induced Encephalopathies

B. Dietzschold

1 Introduction

The mechanism by which virus infections of the central nervous system (CNS) cause neuronal damage are understood in only a few viral CNS diseases in which there is evidence that the virus directly destroys its target cells. In many other cases of virus-induced encephalopathy, the virus does not directly destroy neural tissue but causes indirect damage by altering neuronal functions (Fu et al. 1993; Lipkin et al. 1988a,b) or by triggering cell-mediated responses within the CNS (Byrne and Oldstone 1984; Carbone et al. 1988; Doherty et al. 1976). Soluble factors such as proinflammatory cytokines, proteases, free radicals, and neurotoxins produced by immune cells are thought to play an important role in the process of neuronal destruction (Shankar et al. 1992; Koprowski et al. 1993). It is believed that proinflammatory cytokines in particular are exceptionally important in the process of neuronal destruction (Selmaj 1992). It has been shown that cytokines can have direct cytotoxic effects. For example, the intracisternal challenge of interleukin-1 (IL-1) or tumor necrosis factor (TNF) in rats induced meningitis and blood brain barrier damage (Quagliarello et al. 1991). Recently, attention has focused on the possibility that reactive nitrogen intermediates

Department of Microbiology & Immunology, Center for Neurovirology, J.A.H. # 459, Thomas Jefferson University, 1020 Locust Street, Philadelphia, PA 19107, USA

generated by nitric oxide synthases (NOS) directly damage neural tissue. Both the constitutive form of NOS (cNOS), and the inducible calcium-independent NOS (iNOS) are thought to play a role in the process of neuronal damage (KOPROWSKI et al. 1993; MURPHY et al. 1993). NO generated by cNOS in neurons has been proposed to initiate excitotoxicity through activation of guanylate cyclase (DAWSON et al. 1991). However, the involvement of cyclic GMP in NO-mediated neurotoxicity is questionable since inhibitors of guanylate cyclase failed to protect neuron cultures from the toxic effect of NMDA or the NO donor sodium nitroprusside (LUSTIG et al. 1992). In contrast, increasing evidence suggests a role for activated microglia or infiltrating macrophages in neuronal damage and white matter pathologies by a TNF-and iNOS-dependent process. It has been shown in several neurological diseases that the expression of proinflammatory cytokines, such as IL-1, TNF, and interferon (IFN)-γ, which can prime macrophages and microglial cells for the production of reactive nitrogen intermediates (DING et al. 1988; NATHAN 1992), correlates with the degree of inflammatory lesions in the CNS and with the severity of neurological signs (KOPROWSKI et al. 1993; MUPRHY et al. 1993). Furthermore, the observation that iNOS mRNA is present in the brains of virus-infected rats or rats with experimental allergic encephalitis (EAE) (KOPROWSKI et al. 1993), is very suggestive for the involvement of NO in neuronal damage. To further investigate the potential pathophysiological role of NO in virus-induced neuropathogenesis, we have determined the kinetics of iNOS and cNOS expression in rabies virus and Borna disease virus (BDV)-infected rat brains. Rabies virus and BDV were chosen in this study because both viruses are highly neurotropic. Infection with BDV results in a persistent CNS infection which is characterized by massive infiltration of inflammatory cells. Our results show that iNOS expression which is not detectable in normal brain tissue is strongly up-regulated in the brain of rats infected with BDV or rabies virus and levels of iNOS activity correlate with clinical severity. In contrast, expression of neuronal cNOS, which is detectable in normal brain, is significantly decreased during BDV and rabies virus infection.

2 Kinetics of Expression of Tumor Necrosis Factor-α and Inducible Nitric Oxide Synthase mRNA in Rat Brains Following Borna Disease Virus Infection

Since it has been proposed that TNF-α, TNF-β and IFN-γ can prime macrophages for production of inducible reactive nitrogen intermediates, we compared the kinetics of BDV-induced changes in the mRNA levels of iNOS and TNF-α. The results showed that iNOS and TNF-α mRNAs, normally nondetectable in the brain, were present in the brain after infection with BDV (ZHENG et al. 1993). Although reverse transcriptase-polymerase chain reaction analysis was used, which yields qualitative rather than quantitative results, highest levels of iNOS mRNA were detected 21 days postinfection (p.i.) when animals exhibited neurological signs such as seizures, convulsions, and tremore. Expression of TNF-α mRNA was at a maximum 17 days p.i. and decreased at 21 days p.i. (ZHENG et al. 1993) indicating that the expression of TNF-α precedes that of iNOS. This result is consistent with the notion that certain proinflammatory cytokines such as TNF-α, IL-1 and IFN-γ might effect CNS damage by stimulating inflammatory cells to produce oxygen and nitrogen intermediates.

3 Localization of Inducible Nitric Oxide Synthase Expression in the Borna Disease Virus-Infected Brain

In situ hybridization experiments revealed that iNOS mRNA is expressed in brain areas that are preferentially infected by BDV, such as the hippocampus and basolateral parts of the cortex. However, unlike BDV RNA, which is evenly distributed among neuronal cell layers, iNOS mRNA exhibited a patchy distribution pattern. No iNOS mRNA could be detected in brains of normal rats.

iNOS producing cells in the BDV infected brain were identified by immunohistochemistry using antibodies against iNOS, the macrophage/monocyte marker ED1, and the astroglial marker GFAP. This analysis revealed a generalized astrogliosis accompanied by massive macrophage infiltration in brains of rats with Borna disease (E. Weihe, unpublished). However, cells expressing GFAP and those expressing ED1 were distinctly segregated. The distribution pattern of iNOS-positive cells was very similar to that of ED1-positive cells, and the staining for iNOS was confirmed to the same type of cells as that stained for ED1. Conversely, iNOS-positive cells were absent in areas of intensive GFAP staining. The segregation of areas containing GFAP-stained cells and areas with an accumulation of cells stained for ED1 and iNOS may reflect the damaging effect of NO released by activated macrophages or activated microglia. It is likely that astrocytes, neurons, and oligodendrocytes located in areas where iNOS-positive cells are present are functionally impaired or even destroyed. Furthermore, the observation that cells intensely stained for iNOS are present in areas where neurons were still morphologically intact, whereas iNOS expression is less prominent in areas of massive tissue destruction (ZHENG et al. 1993), suggests that the generation of NO by iNOS is an early event in the process of neurological damage.

4 Effect of Viral Infections on Neuronal Constitutive Nitric Oxide Synthase in the Brain

Excessive amounts of NO in neurons generated by cNOS have been suggested to be involved in neurodegenerative processes such as excitotoxicity (Dawson et al. 1991). However, determination of NOS activities in the brain indicated that activity of cNOS in the cerebrum decreased after infection with BDV (Table 1) or rabies virus (Akaike et al. 1994). In BD, the decrease in cNOS activity which could already be observed 16 days after infection, declined to 30% of the activity found in normal rat brain on day 22 after infection.

Western blot analysis using an antibody specific for cNOS revealed a reduction in neuronal cNOS in the brain after infection with BDV. The observation that cNOS mRNA also decreased in a time-dependent manner after BDV infection indicates that the decrease in neuronal cNOS activity is most likely due to a decrease in the expression of neuronal cNOS mRNA.

The decrease in cNOS expression could be the result of extensive neuronal loss that might occur in virus-infected brains. However, the activity of choline acetyl; transferase remained unchanged after BDV or rabies virus infection. Furthermore, expression of several immediate-early response genes, such as c-*fos*, *jun* B, *egr*-1, and the late response gene encoding proenkephalin is markedly increased in BDV-infected neurons of the cerebral cortex and the hippocampus (Fu et al. 1993). Therefore, it is unlikely that the decrease in cNOS activity is caused by a massive neuronal destruction. Selective damage of nitroxergic neurons is also unlikely because recent observations have shown that, in lesions of patients with Alzheimer's disease or in brains with experimentally induced ischemia, nitroxergic neurons are spared from destruction (Hyman et al. 1992; Uemura et al. 1990). Since BDV or rabies virus can induce the expression of transcription factors, it is possible that the decrease of cNOS expression is the

Table 1. Inducible and Constitutive Nitric Oxide Synthase in Normal and Borna Disease Virus Infected Rat Brain[a]

Days Postinfection	Enzyme activity (μ*M*/20 min/g tissue)	
	iNOS	cNOS
Controls	0	28.9 ± 1.1
16	0.39 ± 0.1	23.4 ± 1.2
18	1.92 ± 1.0	19.2 ± 0.7
19	1.78 ± 0.7	18.8 ± 1.0
20	2.17 ± 0.4	14.7 ± 1.4
22	1.13 ± 0.5	8.7 ± 1.4

[a] iNOS, inducible nitric oxide synthase; cNOS, constitutive nitric oxide synthase.
NOS activity was measured radiochemically as described by Bredt and Snyder (1989). To differentiate the activity of iNOS from cNOS, the calmodulin inhibitor trifluroperazine (final concentration: 100 μ*M*) was added to the assay in vitro

result of the down-regulation of cNOS mRNA transcription by some of these transcription factors, although it cannot be excluded that post-translational modifications of NOS such as phosphorylation (BREDT et al. 1992) or inhibition of NOS by soluble factors such as IL-4 and transforming growth factor (TGF)-β (NATHAN 1992) contribute to the decrease of cNOS activity.

NO generated by neuronal cNOS is thought to be an important neurotransmitter that plays a role in long-term potentiation and long-term depression (BLISS and COLLINGRIDGE 1993). Since BDV and rabies virus replicate preferentially in the hippocampal formation, it is possible that the behavioral abnormalities seen in these virus infections could be partially due to the suppression of neuronal cNOS.

5 Effect of Nitric Oxide Synthase Inhibitors on Borna Disease Virus-Induced Encephalopathy

To determine whether NO, generated by macrophages, plays a major role in the development of neurological symptoms and neuropathology of BDV infection, rats were treated with NOS inhibitors such as L-*N*-monomethylarginine (L-NMMA) and L-nitroargininemethylester (L-NAME) (Table 2). However, none of these inhibitors exerted a therapeutic effect. Treatment with L-NMMA resulted in an exacerbation of clinical signs. These data suggest that NO by itself does not play a pivotal role in the neuropathogenesis of viral CNS infections. On the contrary, the exacerbation of clinical signs after treatment with NOS inhibitors suggests that reduced NO production, caused by the decrease in neuronal cNOS activity in BDV-infected rats, may actually contribute to the disease process.

Table 2. Effect of Nitric Oxide Synthase Inhibitors and Scavenger on Borna Disease Virus Infected Rats

Treatment[a]	Rats with clinical signs 10 days after treatment	Survivor rate	Loss of body weight 25 days after infection (gm)[b]
Control	4/5	5/5	11.26 ± 4.99
L-NAME	5/5	4/5	18.95 ± 5.54
L-NMMA	5/5	4/5	26.17 ± 2.81

[a] Rats were treated beginning at day 15 postinfection (p.i.) for 10 days. All compounds were delivered via i.p. implanted Alzet pumps; L-NAME and L-NMMA at a rate of 140 μ/h/kg. Control animals received PBS via i.p. implanted Alzet osmotic pumps.

[b] Data are mean ± SE of four or five rats that survived 25 days after Borna disease virus infection. Loss of body weight in control and L-NMMA-treated rats differed significantly ($p < 0.05$), two-tailed t-test.

References

Akaike T, Weihe E, Schaeffer MK-H, Fu ZF, Zheng YM, Vogel WH, Schmidt HHHW, Koprowski H, Dietzschold B (1994) Effect of neurotropic virus infection on neuronal and inducible nitric oxide synthase activity in rat brain. J Neurovirol (in press)
Bliss TVP, Collingridge GL (1993) A synaptic model of memory: long-term potentiation in the hippocampus. Nature 361: 31–39
Bredt DS, Snyder SH (1989) Nitric oxide mediates glutamate-linked enhancement of cGMP levels in the cerebellum. Proc Natl Acad Sci USA 86: 9030–9033
Bredt DS, Ferris CD, Snyder SH (1992) Nitric oxide synthase regulation sites: phosphorylation by cyclic AMP-dependent protein kinase, protein kinase C, and calcium/calmodulin protein kinase; identification of flavin and calmodulin binding sites. J Biol Chem 267: 10976–10981
Byrne JA, Oldstone MBA (1984) Biology of cloned cytotoxic T lymphocytes specific for lymphocytic choriomeningitis: clearance of virus in vivo. J Virol 51: 682–686
Carbone KM, Duchala CS, Narayan O (1988) Borna disease: an immunopathologic response to viral infection in the CNS. Ann N.Y Acad Sci 540: 661–662
Dawson VL, Dawson TM, London ED, Bredt DS, Snyder SH (1991) Nitric oxide mediates glutamate neurotoxicity in primary cortical cultures. Proc Natl Acad Sci USA 88: 6368–6371
Ding AH, Nathan CF, Stuehr DJ (1988) Release of reactive nitrogen and oxygen intermediates from mouse peritoneal macrophages. Comparison of activating cytokines and evidence for independent production. J Immunol 141: 2407–2412
Doherty PC, Dunlop MBC, Parish CR, Zinkernagal RM (1976) Inflammatory process in murine lymphocytic choriomeningitis is maximal in H-2K or H-2D compatible interactions. J Immunol 117: 187–190
Fu ZF, Weihe E, Zheng YM, Schäffer MK-H, Sheng H, Corisdeo S, Rauscher FJIII, Koprowski H, Dietzschold B (1993) Differential effects of rabies and Borna disease viruses on immediate-early- and late-response gene expression in brain tissue. J Virol 67: 6674–6681
Hyman BT, Marzloff K, Wenniger JJ, Dawson TM, Bredt DS, Snyder SH (1992) Relative sparing of nitric oxide synthase-containing neurons in the hippocampal formation in Alzheimer's disease. Ann Neurol 32: 818–820
Koprowski H, Zheng YM, Heber-Katz E, Fraser N, Rorke L, Fu ZF, Hanlon CA, Dietzschold B (1993) In vivo expression of inducible nitric oxide synthase in experimentally induced neurologic diseases. Proc Natl Acad Sci USA 90: 3024–3027
Lipkin WI, Battenberg ELF, Bloom FE, Oldstone MBA (1988a) Viral infection of neurons can depress neurotransmitter mRNA levels without histologic injury. Brain Res 451: 333–339
Lipkin WI, Carbone KM, Wilson MC, Duchala CS, Narayan O, Oldstone MBA (1988b) Neurotransmitter abnormalities in Borna disease. Brain Res 475: 336–370
Lustig HS, von Brauchitsch KL, Chan J, Greenberg DA (1992) Cyclic GMP modulators and excitotoxic injury in cerebral cortical cultures. Brain Res 577: 343–346
Murphy S, Simons ML, Agullo L, Garcia A, Feinstein DL, Galea E, Reis DJ, Minc-Golomb D, Schwartz JP (1993) Synthesis of nitric oxide in CNS glial cells. Trends Neurosci 16: 323–328
Nathan C (1992) Nitric oxide as a secretory product of mammalian cells. FASEB J 6: 3051–3064
Quagliarello VJ, Wispelwey B, Long WJ Jr, Scheid WM (1991) Recombinant human interleukin-1 induces meningitis and blood-brain barrier injury in the rat. J Clin Invest 87: 1360–1366
Selmaj KW (1992) The role of cytokines inflammatory conditions of the central nervous system. Semin Neurosci 4: 221–225
Shankar V, Kao M, Hamir AN, Sheng H, Koprowski H, Dietzschold B (1992) Kinetics of virus spread and changes in levels of several cytokine mRNAs in brain after intranasal infection of rats with Borna disease virus. J Virol 66: 992–998
Uemura Y, Kowall NW, Beal MF (1990) Selective sparing of NADPH-diaphorase-somatostatin-neuropeptide Y neurons in ischemic bergil striatum. Ann Neurol 27: 620–625
Zheng YM, Schäffer MK-H, Weihe E, Sheng H, Corisdeo S, Fu ZF, Koprowski H, Dietzschold B (1993) Severity of neurological signs and degree of inflammatory lesions in the brains of rats with Borna disease correlate with the induction of nitric oxide synthase. J Virol 67: 5786–5791

The Role of Peroxynitrite in Nitric Oxide-Mediated Toxicity

J.P. CROW and J.S. BECKMAN

1 Introduction

Nitric oxide is a hydrophobic gas with chemical properties that make it uniquely suited as both an intra- and intercellular messenger. It is a relatively stable, charge-neutral radical with a strong tendency to interact and react with other species possessing unpaired electrons such as superoxide, iron, and molecular oxygen. Thus, it has several potential toxic mechanisms; actual toxicity is highly dependent on nitric oxide concentration and the particular microenvironment in which it is produced.

Nitric oxide has been reported to inhibit critical iron-sulfur-containing enzymes involved in mitochondrial respiration (WELSH et al. 1991), to inhibit ribonucleotide reductase (KWON et al. 1991; LEPOIVRE et al. 1991), and to damage DNA directly (NGUYEN et al. 1992; WINK et al. 1991). In addition, nitric oxide has been implicated in reoxygenation injury following ischemia (JAESCHKE et al. 1992; Masini et al. 1991), glutamate-mediated neuronal toxicity (CAZEVIEILLE et al. 1993;

University of Alabama at Birmingham, Department of Anesthesiology, 619 19th Street South, THT 958, Birmingham, Alabama 35233, USA

DAWSON et al. 1993), inflammation (BERRAZUETA et al. 1990; KRONCKE et al. 1991; MULLIGAN et al. 1991), graft host disease (HOFFMAN et al. 1992; GARSIDE et al. 1992; LANGREHR et al. 1992), and as a major arm of host defense against viruses, bacteria, and other intracellular parasites (GREEN et al. 1990a,b). In many cases these findings were based on the ability of either nitric oxide synthase inhibitors or nitric oxide scavengers like hemoglobin to protect against a particular pathological change or, in the latter case, to inhibit parasite killing by activated macrophages. Conversely, compounds which release nitric oxide typically cause or exacerbate the change. The most reasonable extrapolation from these studies is that nitric oxide is directly toxic.

Many investigators readily accept the notion that nitric oxide is toxic, given its reputation as a reactive radical with a short biological half-life. However, reactive is not synonomous with toxic. Nitric oxide is known to react with molecular oxygen to give nitrogen dioxide (the orange-brown gas in smog), which is a strong oxidant and potent toxin in any biological system, but significant nitrogen dioxide formation occurs only at nitric oxide concentrations far in excess of those seen in vivo. The reaction of nitric oxide with superoxide occurs very rapidly even at physiological concentrations of nitric oxide and offers a more plausible explanation for both the short biological half-life of nitric oxide and its participation in toxic reactions.

An effort will be made in this chapter to examine the chemistry of nitric oxide, with particular attention to reactions which can and do occur in biological systems and those which are unlikely to occur. Major emphasis will be given to the reaction of nitric oxide with superoxide in vivo to form peroxynitrite, a potent oxidant with multiple reaction pathways. A strong case will be made for the participation of peroxynitrite in nitric oxide-mediated toxicity. Existing theories of oxygen radical-mediated toxicity will first be examined and attempts will be made to account for older findings in light of newer hypotheses.

2 The Superoxide Theory of Toxicity

Molecular oxygen is readily reduced by one electron to give superoxide anion. The term "superoxide" was originally coined to describe its unusual electronic configuration rather than its reactivity (PAULING 1979). However, for many years, superoxide was taken to mean "superoxidizing". Reputable biochemistry textbooks still refer to superoxide anion as a "highly reactive and destructive radical". At physiological pH, superoxide has little potential to oxidize other molecules since this would require placing a second negative charge on superoxide (SAWYER and VALENTINE 1981; BAUM 1984). Neutralization of the first negative charge by protonation facilitates superoxide acting as an oxidant; however, the pK_a for superoxide is 4.8 (BIELSKI 1978). Thus, superoxide acts as an oxidant only in a more acidic environment or when the molecule under attack is a good one-electron donor such as reduced iron in Fe (II)-sulfur centers. Iron (III) is readily reduced by superoxide; superoxides' ability to carry out the one-electron reduction of iron (III) in cytochrome c serves as the basis for a frequently used assay (McCord and Fridovich 1969).

Superoxide dismutase (SOD) affords protection against oxidant stress and ischemic injury in many model systems (Flaherty and Weisfeldt 1988; Zweier et al. 1987). The mechanism by which SOD protects remains unclear since few specific biological targets for superoxide have been identified. Superoxide appears to reversibly inactivate mitochondrial Fe (II)-sulfur enzymes in bacteria (Gardner and Fridovich 1991a,b); the percent of dynamically inactivated aconitase has seen used to estimate the steady-state concentration of superoxide (Gardner and Fridovich 1992). However, the contribution of this ongoing inactivation/ reactivation to overt cellular toxicity is not clear. To account for the effects of SOD in the absence of clear direct toxicity of superoxide, the superoxide-driven Fenten reaction was proposed whereby superoxide reduces Fe^{+3} to Fe^{+2}, which, in turn, reduces hydrogen peroxide resulting in the formation of the strongly oxidizing hydroxyl radical (see below). Numerous problems exist with this proposal, not the least of which is its inability to account for nitric oxide-dependent toxicity.

3 Hydroxyl Radical Is Too Reactive To Be Highly Toxic

For many years, the accepted explanation for superoxide-derived toxicity has been hydroxyl radical production via the iron-catalyzed Haber-Weiss reaction (Haber and Weiss 1934):

$$O_2^- + Fe^{3+} \rightarrow O_2 + Fe^{2+}$$

$$Fe^{2+} + H_2O_2 \rightarrow \cdot OH + {}^-OH + Fe^{3+}$$

In this scheme, superoxide serves primarily as the reductant for iron. This seems unlikely when considering that the steady-state superoxide concentration is in the range of 10–100p*M* (Gardner and Fridovich 1992) while other reductants like ascorbate are present in much higher concentrations in vivo and also reduce Fe^{3+} to Fe^{2+} quite effectively. On the other hand, if superoxide served primarily as the source of hydrogen peroxide, then SOD would not be protective (since hydrogen peroxide is formed from the reaction of superoxide with SOD). Superoxide may also serve to increase the pool of free iron by releasing it from ferritin (Biemond et al. 1984; Thomas et al. 1985). However, even with a relatively large pool of Fe^{2+}, the rate at which it reduces hydrogen peroxide to produce hydroxyl radical is quite slow (10^3–10^4 $M^{-1}s^{-1}$) (Rush and Koppenol 1989). Thus, the iron-catalyzed Haber-Weiss reaction requires the interaction of three species present in low concentrations in vivo due to efficient scavenging systems (SOD, catalase, ferritin), making the rate of hydroxyl radical formation dependent on the product of three minuscule concentrations which react at relatively low rates.

ydroxyl radical were efficiently formed in vivo, it is simply too reactive ific damage. The ability of hydroxyl radical to react at near diffusion- with virtually all biological molecules limits its effective diffusion to only a few angstroms. Thus, hydroxyl radical would have to be formed directly at critical sites (e.g., active sites of enzymes, bases of DNA, within membranes) to affect significant injury. Many such critical targets are capable of binding iron in vitro thereby promoting hydroxyl radical formation in close proximity. However, the rates of hydrogen peroxide reduction by bound iron can be even lower than with free iron due to nonoptimal coordination geometries. Overall, the slow rates of reactions leading to hydroxyl radical production and the low concentrations of the reactants suggest that other reactions may be important for understanding superoxide toxicity.

4 The New Radical Era—Nitric Oxide as Cellular Messenger

The discoveries of nitric oxide synthases and nitric oxide production in biological systems ushered in a new era of oxygen radical enthusiasm reminiscent of that seen previously with superoxide. A major biological difference between superoxide and nitric oxide is that cells have adapted to tolerate the inescapable production of superoxide, while a variety of isoenzymes have evolved to deliberately and specially produce nitric oxide. At present, there is no evidence for "a nitric oxide dismutase" to eliminate this newly discovered reactive radical. Important questions are naturally raised: what determines the half-life of nitric oxide in biological systems? How are its physiological effects terminated and how is toxicity avoided or minimized? Earlier data suggested that superoxide might be involved.

5 Superoxide Dismutase Increases the Biological Half-Life of Nitric Oxide

Prior to conclusive demonstration that endothelium-derived relaxing factor (EDRF) was indeed nitric oxide, SOD had been found to potentiate or prolong the vasorelaxant effects of EDRF (Ignarro et al. 1989; Hutchinson et al. 1987). This led to the concept of superoxide as an endothelial-dependent contracting factor (EDCF) and later to the notion that superoxide simply detoxifies nitric oxide by a direct radical-radical annihiliation reaction. The concept of detoxification will be discussed later. Here, it is noteworthy that the reaction of superoxide with nitric oxide (Huie and Padmaja 1993) is three fold faster than the reaction of superoxide with SOD (Klug et al. 1972) and 30-fold faster than the reaction of superoxide with the Escherichia coli iron-sulfur enzyme 6-phosphogluconate dehydratase

(GARDNER and FRIDOVICH 1991b). Thus, to date, nitric oxide is the only identified target for superoxide which reacts fast enough and can exist at sufficient concentrations to effectively compete with SOD. The ability of SOD to affect the biological half-life of nitric oxide provides strong evidence that the reaction of nitric oxide with superoxide is significant in vivo.

The reaction between nitric oxide and superoxide is directly analogous to a binary chemical bomb, in which two fairly innocuous substances are mixed to produce a potent toxin. Macrophages and neutrophils, cell types which simultaneously produce nitric oxide and superoxide, appear to utilize this strategy to kill invading parasites; accompanying inflammation of host tissues may also be a direct result of this programmed oxidant attack. Peroxynitrite is a strong oxidant and nitrating agent capable of producing tissue injury by targeting key cellular components. Formation of peroxynitrite in cell systems is inhibited by SOD in a dose-dependent manner and by inhibitors of nitric oxide synthase (ISCHIROPOULOS et al. 1992a). Tissues containing constitutive nitric oxide synthase would be primed during ischemic periods to produce both nitric oxide and superoxide during reoxygenation. Hyperoxic conditions would also promote sumultaneous production. Thus, many of the findings related to tissue injury (and protection) under conditions of oxidant stress, ischemia/reperfusion, and inflammation can be explained on the basis of peroxynitrite formation, without the need to invoke hydroxyl radical formation via sequential, concerted reactions that occur at relatively low rates.

6 The Biological Chemistry of Nitric Oxide

Nitric oxide is a radical by virtue of having one unpaired outer shell electron. However, this is not an inherently unstable state since nitrogen normally possesses an odd number of valence electrons (which is why the nitric oxide radical has no formal charge). Nitric oxide can undergo a one-electron reduction to give nitroxyl anion (NO^-) which can also produce peroxynitrite via a direct reaction with molecular oxygen. However, it seems more likely that nitric oxide would react directly with other radicals or one-electron donors/acceptors such as the iron in heme, particularly considering the abundance of heme-containing proteins. Nitric oxide is unusual in that it reacts with both ferrous and ferric forms of heme iron. Signal transduction is accomplished by the binding of nitric oxide to the heme prosthetic group of guanylate cyclase (reaction 1) which activates the enzyme to produce cGMP

$$\text{GC-heme-Fe}^{+2} + \cdot\text{NO} \rightarrow \text{GC-heme-Fe}^{+2}\text{– NO.} \qquad \text{(reaction 1)}$$

The major sink for nitric oxide which diffuses to the vasculature is oxyhemoglobin. This results in the formation of nitrate and methemoglobin (reaction 2)

$$\text{Hb-Fe}^{+2}\text{—}O_2 + \cdot\text{NO} \rightarrow \text{Hb-Fe}^{+2}\text{—OONO} \rightarrow \text{Hb-Fe}^{+3} + NO_3^-. \qquad \text{(reaction 2)}$$

The most abundant, hydrophobic, small molecule containing an unpaired electron with which nitric oxide can readily react is oxygen. Nitric oxide reacts in a 2:1 molar ratio with molecular oxygen to form nitrogen dioxide (reaction 3), the orange gas seen when concentrated nitric oxide is released from a cylinder

$$2 \cdot NO + O_2 \rightarrow 2 \cdot NO_2 \,. \qquad \text{(reaction 3)}$$

The rate of nitrogen dioxide formation via this reaction is $2 \times 10^6\ M^{-2}s^{-1}$ at 25 °C (FORD et al. 1993) and since it requires the interaction of three molecules, the rate is dependent on the square of nitric oxide concentration (reaction 4). This has important biological implications

$$d[NO]/dt = k_3\,[NO]^2\,[O_2] \qquad \text{where } k_3 = 4 \times 10^6\ M^{-2}s^{-1} \,. \qquad \text{(reaction 4)}$$

Concentrations of nitric oxide produced in vivo (~100 n*M*) (SHIBUKI 1990) are 10 000-fold lower than the local concentrations at the nozzle of a gas cylinder. The rate of reaction between nitric oxide and oxygen at their respective physiological concentrations would, therefore, be 100 000 000 times slower. Calculations based solely on reaction with oxygen reveal that 0.7 h would be required for 100 n*M* nitric oxide to decrease to 50 n*M* (i.e., first half-life), 1.4 h to decrease from 50 n*M* to 25 n*M* and so on. Clearly, the reaction of nitric oxide with molecular oxygen cannot account for its short (1–10 s) biological half-life. The rate constant for the reaction of nitric oxide with superoxide is $6.7 \times 10^9\ M^{-1}s^{-1}$ (HUIE and PADMAJA 1993) and, since it is bimolecular, the rate is not dependent on the square of nitric oxide concentration (reaction 5)

$$\cdot NO + O_2^{\cdot -} \rightarrow ONOO^- \,. \qquad \text{(reaction 5)}$$

The fact that SOD can potentiate or prolong the biological effects of nitric oxide provides strong evidence that reaction with superoxide is a major pathway for termination of nitric oxide.

7 Formation, Decomposition, and Reactivity of Peroxynitrite

For in vitro use, peroxynitrite is readily synthesized from hydrogen peroxide and nitrous acid; the reaction mixture is rapidly quenching with sodium hydroxide to yield peroxynitrite anion (REED et al. 1974). Peroxynitrite is stable in alkaline solutions for several weeks. Dilution of the alkaline stock solution into pH 7.4 potassium phosphate results in protonation to give peroxynitrous acid (pK_a 6.8) which rapidly decomposes to yield nitrate; in the presence of other target molecules, multiple oxidation products are seen (see below). Peroxynitrite formation in vivo is almost certainly due to the reaction of nitric oxide with superoxide. This rate of formation is near the diffusion limit, which means that virtually every collision between molecules of nitric oxide and superoxide will result in a reaction to produce peroxynitrite.

At pH 7.4 and 37 °C the half-life of peroxynitrite is 1–2 s (BECKMAN et al. 1990). Peroxynitrous acid can decompose via three major pathways. Reaction of peroxynitrite with target molecules results in products characteristic of both nitrogen dioxide and hydroxyl radical (reaction 6). However, kinetic data and thermodynamic calculations indicate that no free hydroxyl radical is produced (KOPPENOL et al. 1992). We have proposed that peroxynitrous acid reacts as a vibrationally activated complex rather than via homolytic fission of peroxynitrous acid as depicted in reaction 6 (CROW et al. 1994)

$$ONOO^- + H^+ \rightleftarrows ONOOH \rightarrow \text{"}\cdot NO_2 + \cdot OH\text{"} . \qquad \text{(reaction 6)}$$

Peroxynitrite will directly nitrate phenolic compounds (HALFPENNY and ROBINSON 1952; ISCHIROPOULOS et al. 1992b; BECKMAN et al. 1992). Nitration is catalyzed by transition metals and by Cu,Zn-SOD and is thought to occur via formation of a reactive nitronium ion (reaction 7) (ISCHIROPOULOS et al. 1992b; SMITH et al. 1992). The primary consequence of this reaction in vivo is the modification of tyrosine residues in proteins, which we have detected using an antibody raised to peroxynitrite-treated keyhole limpet hemocyann (KLH, see below)

$$ONOO^- + Metal^{2+} \rightarrow NO_2^+ \ldots O^- \text{—} Metal^{1+} \rightarrow$$
$$protein\text{—}tyr\text{—}NO_2 + HO\text{-}Metal^{1+} + Metal^{2+} + H_2O . \qquad \text{(reaction 7)}$$

In the absence of target molecules, the primary decomposition product of peroxynitrite at neutral pH is nitrate (reaction 8)

$$ONOO^- + H^+ \rightleftarrows ONOOH \rightarrow HNO_3 \rightleftarrows NO_3^- + H^+ . \qquad \text{(reaction 8)}$$

Reactions with target molecules compete with this relatively innocuous route of decomposition. Thus the variety and yield of products from peroxynitrite will vary depending on the relative abundance of specific targets and the presence or absence of transition metals or metalloproteins.

8 Peroxynitrite: The Smart Bomb of Biological Oxidants

Transition metals are not required for either the formation or the decomposition of peroxynitrite although metals like iron and copper will catalyze nitration reactions. Decomposition of one molecule of peroxynitrous acid yields the equivalent of two potent one-electron oxidants each capable of participating in a variety of reactions. Peroxynitrous acid can nitrate phenolic rings (HALFPENNY and ROBINSON 1952; ISCHIROPOULOS et al. 1992b; BECKMAN et al. 1992), hydroxylate aromatic rings (HALFPENNY and ROBINSON 1952; HOGG et al. 1992), and oxidize lipids (RADI et al. 1991a), proteins (MORENO and PRYOR 1992), and DNA (KING et al. 1992). Under physiological conditions, peroxynitrite is sufficiently stable to diffuse several cell diameters to reach critical cellular targets before becoming protonated and decomposing. We have also seen that it can cross lipid membranes, probably in the protonated (acid) form (J. Chen and J.S. Beckman, unpublished observations). While most oxidative species are derived from peroxynitrous acid, direct reactions of the otherwise stable peroxynitrite anion occur with nucleophiles like thiols (RADI et al. 1991b) which effectively targets its oxidative capabilities. Thus, peroxynitrite may be considered the "smart bomb" of biological oxidants in that it is formed from two weak oxidants in a metal-independent manner, it can deliver multiple reactivites (including hydroxyl radical-like reactivity) over a relatively long distance, and the anionic form is stable except for its reactivity toward thiols.

9 The Dark Side of Superoxide Dismutase—Catalysis of Nitration by Peroxynitrite

Nitration of phenolic compounds by peroxynitrite is readily catalyzed by Cu, Zn-SOD (ISCHIROPOULOS et al. 1992b). Removal of the copper atom from the active site abolishes both nitrating and dismutase activities and activities are restored by copper replacement (SMITH et al. 1992). Catalysis of nitration by SOD is faster and more efficient than with either free copper or EDDA-chelated copper, suggesting that the substrate binding region of SOD recognizes peroxynitrite. This seems reasonable based on the structural similarity between peroxynitrite and superoxide and fact that both are negatively charged. The ability of SOD to enchance nitration is modest in simple in vitro systems. However, in complex media such as brain homogenates or plasma, where competing side reactions would tend to decrease total nitration, the enhancement by SOD is pronounced and consistent with a direct reaction between peroxynitrite and SOD. Again, this difference suggests that the electrostatic field of SOD, which serves to attract the negatively charged superoxide anion, may also attract peroxynitrite anion.

In the absence of suitable targets compounds (e.g., low molecular weight phenolics or tyrosine residues in proteins), bovine Cu,Zn SOD will catalyze nitration of its own single tyrosine residue. The distance between the catalytic site and the affected tyrosine residue suggests that nitration is inter- rather than intra-molecular. Nitration renders the protein yellow at neutral pH but has no effect on dismutating or subsequent nitrating activity (SMITH et al. 1992).

10 Biological Relevance: Evidence for Peroxynitrite Formation In Vivo

Superoxide dismutase prolongs the biological half-life of either endogenous nitric oxide or authentic nitric oxide added to a tissue bath (IGNARRO et al. 1987, 1989; KELM and SCHRADER 1990; BATES et al. 1991; CHEN and LEE 1993). The most reasonable interpretation of this prolongation is that SOD scavenges superoxide, thereby preventing its reaction with nitric oxide. The transient, intermediate form of SOD, Cu^{+1}-SOD, has been reported to directly reduce nitric oxide to nitroxyl anion under anaerobic conditions (MURPHY and SIES 1991). However, it is not clear whether this would occur to any extent under physiological (aerobic) conditions in which the sole source of reductant for SOD is superoxide itself. Provided that no direct reaction occurs between nitric oxide and SOD, the ability of SOD to prolong the half-life of endogenous nitric oxide is consistent with ongoing formation of peroxynitrite even at low physiological concentrations of superoxide and nitric oxide.

Using nitration of a more water-soluble tyrosine analogue (*p*-hydroxyphenyl-acetic acid) as an index, we have shown that virtually all of the nitric oxide produced by activated macrophages is converted to peroxynitrite (ISCHIROPOULOS et al. 1992a). Peroxynitrite production by inflammatory cells may represent a programmed killing mechanism whereby two relatively non-toxic radical species are simultaneously and deliberately produced to form a potent oxidant directed against an invading microbe. Peroxynitrite formed outside the inflammatory cell would set up a concentration gradient favoring diffusion away from (and thereby protection of) the producing cell and toward the invader. The presence of SOD inside the microbial cell would offer no protection against preformed peroxynitrite and might actually serve to mediate injury by catalyzing nitration of microbial proteins.

Under normal physiological conditions the concentration of SOD is 100- to 1000-fold higher than nitric oxide and the rate constant for SOD's reaction with superoxide is only three fold lower (KLUG et al. 1972) than the rate constant for nitric oxide and superoxide (HUIE and PADMAJA 1993). Therefore, SOD effectively outcompetes nitric oxide for the available superoxide by approximately 30-fold and thereby provides the primary defense against host cell injury by limiting peroxynitrite formation. However, in inflammatory conditions (e.g., sepsis), in which nitric oxide production is increased by induction of a constitutively active isoenzyme of nitric oxide synthase, peroxynitrite formation would predominate and the role SOD may shift from prevention to promotion of injury by catalyzing nitration. This could account for the exacerbation of injury following exogenous administration of high doses of SOD in the same model systems in which lower doses of SOD were protective (PARKS et al. 1982; MATSUDA et al. 1991; OMAR et al. 1990).

Peroxynitrite readily nitrates tyrosine present as residues in proteins or as the free amino acid. SOD-catalyzed nitration by peroxynitrite occurs at a rate of $10^5\ M^{-1}s^{-1}$ (BECKMAN et al. 1992). Nitrogen dioxide will nitrate tyrosine at a similar rate ($3.2 \times 10^5\ M^{-1}s^{-1}$) (PRUTZ et al. 1985), but relatively high concentrations are required and the overall reaction efficiency is at least 100-fold lower than for peroxynitrite (Crow and Beckman, unpublished observations). Nitrotyrosine has been measured in human urine (OHSHIMA et al. 1990, 1991), indicating that it is stable as the free amino acid and suggesting that the nitro group is not subject to significant metabolic transformation. Thus, nitration of the phenolic ring of tryosine residues in proteins is a stable covalent modification which appears to be a specific biological marker of peroxynitrite.

We have raised an antibody in rabbits that recognizes a number of proteins treated with peroxynitrite in vitro at a titer of 1/256,000. The antibodies do not cross-react with native albumin, hemoglobin, SOD, histone, lysozyme, lung surfactant proteins, rat brain homogenates or human CSF, but recognize such proteins after treatment with peroxynitrite in both ELISA assays or western blots. The presence of nitrotyrosine has been confirmed by amino acid analysis of hydrolyzed proteins (Crow and Beckman, unpublished). Preincubation of the antibody with 10 m*M* nitrotyrosine blocks subsequent binding to nitrated proteins whereas aminotyrosine and phosphotyrosine do not. Treatment of nitrated proteins or tissue sections with dithionite, which reduces nitrotyrosine to aminotyrosine, (Fig. 1) also blocks binding.

We present immunohistochemical evidence (using this antibody) that lung tissue from a patient with adult respiratory distress syndrome (ARDS) (Figs. 2A, 3A) and a lesion from an atherosclerotic human artery show discrete areas of heavy staining for nitrated proteins (Fig. 2B). In lung, staining appears to be most intense in regions immediately surrounding inflammatory cells but is not restricted to those areas; note that erythrocytes immediately adjacent to inflammatory cells do not stain. In the atheroscelerotic vessel, the endothelial lining stains heavily as does the plaque itself, where evidence of inflammatory cell infiltration is seen. Staining of lung is blocked by preincubation of the antibody with 10 m*M* nitrotyrosine (Fig. 3B) or by pretreatment of the tissue with dithionite (not shown). Using preimmune sera, a pattern of very light, homogeneous staining was seen which is essentially identical to that seen when the primary antibody is preincubated with nitrotyrosine (Fig. 3B).

The immunohistochemical results strongly indicate enhanced peroxynitrite formation in inflammatory conditions. We have shown previously that activated macrophages produce relatively large amounts of peroxynitrite under conditions of simultaneous superoxide and nitric oxide production (Ischiropoulos et al. 1992a). Thus, in the case of lung and atheroma macrophages and neutrophils are a likely source of peroxynitrite.

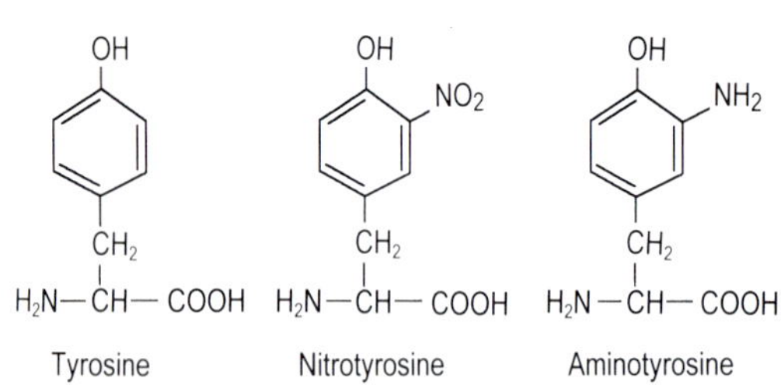

Fig. 1. Structures of the naturally occurring amino acid tyrosine and its reaction product with peroxynitrite, 3-nitrotyrosine. Nitrotyrosine, either as the free amino acid or within a protein, can be chemically reduced with dithionite (sodium hydrosulfite) to 3-aminotyrosine. Under alkaline conditions, nitrotyrosine absorbs at 420 nm whereas aminotyrosine does not. Thus, reduction can be monitored spectrally at 420 nm. This procedure is used to tentatively identify and quantify nitrotyrosine in various assays

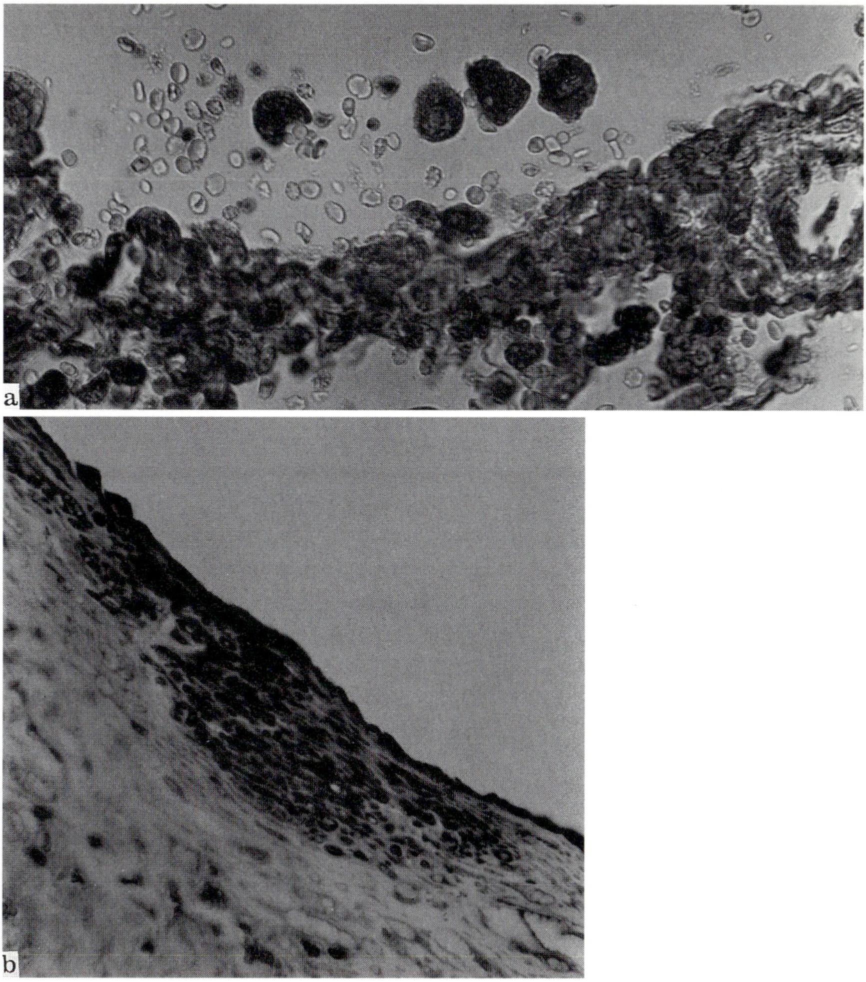

Fig. 2. a Lung tissue from a human patient with ARDS. **b** Atheromatous plaque in large vessel from human patient. Fixed tissues sections were treated with rabbit polyclonal antibody raised to peroxynitrite-treated KLH (1:1000), washed, treated with anti-rabbit globulin conjugated to biotin, washed again, and treated with avidin-conjugated horse-radish peroxidase. Peroxidase substrate was added and allowed to react. The section was then washed and counterstained with hematoxylin and eosin

11 Possible Mechanisms of Peroxynitrite-Mediated Injury

Nitration of proteins by peroxynitrite may be injurious via multiple mechanisms; alteration of tyrosine phosphorylation, altered protein function, increased protein turnover due to enhanced proteolysis, and initiation of autoimmune reactions

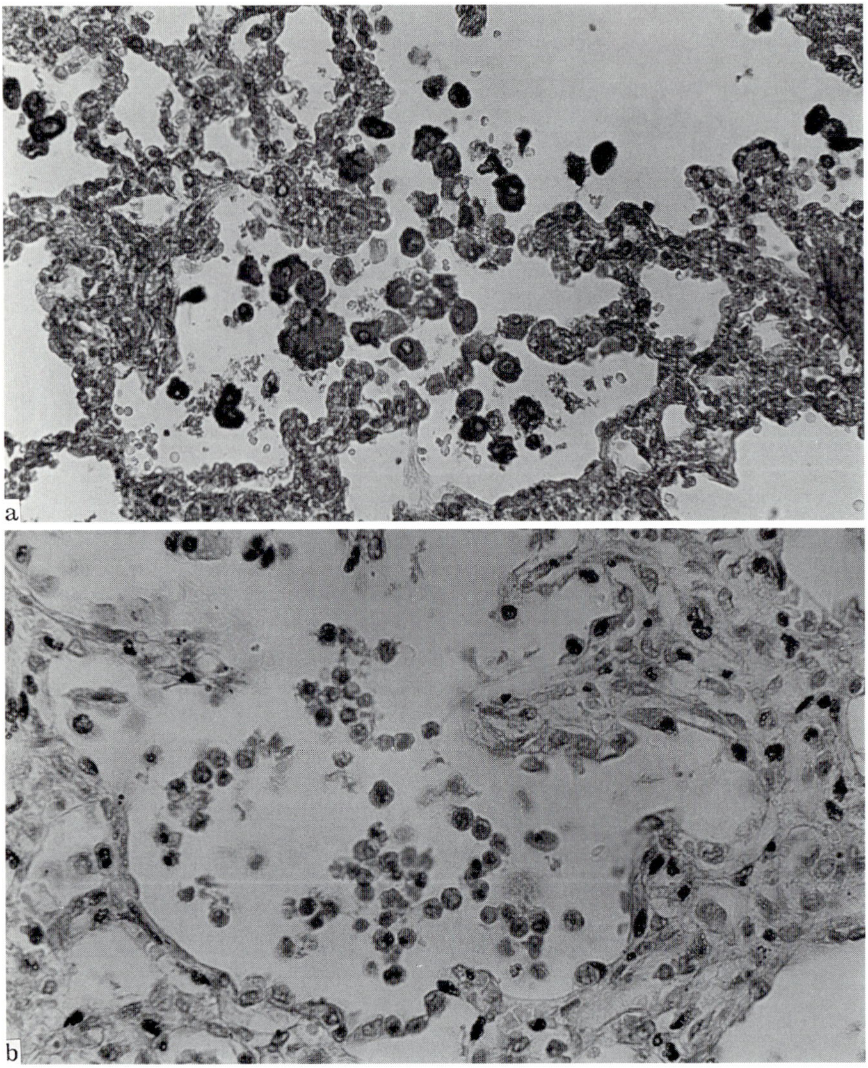

Fig. 3. a Lung section (low power) from same patient as in Fig. 2A stained with anti-nitrotyrosine antibody as in Fig. 2. **b** Same as **a** except that primary antibody was preincubated with 10 m*M* nitrotyrosine

to antigenic haptens created by protein nitration. Nitration can alter protein conformation by decreasing the hydrophobicity of tyrosine and/or by introducing a negative charge into a previously charge-neutral region of the protein. The ability of nitration of mimic phosphorylation is currently under investigation; however, the ability of nitration to inhibit tyrosine phosphorylation has been demonstrated (MARTIN et al. 1990).

Tetranitromethane has been used for many years by biochemists to nitrate tyrosines in isolated proteins to investigate their role in protein or enzyme function. Most enzymes do not depend on tyrosines for catalytic activity and many tyrosines can be nitrated without dramatically affecting activity. However, tetranitromethane inactivates the C1q binding capacity of human IgG (McCall and Easterbrook-Smith 1989), abolishes the inhibitory activity of α-1-proteinase inhibitor toward elastase (Moreno and Pryor 1992), inhibits the activity of cytochrome P450 (Janig et al. 1987, 1988), a-thrombin (Scully et al. 1992), and mitochondrial ATPase (Wu and Fisher 1982; Guerrieri et al. 1984) and inhibits the binding of human high density lipoprotein to liver plasma membranes (Darley-Usmar et al. 1992; Grahm et al. 1993). Even if the activity of a given protein were unaffected by nitration, it could alter protein conformation sufficiently to "tag" it for proteolysis or cause it to be recognized as foreign by the immune system.

The technique of chemically linking dinitrophenol (which resembles nitrotyrosine) to proteins to make them more antigenic has been used for many years. This fact, together with our own experience of making an antibody to nitrated KLH protein, suggests that proteins nitrated in vivo may be highly antigenic and serve as the initiating stimulus for an autoimmune response.

12 Nitric Oxide—The Molecular Chameleon

The availability and widespread use of potent inhibitors of nitric oxide synthase has firmly established the fundamental role of nitric oxide in a wide range of physiological processes. Yet, the ability of these inhibitors to reduce or prevent injury in conditions normally associated with increased nitric oxide production (e.g., sepsis, ischemia/reperfusion) indicates that nitric oxide also has significant toxicities associated with its production. The relative lack of direct toxicity in most systems suggests that some reaction product of nitric oxide is responsible. The ability of nitric oxide to react instantaneously with superoxide, coupled with the ability of the reaction product, peroxynitrite, to permanently modify a variety of biological molecules, provides a plausible toxicological explanation and is totally consistent with the ability of SOD to prolong the biological half-life of nitric oxide and to protect against injury in response to oxidative stress. Indeed, the footprint of peroxynitrite, nitrotyrosine, is revealed in pathologic tissues by immunohistological techniques and confirmed be chemical analysis.

A current debate centers around whether the predominant effect of nitric oxide is to cause injury or to prevent it. This is particularly germane given the clinical potential of inhaled nitric oxide as therapy for pulmonary edema and possibly ARDS. This question can only be approached by considering specific conditions in which the chemistry of nitric oxide may provide some clues.

Concentrations of 10–100 n*M* nitric oxide are produced in vivo for signal transduction; this obviously poses no real danger to humans since we produce it continuously for 70–80 years or longer. At these levels SOD would out-compete nitric oxide for superoxide by 340-fold (based on relative concentration times reaction rate), thereby limiting peroxynitrite formation. At higher nitric oxide concentrations such as seen in ischemic brain (2–4 μ*M*) and in the vicinity of macrophages (1–10 μ*M*) (Ischiropoulos et al. 1992a), the ability of SOD to out-compete nitric oxide for superoxide would be greatly diminished. However, tipping the balance in favor of nitric oxide toxicity almost certainly involves other mechanisms. For example, in studies involving ischemia/reperfusion, the ability of nitric oxide to dilate vessels and thereby increase colateral bloodflow may more than offset any toxicity related to nitric oxide overproduction per se. Within the vasculature, nitric oxide could reduce injury by inhibiting both neutrophil adhesion (Kubes et al. 1991) and platelet aggregation (Macdonald et al. 1988).

Other factors which may contribute to the observed toxicity/protection by nitric oxide relate to its site of production. The constitutive nitric oxide synthase in endothelial cells appears to be bound to the plasma membrane (Pollock et al. 1991). The hydrophobic character of nitric oxide would favor its partitioning into the lipid membrane theraby preventing its reaction with the charged, hydrophilic superoxide molecule. Nitric oxide within the lipid bilayer would be positioned to react with lipid peroxyl radicals (rate constant = $3 \times 10^9\ M^{-1}s^{-1}$) (Padmaja and Huie 1993). Formation of a stable organic nitrate ($R\text{-}NO_3$) could effectively terminate lipid peroxidation and thereby be protective.

We have recently investigated the reaction between nitric oxide and peroxy nitrite. The products of this reaction have not been characterized, but the reaction is quite rapid and appears to destroy nitric oxide while enhancing the nitration yield from peroxynitrite (Crow and Beckman, manuscript in preparation).

It has been suggested that since peroxynitrite decomposes to innocuous nitrate, the reaction of nitric oxide with superoxide to give peroxynitrite represents a detoxification mechanism. Two crucial points must be considered. (1) Peroxynitrite is a much stronger oxidant than is nitric oxide and is more damaging in every system which has been examined; thus, the idea of forming a potent oxidant to detoxify a weaker oxidant is unsound. (2) The products of peroxynitrite reactions are dependent on the targets present. Oxidative, hydroxylating, and nitrating reactions of peroxynitrite with biological molecules compete with the decomposition pathway (to give nitrate); the primary determinants of product distribution are the relative concentrations of the reactants and the relative rate constants for the respective reactions. In a simple system in which the total concentration of all target molecules is extremely low, the predominant product of peroxynitrite would be nitrate simply because there are no other possible reaction pathways. This is analogous to comparing the results of an explosion which occurred in a crowded room with the same explosion in a empty room. Clearly, the total target concentration would be high inside or on the surface of a cell, where nitric oxide is most likely to encounter superoxide and form peroxynitrite. Thus, it is imperative that toxicity arguments be based on realistic physiological milieu and known reactivities of the species involved.

References

Bates JN, Harrison DG, Myers RR, Minor RL (1991) EDRF: nitrosylated compound or authentic nitric oxide. Basic Res Cardiol 86 [Suppl 2]: 17–26

Baum RM (1984) Superoxide theory of oxygen toxicity is center of heated debate. Chem. Engin. News 9: 20–28

Beckman JS, Beckman TW, Chen J, Marshall PA, Freeman BA (1990) Apparent hydroxyl radical production by peroxynitrite: immplications for endothelial injury from nitric oxide and superoxide. Proc Natl Acad Sci USA 87: 1620–1624

Beckman JS, Ischiropoulos H, Zhu L, van der Woerd M, Smith C, Chen J, Harrison J, Martin JC, Tsai M (1992) Kinetics of superoxide dismutase- and iron-catalyzed nitration of phenolics by peroxynitrite. Arch Biochem Biophys 298: 438–445

Berrazue JR, Lopez-Jaramillo P, Moncada S (1990) Nitric oxide: from endogenous vasodilator to biologic mediator (Review in Spanish). Rev Esp Cardiol 43: 421–431

Bielski BHJ (1978) Re-evaluation of the spectral and kinetic properties of HO_2 and O_2 free radicals. Photochem Photobiol 28: 645–649

Biemond P, van Eijk HG, Swaak AJG, Koster JF (1984) Iron mobilization from ferritin by superoxide derived from phagocytosing polymorphonuclear leucocytes. Possible mechanism in inflammation diseases. J Clin Invest 73: 1576–1579

Cazevieille C, Muller A, Meynier F, Bonne C (1993) Superoxide and nitric oxide cooperation in hypoxia/reoxygenation-induced neuron injury. Free Radic Biol Med 14: 389–395

Chen FY, Lee TJ (1993) Role of nitric oxide in neurogenic vasodilation of porcine cerebral artery. Pharmacol Exp Ther 265: 339–345

Crow JP, Spruell C, Chen J, Gunn C, Ischiropoulos H, Tsai M, Smith CD, Radi R, Koenol WH, Beckman JS (1994) On the pH-dependent yield of hydroxyl radical products from peroxynitrite. Free radical Biol Med (in press)

Darley-Usmar VM, Hogg N, O'Leary VJ, Wilson MT, Moncada S (1992) The simultaneous generation of superoxide and nitric oxide can initiate lipid peroxidation in human low density lipoprotein. Free Radic Res Commun 17: 9–20

Dawson VL, Dawson TM, Bartley DA, Uhl GR, Snyder SH (1993) Mechanisms of nitric oxide-mediated netrotoxicity in primary brain cultures. J Neurosci 13: 2651–2661

Flaherty JT, Weisfeldt ML (1988) Reperfusion injury. Free Radic Biol Med 5: 409-419

Ford PC, Wink DA, Stanbury DM (1993) Autoxidation kinetics of aqueous nitric oxide (review). FEBS Lett 326: 1–3

Gardner PR, Fridovich I (1991a) superoxide sensitivity of the *Escherichia coli* aconitase. J Biol Chem 266: 19328–19333

Gardner PR, Fridovich I (1991b) Superoxide sensitivity of the *Escherichia coli* 6-phosphogluconate dehydratase. J Biol Chem 266: 1478–1483

Gardner PR, Fridovich I (1992) Inactivation-reactivation of aconitase in *Escherichia coli*. A sensitive measure of superoxide radical. J Biol Chem 267: 8757–8763

Garside P, Hutton AK, Severn A, Liew FY, Mowat AM (1992) Nitric oxide mediates intestinal pathology in graft-vs.-host disease. Eur J Immunol 22: 2141–2145

Graham A, Hogg N, Kalyanaraman B, O'Leary V, Darley-Usmar V, Moncada S (1993) Peroxynitrite modification of low-density lipoprotein leads to recognition by the macrophage scavenger receptor. FEBS Lett 330: 181–185

Green SJ, Mellouk S, Hoffman SL, Meltzer MS, Nacy CA (1990a) Cellular mechanisms of nonspecific immunity to intracellular infection: cytokine-induced synthesis of toxic nitrogen oxides from L-arginine by macrophages and hepatocytes. Immunol Lett 25: 15–19

Green SJ, Meltzer MS, Hibbs JB, Jr Nacy CA (1990b) Activated macrophages destory intracellular *Leishmania major* amastigotes by an L-arginine-dependent killing mechanism. J Immunol 144: 278–283

Guerrieri F, Yagi A, Yagi T, Papa S (1984) On the mechanism of H^+ translocation by mitochondrial H+ -ATPase. Studies with chemical modifier of tyrosine residues. J Bioenerg Biomembr 16: 251–262

Haber F, Weiss J (1934) The catalytic decomposition of hydrogen peroxide by iron salts. Proc R Soc 147: 332–351

Halfpenny E, Robinson PL (1952) The nitration and hydroxylation of aromatic compounds by pernitrous acid. J Chem Soc 939–946

Hoffman RA, Langrehr JM, Simmons RL (1992) The role of inducible nitric oxide synthetase during graft-versus-host disease. Transplant Proc 24: 2856

Hogg N, Darley-Usmar VM, Wilson MT, Moncada S (1992) Production of hydroxyl radicals from the simultaneous generation of superoxide and nitric oxide. Biochem J 281: 419–424

Huie RE, Padmaja S (1993) The reaction of NO with superoxide. Free Radic Res Commun 18: 195–199

Hutchinson PJ, Palmer RM, Moncada S (1987) Comparative pharmacology of EDRF and nitric oxide on vascular strips. Eur J Pharmacol 141: 445–451

Ignarro LJ, Buga GM, Wood KS, Byrns RE, Chaudhuri G (1987) Endothelium-derived relaxing factor produced and released from artery and vein is nitric oxide. Proc Natl Acad Sci USA 84: 9265–9269

Ignarro LJ, Gold ME, Buga GM, Byrns RE, Wood KS, Chaudhuri G, Frank G (1989) Basic polyamino acids rich in arginine, lysine, or ornithine cause both enhancement of and refractoriness to formation of endothelium-derived nitric oxide in pulmonary artery and vein. Circ Res 64: 315–329

Ischiropoulos H, Zhu L, Beckman JS (1992a) Peroxynitrite formation from macrophage-derived nitric oxide. Arch Bioch Biophys 298: 446–451

Ischiropoulos H, Zhu L, Chen J, Tsai M, Martin JC, Smith CD, Beckman JS (1992b) Peroxynitrite-mediated tyrosine nitration catalyzed by superoxide dismutase. Arch Biochem Biophys 298: 431–437

Jaeschke H, Schini VB, Farhood A (1992) Role of nitric oxide in the oxidant stress during ischemia/reperfusion injury of the liver. Life Sci 50: 1797–1804

Janig GR, Kraft R, Blanck J, Ristau O, Rabe H, Ruckpaul K (1987) Chemical modification of cytochrome P-250 LM4. Identification of functionally linked tyrosine residues. Biochim Biophys Acta 91: 512–523

Janig GR, Kraft R, Rabe H, Makower A, Ruckpaul K (1988) Comparative studies on the accessibility and functional importance of tyrosine residues in cytochrome P-450 isozymes. Biomed Biochim Acta 47: 565–579

Kelm M, Schrader J (1990) Control of coronary vascular tone by nitric oxide. Circ Res 66: 1561–1575

King PA, Anderson VE, Edwards JO, Gustafson G, Plumb RC, Suggs JW (1992) A stable solid that generates hydroxyl radical upon dissolution in aqueous solution: Reaction with proteins and nucleic acid. J Am Chem Soc 114: 5430–5432

Klug D, Rabani J, Fridovich I (1972) A direct demonstration of the catalytic action of superoxide dismutase through the issue of pulse radiolysis. J Biol Chem 247: 4839–4842

Koppenol WH, Moreno JJ, Pryor WA, Ischiropoulos H, Beckman JS (1992) Peroxynitrite, a cloaked oxidant formed by nitric oxide and superoxide. Chem Res Toxicol 5: 834–842

Kroncke KD, Kolb-Bachofen V, Berschick B, Burkart V, Kolb H (1991) Activated macrophages kill pancreatic syngeneic islet cells via arginine-dependent nitric oxide generation. Biochem Biophys Res Commun 175: 752–758

Kubes P, Suzuki M, Granger DN (1991) Nitric oxide: an endogenous modulator of leukocyte adhesion. Proc Nat Acad Sci USA 88: 4651–4655

Kwon NS, Stuehr DJ, Nathan CF (1991) Inhibition of tumor cell ribonucleotide reductase by macrophage-derived nitric oxide. J Exp Med 174: 761–767

Langrehr JM, Murase N, Markus PM, Cai X, Neuhaus P, Schraut W, Simmons RL, Hoffman RA (1992) Nitric oxide production in host-versus-graft and graft-versus-host reactions in the rat. J Clin Invest 90: 679–683

Lepoivre M, Fieschi F, Coves J, Thelander L, Fontecave M (1991) Inactivation of ribonucleotide reductase by nitric oxide. Biochem Biophys Res Commun 179: 442–448

Macdonald PS, Read MA, Dusting GJ (1988) Synergistic inhibition of platelet aggregation by endothelium-derived relaxing factor and prostacyclin. Thromb Res 49: 437–449

Martin BL, Wu D, Jakes S, Graves DJ (1990) Chemical influences on the specificity of tyrosine phosphorylation. J Biol Chem 265: 7108–7111

Masini E, Bianchi S, Mugai L, Gambassi F, Lupini M, Pistelli A, Mannaioni PF (1991) The effect of nitric oxide generators on ischemia reperfusion injury and histamine release in isolated perfused guinea-pig heart. Agents Actions 33: 53–56

Matsuda M, Fujiwara H, Kawamura A, Ishida M, Takemura G, Kida M, Uegaito T, Fujiwara Y, Fujiwara T, Kawai C (1991) Failure to reduce infarct size by intracoronary infusion of recombinant human superoxide dismutase at reperfusion in the procine heart—immunohistochemical and histological analysis. J Mol Cell Cardiol 23: 1287–1296

McCall MN, Easterbrook-Smith SB (1989) Comparison of the role of tyrosine residues in human IgG and rabbit IgG in binding of complement subcomponent C1q. Biochem J 257: 845–851

McCord JM, Fridovich I (1969) Superoxide dismutase and enzymic function for erythrocuperein (hemocuprein). J Biol Chem 244: 6049–6055

Moreno JJ, Pryor WA (1992) Inactivation of alpha 1-proteinase inhibitor by peroxynitrate. Chem Res Toxicol 5: 425–431

Mulligan MS, Hevel JM, Marletta MA, Ward PA (1991) Tissue injury caused by deposition of immune complexes is L-arginine dependent. Proc Natl Acad Sci USA 88: 6338–6342

Murphy ME, Sies H (1991) Reversible conversion of nitroxyl anion to nitric oxide by superoxide dismutase. Proc Natl Acad Sci USA 88: 10860–10864

Nguyen T, Brunson D, Crespi CL, Penman W, Wishnok JS, Tannebaum, SR (1992) DNA damage and mutation in human cells exposed to nitric oxide in vitro. Proc Nat Acad Sci USA 89: 3030–3034

Ohshima H, Friesen M, Brouet I, Bartsch H (1990) Nitrotyrosine as a new marker for endogenous nitrosation and nitration of proteins. Food Chem Toxicol 28: 647–652

Ohshima H, Brouet I, Friesen M, Bartsch H (1991) Nitrotyrosine as a new marker for endogenous nitrosation and nitration. IARC Scientific Publications, Lyon, 443–448

Omar BA, Gad NM, Jordan MC, Striplin SP, Russell WJ, Downey JM, McCord JM (1990) Cardioprotection by Cu,Zn-superoxide dismutase is lost at high doses in the reoxygenated heart. Free Radic Biol Med 9: 465–471

Padmaja S, Huie RE (1993) The reaction of nitric oxide with organic peroxyl radicals. Biochem Biophys Res Commun 195: 539–544

Parks DA, Bulkley GB, Granger DN, Hamilton SR, McCord JM (1982) Ischemic injury in the cat small intestine: role of superoxide radicals. Gastroenterology 82: 9–15

Pauling L (1979) The discovery of the superoxide radical. Trends Biochem Sci 4: N270–N271

Pollock JS, Forstermann U, Mitchell JA, Warner TD, Schmidt HHHW, Nakane M, Murad F (1991) Purification and characterization of particulate endothelium-derived relaxing factor synthase from cultured and native bovine aortic endothelial cells. Proc Natl Acad Sci USA 88: 10480–10484

Prutz WA, Monig H, Butlker J, Land EJ (1985) Reactions of nitrogen dioxide in aqueous model systems: oxidation of tyrosine units in peptides and proteins. Arch ;Biochem Biophys 243: 125–134

Radi R, Beckman JS, Bush KM, Freeman BA (1991a) Peroxynitrite-induced membrane lipid peroxidation: the cytotoxic potential of superoxide and nitric oxide. Arch Biochem Biophys 288: 481–487

Radi R, Beckman JS, Bush KM, Freeman BA (1991b) Peroxynitrite oxidation of sulfhydryls. The cytotoxic potential of superoxide and nitirc oxide. J Biol Chem 266: 4244–4250

Reed JW, Ho HH, Jolly WL (1974) Chemical synthesis with a quenched flow reactor. Hydroxytrihydroborate and peroxynitrite. JACS 96: 1248–1249

Rush JD, Koppenol WH (1989) Reactive intermediates formed by the interaction of hydrogen peroxide and ferrous complexes. In: Beaumont P, Deeble D, Parsons B, Rice-Evans C (eds) Free radicals, metal ions and biopolymers. Richelieu, London, pp 33–44

Saran M, Michel C, Bors W (1990) Reaction of NO with O_{2-}. implications for the action of endothelium-derived relaxing factor (EDRF). Free Radic Res Commun 10: 221–226

Sawyer DT, Valentine JS (1981) How super is superoxide? Accts Chem Res 14: 393–400

Scully MF, Kakkar VV, Goodwin CA (1992) Non-specific influence of chemical modification upon the properties of antithrombin III: modification of carboxyl groups. Thromb Res 67: 447–456

Shibuki K (1990) An electrochemical microprobe for detecting nitric oxide release in brain tissue. Neurosci Res 9: 69–76

Smith CD, Carson M, van der Woerd M, Chen J, Ischiropoulos H, Beckman JS (1992) Crystal structure of peroxynitrite-modified bovine Cu,Zn-superoxide dismutase. Arch Biochem Biophys 299: 350–355

Thomas EE, Morehouse LA, Aust SD (1985) Ferritin and superoxide-dependent lipid peroxidation. J Biol Chem 260: 3275–3280

Welsh N, Eizirik DL, Bendtzen K, Sandler S (1991) Interleukin-1 beta-induced nitric oxide production in isolated rat pancreatic islets requires gene transcription and may lead to inhibition of the Krebs cycle enzyme aconitase. Endocrinology 129: 3167–3173

Wink DA, Kasprzak KD, Maragos CM, Elespuru RK, Misra M, Dunams TM, Cebula TA, Koch WH, Andrews AW, Allen JS et al. (1991) DNA deaminating ability and genotoxicity of nitric oxide and its progenitors. Science 254: 1001–1003

Wu LN, Fisher RR (1982) Modification of bovine heart mitochondrial transhydrogenase with tetranitromethane. Biochim Biophys Acta 681: 388–396

Zweier JL, Rayburn BK, Flaherty JT, Weisfeldt ML. (1987) Recombinant superoxide dismutase reduces oxygen free radical concentrations in reperfused myocardium. J Clin Invest 80(6): 1728–1734

Regulation and Dysregulation of Constitutive Nitric Oxide Synthases Types I and III

H.H.H.W. SCHMIDT, H. HOFMANN, and P. OGILVIE

1 Introduction

Mammalian cells synthesize nitric oxide (NO) from L-arginine by NO synthases (NOS), members of the cytochrome P_{450} gene family. NO represents the best understood member of a novel class of redox-active transmembrane and intracellular signaling molecules that may also include CO and oxygen radicals. When formed and released in nanomolar concentrations, the main biological effect of NO is to trigger the guanosine cyclic 3',5'-monophosphate signalling cascade by activating the cGMP-generating enzyme soluble guanylyl cyclase (SCHMIDT et al. 1993).

2 Nitric Oxide Synthase Isozymes

In mammalian cells, NO is enzymatically formed from a terminal guanidino-nitrogen of L-arginine (HIBBS et al. 1987; PALMER et al. 1988; SCHMIDT et al. 1988 a,b) by a family of at least three distinct NOS isozymes (E.C. 1.14.23) (SCHMIDT et al. 1991). All described NOSs yield L-citrulline as a coproduct of this reaction.

Department of Clinical Biochemistry and Pathobiochemistry, Medical University Clinic, Versbacher Strasse 5, 97078 Würzburg, Germany

Table 1. Structural and Kinetic characteristics of nitric oxide synthase

Parameter binding site	Type I (brain)	Type II (macrophage)	Type III (endothelial)
Structure			
Native	Homodimer	Homodimer	?
Denatured, kDa	160	130	133
Calmodulin			
Binds	Reversible	Copurifies	Reversible
Essential	Yes	?	Yes
Regulation	Free Ca^{2+}	Constitutive	Free Ca^{2+}
L-Arginine			
K_m (μm)	2–7	3–32	3
V_{max}, (nmol/mg per min)	≥ 300	≤ 1600	15
Co/posttranslational modifications			
Phosphorylation	Yes	?	Yes
Myristoylation	No	No	Yes
Subcellular localization (100 000 × g)	Soluble	Soluble	Posttranslationally regulated

The original classification of NOS into NOS-I, II, and III (Table 1) was based on the physical and biochemical characteristics of the purified enzymes, i.e., subcellular location (soluble vs particulate fraction) and regulation by the free Ca^{2+} concentration (SCHMIDT et al. 1991), and has been confirmed by the recent molecular cloning and expression studies of three corresponding genes (BREDT et al. 1991; MARSDEN et al. 1992; XIE et al. 1992; NAKANE et al. 1993). Other current classifications group NOS isoforms according to their: (1) regulation of expression (constitutive and inducible NOS), (2) regulation by free Ca^{2+} (Ca^{2+}-dependent and Ca^{2+}-independent NOS; (NATHAN 1992)), or (3) primary source for purification (brain-, macrophage-, and endothelial-type NOS) (SNYDER and BREDT 1991; MARSDEN et al. 1992; NATHAN 1992; SESSA et al. 1992). However, NOS-I, for example, is not exlusively expressed in brain but also in the peripheral nervous system and in nonneuronal cells (SCHMIDT et al. 1992e; WILCOX et al. 1992), and its expression is not constitutive under all conditions (see below). Moreover, the terms "constitutive" or "Ca^{2+}-dependent" NOS do not differentiate between type I and type III, which derive from two distinct genes (see below).

3 Nitric Oxide Synthase Structure

Nitric oxide synthases are homodimers (NOS_2) (SCHMIDT et al. 1991) of subunits which range between 130 and 160 kDa (BREDT et al. 1991; JANSSENS et al. 1992; LOWENSTEIN et al. 1992; LYSONS et al. 1992; MARSDEN et al. 1992; SESSA et al. 1992; XIE et al. 1992; NAKANE et al. 1993). The COOH-terminal half of all NOS

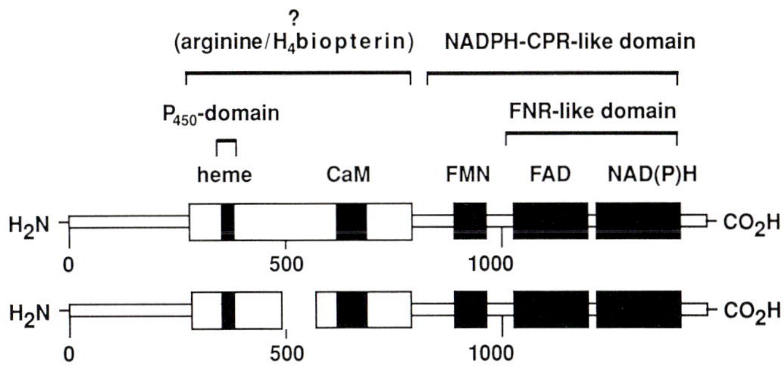

Fig. 1. Consensus sequences for binding sites and protein domains of nitric oxide synthases (NOSs). Catalytic sites (*filled boxes*) include the cytochrome P_{450}-like oxidase (heme binding site) and the P_{450} reductase (*CPR*)-like reductase domain (FMN, FAD, NADPH binding sites). Regulatory domains include the calmodulin binding site; several potential phosphorylation sites exist (not shown). The L-arginine and tetrahydrobiopterin (H_4biopterin) binding sites may locate to a conserved region in the NH_2-terminal half (*open boxes*). An in-frame deletion of this region and an NOS-I greatly reduced in activity and smaller by about 10 kDa has been described

contains bindings motifs for NADPH, FAD and FMN, identical to NADPH-cytochrome P_{450} reductase (CPR), one other flavin nucleotide reductase that contains both flavin nucleotides, FMN and FAD (Fig.1). In addition, the NOS heme moiety presents the typical CO difference spectrum of P_{450}-type enzymes. Due to its CPR-like domain NOS can, thus, be viewed as a mechanistically self-contained cytochrome P_{450}/CPR chimeric protein which combines both an oxidase and a reductase domain in a single polypeptide. All NOS bind calmodulin in a process which is either Ca^{2+}-dependent (NOS-I and III) or Ca^{2+}-independent (NOS-II) (CHO et al. 1992; SCHMIDT et al. 1992b). In the case of NOS-II, calmodulin and enzyme copurify, suggesting a native heterotetrameric structure (NOS_2/calmodulin$_2$) of this isoform. Calmodulin is the key regulator of electron flow within NOS (HEINZEL et al. 1992; KLATT et al. 1992). Additional binding sites within the NOS polypeptide must be postulated for tetrahydrobiopterin (BH_4) and L-arginine, but have not been identified. BH_4 is utilized as a cofactor by several other amino acid hydroxylases. However, no other BH_4 utilizing enzyme but NOS binds BH_4 with an affinity that permits copurification of enzyme and cofactor (up to 2 mol per dimer). This fact and some of the effects of BH_4 on NO catalysis (see below) suggest a novel BH_4 binding site in NOS. Moreover, dimerization of NOS is a prerequisite for L-arginine turnover and BH_4-dependent (STUEHR et al. 1991). The NH_2-terminal region C of NOS (Fig.2) is very likely to contain the domains for BH_4 binding, L-arginine binding and possibly subunit dimerization. Hydrodynamic characterization of NOS suggests not only a homodimeric structure, but revealed also a frictional ratio indicative of an elongated tertiary structure hindering intramolecular contact between the oxidase domain (NH_2-terminal) and the oxidase domain (COOH-terminal). A yet hypothetical head-to-tail arrangement of NOS at the C region would allow for intermolecular contact and electron flow between the oxidase domain of one monomer and the reductase domain of another monomer (Fig. 2).

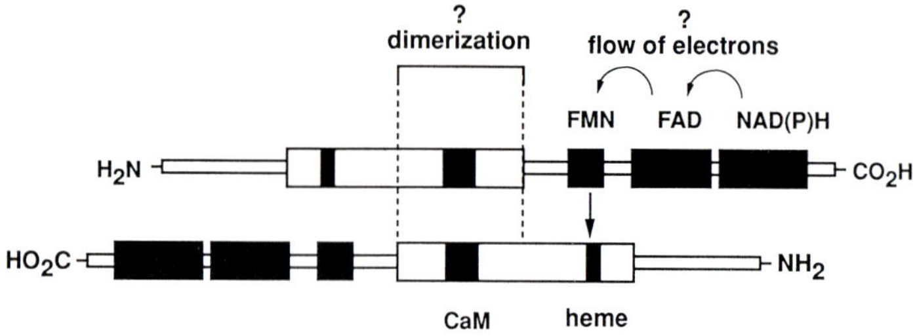

Fig. 2. Hypothetical model for nitric oxide synthase (NOS) dimerization and electron flow. Native NOS monomers have the hydrodynamic characteristics of an elongated polypeptide. Dimerization in a hypothetical head-to-tail orientation would allow for electron flow between the oxidase and reductase domains. This tertiary structure of NOS may be regulated by calmodulin and region C (*white box*), which is conserved between all NOSs

Interestingly a mutant of NOS-I mRNA bearing an in-frame deletion in region C has recently been reported in different neuroblastoma cell lines. A NOS-I truncated by about 10 kDa, but not originating from the same in-frame deletion, is expressed in embryonal rat brain and appears to have a greatly reduced specific activity.

4 Mechanisms of Nitric Oxide Synthesis

The oxidation of a terminal guanidino nitrogen of L-arginine to NO is broadly accepted as a working hypothesis to explain the mechanism of NOS catalysis (HEVEL et al. 1991; MAYER et al. 1992; SCHMIDT et al. 1992b; STUEHR and GRIFFITH 1992), whereas considerable debate concerns additional reactions NO may undergo postsynthesis, leading to storage or transport forms of this messenger molecule. Three, possibly four, cofactors (heme, FMN, FAD and H_4 biopterin) and two cosubstrates (O_2 and NADPH) participate in enzymatic NO formation. Electron flow is regulated by calmodulin binding. Dissociation of traces of OHArg from NOS were identified by HPLC and GC/MS. EPR data also suggest the formation of a N^{ω}-hydroxy-L-arginine (OHArg) cation radical (PRÓNAI et al. 1991). Since OHArg is utilized as an alternative substrate, it is believed to represent an intermediate of the NOS reaction. The stoichiometry of the electron transfer can fully be explained by the use of 1.5 mol NADPH per mol NO formed. It is not clear whether BH_4 is required as an additional cofactor or NADPH-dependently recycled. In all other known BH_4-utilizing enzymes, the quinoid form of dihydrobiopterin (q-BH_2) is the product of normal catalysis and is regenerated to BH_4 by a NADH-dependent q-BH_2 reductase (dihydropteridine reductase). Recyclable q-BH_2 was not detected with catalytically active NOS. However, NOS copurifies with BH_4 indicative of an inherent, NADPH-dependent biopterin reductase activity (SCHMIDT et al. 1992b). Alternatively, it was suggested that BH_4 is not metabolized during normal catalysis but acts as an allosteric activator or stabilizer of the active center (GIOVANELLI et al. 1991). The BH_4 binding site of NOS also appears to be different from that in other known BH_4-metabolizing enzymes as indicated by: (1) a uniquely small K_a value (CRAINE et al. 1972; FIRGAIRA et al. 1981), (2) high substrate specificity (KWON et al. 1989; TAYEH and MARLETTA 1989; GIOVANELLI et al. 1991) of NOS for BH_4, and (3) the resistance of NOS to inhibition even by high concentrations of methotrexate, which inhibits dihydrofolate and dihydropteridine reductases (GIOVANELLI et al. 1991).

The oxidase and reductase activities of NOS extend to other substrates (Table 2). Many of the reductions catalyzed by NOS are likely to be related to its CPR-like domain and generate an electron pressure towards the oxidase domain. As expected from the close similarity of NOS to CPR, NOS reduces cytochrome c in vitro. The mechanism of cytochrome c reduction by NOS is unclear. NOS forms superoxide anions (see above), as does CPR, but also interacts directly with cytochrome c by a highly affinity protein–protein interaction, whereas CPR does not bind to cytochrome c (KLATT et al. 1992). Furthermore, NOS also reduces cytochrome P_{450}, as indicated by the fact that, in vitro, NOS supports the hydroxylation of *N*-ethylmorphine by cytochrome P_{450}. This suggests that NOS may participate in similar electron transfer processes in vivo (KLATT et al. 1992) and NOSs may be viewed rather as isoforms of CPR, with L-arginine turnover being merely a side effect of their catalytic activities. The close similarity between NOS and the CPR/cytochrome P_{450} system is underscored by the fact that cytochrome P_{450} catalyzes the oxidation of OHArg to NO.

While reduction of cytochromes by NOS is strictly calmodulin-dependent, several artificial electron acceptors, e.g., nitroblue tetrazolium (NBT) and dichlorphenon-indophenol (DCPIP), are calmodulin-*independently* reduced by NOS. This suggests that NOS has: (a) a single reductase domain that is partially dependent on calmodulin binding, depending on the substrate; (b) two reductase domains, one calmodulin-dependent (oxygen, cytochromes) and one calmodulin-independent (biopterin, NBT, DCPIP). The electron acceptor NBT is converted to blue diformazan by NOS and many other dehydrogenases and reductases. Only in the case of NOS, however, is this so-called NADPH-diaphorase activity of NOS remarkably resistant to commonly used protein fixatives, thus enabling the convenient histochemical localization of NOS in paraformaldehyde-treated tissue sections (HOPE et al. 1991; SCHMIDT et al. 1992c). The utilization of NBT by NOS makes NBT also a potent noncompetitive NOS inhibitor, presumably by competing with molecular oxygen for reducing equivalents (SCHMIDT et al. 1992a,b) or by reacting with intermediate superoxide radicals. The aforementioned putative biopterin reductase activity of NOS may be closely related to the NADPH-diaphorase domain of NOS (HOPE et al. 1991).

Table 2. Reactions catalyzed by nitric oxide synthase

Type	Substrate	Cosubstrate	Regulation	Product	Reaction name
Oxidase	O_2	NADPH	Calmodulin	Superoxide, peroxide	NADPH Oxidase
Oxidoreductase	L-Arginine	NADPH, O_2, (BH_4)	Calmodulin	Nitric oxide	Nitric Oxide Synthase
Reductase C	Cyt P_{450}	NADPH, O_2	Calmodulin	Reduced Cyt P-450	CPR
Reductase C	Cyt c	NADPH (direct)	Calmodulin	Reduced Cyt c	Cyt c reductase
Reductase C, I	NBT	NADPH	Partially by calmodulin	Formazan	NADPH diaphorase
Reductase I	DCPIP	NADPH	Independent of calmodulin		

Cyt, Cytochrome; CPR, Cyt P_{450} reductase; BH_4, tetrahydrobiopterin; NBT, nitroblue tetrazolium; DCPIP, dichlorphenol-indophenol.

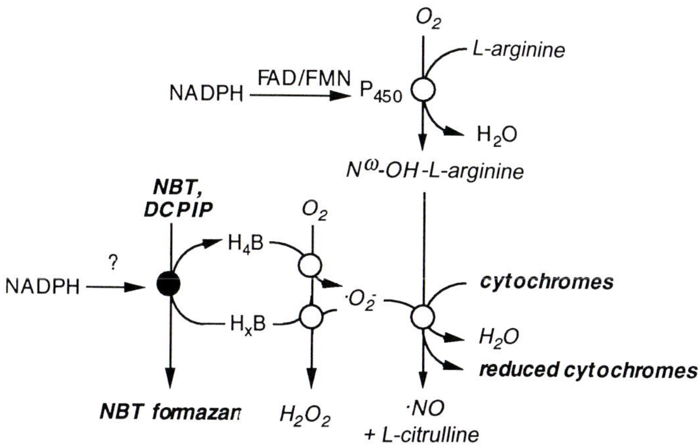

Fig. 3. Proposed reaction mechanisms for nitric oxide (*NO*) synthase. Molecular oxygen is incorporated into both NO and L-citrulline. The flavin cofactors FAD and FMN and the cytochrome P_{450} domain of NOS mediate part of or the entire electron transfer from NADPH to molecular oxygen (oxidase domain). Tetrahydrobiopterin (H_4B) may be involved in this or regulate NOS allosterically. L-Arginine is hydroxylated to the intermediate N^{ω}-hydroxy-L-arg *OHArg*. In the calcium-regulated NOS-I, two reductase activities can be distinguished by their mode of regulation: one that transfers electrons independently of calmodulin (*filled circle*) to artificial acceptors nitro blue tetrazolium (*NBT*) and dichlorphenol-indophenol (*DCPIP*) and may represent a putative biopterin reductase domain of NOS, and another one that transfers electrons in a calmodulin-dependent manner (*open circles*) to cytochromes either directly or via superoxide anions

5 Regulation of Nitric Oxide Synthase Expression and Posttranslational Modification

Types I and III NOS are constitutively expressed and in a very cell-specific manner (Table 3), whereas NOS-II is not constitutive but inducible in virtually every cell after appropriate immunological activation with different cytokines or endotoxins

Table 3. Cells and tissues expressing nitric oxide synthase or nitric oxide synthase immunoreactivity

Type I	Type II	Type III
Neurons	Macrophages	Endothelium
Macula densa	Cardiomyocytes	
Bronchial epithelium	Vascular smooth muscle	
Gastric epithelium	Hepatocytes (Calmodulin-dependent)	
HIT-T15 pancreatic B cells	Intestinal epithelium	
Mast cells	Megakaryocytes	
Human skeletal muscle	Keratinocytes	
Endothelium		
Testis		
Photoreceptors		

Not identified: Polymorphonuclear-leukocytes, platelets, bone, retina, olfactory epithelium.

Table 4. Physiological and pathophysiological regulation of nitric oxide synthase expression and activity

Level	Type I	Type II	Type III
Expression			
Constitutive	Yes	No	Yes
Increased by	Inflammation	IFN-γ, LPS, IL-1,2	Shear stress
Decreased by	IFN-γ, LPS	glucocorticoids, TGF-β, IL-4, 10	TNF-α
Mutations	In-frame deletion	?	?
Posttranslational modifications			
Dimerization	Yes	Yes (cofactor-dependent)	?
Phosphorylation	Yes	?	Yes
Myristoylation	No	No	Yes
Calmodulin binding	Reversible (Ca^{2+})	Irreversible	Reversible (Ca^{2+})
Activity			
Elevated free Ca^{2+}	Essential	No effect	Essential
Endogenous MeArg	Inhibit	Inhibit	Inhibit
Autoinhibition (NO)	No/Yes	No	?
Intracellular pH	?	?	Modulates

IFN-γ, interferon-γ, LPS, lipopolysaccharide; IL, interleukin, TGF-β, transforming growth factor-β; TNF-α, tumor necrosis factor; MeArg, methyl-L-arginine; NO, nitric oxide.

(Table 4). Once expressed, NOS-II binds calmodulin irreversibly, independently of Ca^{2+} and stays maximally active irrespective of the free Ca^{2+} concentration. The induction of NOS-II by cytokines is transcriptionally based. In control cells, mRNA for NOS-II is not detectable.

Modulation of basal NOS-I and III expression may also take place. For example, the same cytokines which induce NOS-II in RAW macrophages have the reverse effect on basal expression of the constitutive type I NOS in this cell line. The mechanisms which lead to reduced expression and the pathophysiological significance of this bidirectional, immunological regulation of NOS expression are unclear (SCHMIDT et al. 1992d; FÖRSTERMANN et al. 1993). Moreover, under physiological conditions, NOS-I expression is also not constant. We were recently able to show that during brain development enzymatically competent NOS-I is only expressed with the onset of synaptogenesis. Here, expression is dramatically increased within 1–2 days and returns to a lower steady state level at the end of synaptogenesis. The peak of expression is also different in different brain regions in pre- but not postnatal brain tissue. No NOS activity but a 150 kDa NOS-I immunoreactive protein was detected. Since the antibody that was used is monospecific for type I NOS, it is likely that this protein represents an embryonal NOS-I deletion mutant (Ogilvie et al., unpublished).

Further regulation of NOS expression takes place at the co- and posttranslational level. As mentioned before, the subcellular distribution of the various NOS isoforms is different. Even the same isoform may be distributed at varying ratios between soluble and particulate cell fractions in different species (SCHMIDT et al. 1992d). However, none of the described NOSs contain an amino acid sequence suggestive of a transmembrane domain (BREDT et al. 1991; LYONS et al. 1992). Therefore, co- or posttranslational modifications such as myristoylation (POLLOCK et al. 1992) and phosphorylation (MICHEL et al. 1993) are very likely to regulate the subcellular localization of NOS. These modifications may also represent an additional means to regulate enzyme activity (SCHMIDT et al. 1992a). The membrane-bound NOS-III is not only myristoylated but has also the lowest specific activity of all known NOSs (POLLOCK et al. 1991). At least for the types I and II isoforms there is considerable evidence that dimerization or an essential conformational change, as detected by gel permeation chromatography, depends on saturating concentrations of BH_4 and possibly also of L-arginine and heme. NOS monomers appear to be inactive.

6 Regulation of Nitric Oxide Synthase Activity

Once expressed, binding of the Ca^{2+}-binding protein calmodulin or phosphorylation is an established mechanism by which enzyme activity of different isoforms of NOS is regulated (Table 1). All NOSs bind calmodulin and have conserved consensus sequences for calmodulin binding. In the case of NOS-I and III, interaction with calmodulin depends on elevated intracellular free Ca^{2+} concentrations ($[Ca^{2+}]_i$). At resting $[Ca^{2+}]$ (< 100 n*M*), these NOSs are inactive. They bind calmodulin and become fully active at increased $[Ca^{2+}]_i$ (≥ 500 n*M*). Calmodulin antagonists, e.g., calmidazolium and trifluoperazine, block calmodulin binding and inhibit Ca^{2+}-induced NO formation. Both compounds more potently inhibit NOS-I than NOS-III, which probably reflects differences in the respective calmodulin binding domains. In various systems, transmembrane Ca^{2+} flux is initiated by the binding of a receptor agonist to its membrane receptor. Thus, receptor occupation and increased $[Ca^{2+}]_i$ can be linked to increased NO formation in cells.

Two forms of NOS-II have been isolated, one from liver (EVANS et al. 1992) and one from a macrophage cell line (HEVEL et al. 1991; STUEHR et al. 1991). At < 100 n*M* free Ca^{2+}, the liver NOS-II is calmodulin-free and inactive. Similar to NOS-I and III, activity increases upon reconstitution of liver NOS-II with calmodulin, but, unlike NOS-I and III, this activation is Ca^{2+}-independent. Macrophage NOS-II has calmodulin constitutively bound and is constitutively active. Three other proteins bind calmodulin also constitutively, i.e., in an apparently Ca^{2+}-independent manner: phosphorylase kinase, a cyclic nucleotide phosphodiesterase, and *Bordetella pertussis* adenylyl cyclase (CHO et al. 1992). Thus, calmodulin binding represents a common activation principle for all NOSs, whereas their dependency on the free intracellular Ca^{2+} concentration distinguishes them.

The predicted amino acid sequence of NOS-I contains consensus sites for phosphorylation by cAMP-dependent protein kinases (BREDT et al. 1991). Forskolin-induced increases in intracellular cAMP levels (SCHMIDT et al.1992d) and cAMP-dependent protein kinases (BRÜNE and LAPETINA 1991) do not regulate NOS, whereas protein kinase C, Ca^{2+}/calmodulin-dependent kinase (NAKANE et al. 1991) and the phosphatase inhibitor okadaic acid (SCHMIDT et al. 1992d) do. In vitro, Ca^{2+}/calmodulin-dependent protein kinase II phosphorylates NOS on both serine and threonine (NAKANE et al. 1991). Phosphorylation is Ca^{2+}- and calmodulin-dependent and results in a marked decrease of NOS activity. Thus, frequent increases in $[Ca^{2+}]_i$ and phosphorylation of NOS by Ca^{2+}/calmodulin-dependent kinase II may represent a negative feedback regulation of NOS activity. Since Ca^{2+}/calmodulin kinase II becomes Ca^{2+}-independent upon autophosphorylation, it is conceivable that phosphorylation and inactivation of NOS proceed even after $[Ca^{2+}]_i$ has returned to basal levels. However, these in vitro observations need to be confirmed at the in vivo level.

Recent work has also focused on autoinhibition of NOS activity. Namely, several laboratories have identified circulating endogenous N^{ω}-methyl-L-arginine (MeArg) and Me_2Arg, which accumulate during renal failure and will systematically inhibit NO biosynthesis, possibly explaining some of the immunological dysfunctions observed under this condition. Whether under other conditions tissue levels are high enough to affect NOSs remains to be established. Moreover, several laboratories have suggested that NO acts back on NOS in a negative autofeedback loop. At least for NOS-I, we were unable to observe effects on citrulline formation either by heme, a NO scavenger, or superoxide dismutase, which should increase the apparent half-life of NO and potentiate auto-inhibition (Hofmann and Schmidt, unpublished). Thus, modulation of NO biosynthesis may occur via endogenous arginine derivatives but possibly not through NO-mediated autoinhibition.

References

Bredt DS, Hwang PM, Glatt CE, Lowenstein C, Reed RR, Snyder SH (1991) Cloned and expressed nitric oxide synthase structurally resembles cytochrome P-450 reductase. Nature 351: 714–718

Brüne B, Lapetina EG, (1991) Phosphorylation of nitric oxide synthase by protein kinase. Biochem Biophys Res Commun 181: 921–926

Cho HJ, Xie Q-W, Calaycay J, Mumford RA, Swiderek KM, Lee TD, Nathan C (1992) Calmodulin is a subunit of nitric oxide synthase from macrophages. J Exp Med 176: 599–604

Craine JE, Hall ES, Kaufman S (1972) The isolation and characterization of dihydropteridine reductase from sheep liver. J Biol Chem 247: 6082–6091

Evans T, Carpenter A, Cohen J (1992) Purification of a distinct isoform of endotoxin-induced nitric oxide synthase from rat liver. Proc Natl Acad Sci USA 89: 5361–5365

Firgaira FA, Cotton RGH, Danks DM (1981) Isolation and characterization of dihydropteridine reductase from human liver. Biochem J 197: 31–43

Förstermann U, Kuk J, Nakane M, Pollock J (1993) Tumor necrosis factor (TNF-α) downregulates the expression of endothelial nitric oxide synthase. Naunyn Schmiedebergs Arch Pharmacol 347 [Suppl]: R64

Giovanelli J, Campos KL, Kaufman S (1991) Tetrahydrobiopterin, a cofactor for rat cerebellar nitric oxide

synthase, does not function as a reactant in the oxygenation of arginine. Proc Natl Acad Sci USA 88: 7091–7095
Heinzel B, John M, Klatt P, Böhme E, Mayer B (1992) Ca^{2+}/calmodulin-dependent formation of hydrogen peroxide by brain nitric oxide synthase. Biochem J 281: 627–630
Hevel JM, White KA, Marletta MA (1991) Purification of the inducible murine macrophage nitric oxide synthase. Identification as a flavoprotein. J Biol Chem 266: 22789–22791
Hibbs JB Jr, Taintor RR, Vavrin Z (1987) Macrophage cytotoxicity: role for L-arginine deiminase and imino nitrogen oxidation to nitrite. Science 235: 473
Hope BT, Michael GJ, Knigge KM, Vincent SR (1991) Neuronal NADPH diaphorase is a nitric oxide synthase. Proc Natl Acad Sci USA 88: 2811–2814
Janssens SP, Shimouchi A, Quertermous T, Bloch DB, Bloch KD (1992) Cloning and expression of a cDNA encoding human endothelium-derived relaxing factor/nitric oxide synthase. J Biol Chem 267: 14519–14522
Klatt P, Heinzel B, John M, Kastner M, Böhme E, Mayer B (1992) Ca^{2+}/calmodulin-dependent cytochrome c reductase activity of brain nitric oxide synthase. J Biol Chem 267: 11374–11378
Kwon NS, Nathan CF, Stuehr, DJ (1989) Reduced biopterin as a cofactor in the generation of nitrogen oxides by murine macrophages. J Biol Chem 264: 20496–20501
Lowenstein C, Glatt CS, Bredt DS, Snyder SH (1992) Cloned and expressed macrophage nitric oxide synthase contrasts with the brain enzyme. Proc Natl Acad Sci USA 89: 6711–6715
Lyons CR, Orloff GJ, Cunningham JM (1992) molecular cloning and functional expression of an inducible nitric oxide synthase from a murine macrophage cell line. J Biol Chem 267: 6370–6374
Marsden PA, Schappert KT, Chen HS, Flowers M, Sundell CL, Wilcox JN, Lamas S, Michel T (1992) Molecular cloning and characterization of human endothelial nitric oxide synthase. FEBS Lett 307: 287–293
Mayer B, Heinzel B, Klatt P, John M, Schmidt K, Böhme E (1992) Nitric oxide synthase-catalyzed activation of oxygen and reduction of cytochromes: reaction mechanisms and possible physiological implications. J Cardiovasc Pharmacol 20 [Suppl 12]: S54–S56
Michel T, Li GK, Busconi L (1993) Phosphorylation and subcellular translocation of endothelial nitric oxide synthase. Proc Natl Acad Sci USA 90: 6252–6256
Nakane M, Mitchell J, Förstermann U, Murad F (1991) Phosphorylation by calcium calmodulin-dependent protein kinase II and protein kinase C modulates the activity of nitric oxide synthase. Biochem Biophys Res Commun 180: 1396–1402
Nakane M, Schmidt HHHW, Pollock JS, Förstermann U, Murad F (1993) Cloned human brain nitric oxide synthase is highly expressed in skeletal muscle. FEBS Lett 316: 175–180
Nathan C (1992) Nitric oxides as a secretory product of mammalian cells. FASEB J 6: 3051–3064
Palmer RMJ, Ashton DS, Moncada S (1988) Vascular endothelial cells synthesize nitric oxide from L-arginine. Nature 333: 664–666
Pollock JS, Förstermann U, Mitchell JA, Warner TD, Schmidt HHHW, Nakane M, Murad F (1991) Purification and characterization of particulate endothelium-derived relaxing factor synthase from cultured and native bovine aortic endothelial cells. Proc Natl Acad Sci USA 88: 10480–10484
Pollock J, Klinghofer V, Förstermann U, Murad F (1992) Endothelial nitric oxide synthase is myristilated. FEBS Lett 309: 402–404
Prónai L, Ichimori K, Nozaki H, Nakazawa H, Okino H, Carmichael AJ, Arroyo CM (1991) Investigation of the existence and biological role of L-arginine/nitric oxide pathway in human platelets by spin trapping/EPR studies. Eur J Biochem 202: 923–930
Schmidt HHHW, Klein MM, Niroomand F, Böhme E (1988a) Is arginine a physiological precursor of endothelium-derived nitric oxide? Eur J Pharmacol 148: 293–295
Schmidt HHHW, Nau H, Wittfoht W, Gerlach J, Prescher K-E, Klein MM, Niroomand F, Böhme E (1988b) Arginine is a physiological precursor of endothelium-derived nitric oxide. Eur J Pharmacol 154: 213–216
Schmidt HHHW, Pollock JS, Nakane M, Gorsky LD, Förstermann U, Murad F (1991) Purification of a soluble isoform of guanylyl cyclase-activating-factor synthase. Proc Natl Acad Sci USA 88: 365–369
Schmidt HHHW, Pollock JSP, Nakane M, Förstermann U, Murad F (1992a) Ca^{2+}/calmodulin-regulated nitric oxide synthases. Cell Calcium 13: 427–434
Schmidt HHHW, Smith RM, Nakane M, Murad F (1992b) Ca^{2+}/Calmodulin-dependent NO synthase type I: a biopteroflavoprotein with Ca^{2+}/calmodulin-independent diaphorase and reductase activities. Biochemistry 31: 3243–3249
Schmidt HHHW, Warner TD, Ishii K, Sheng H, Murad F (1992c) Insulin-secretion from pancreatic B-cells caused by L-arginine-derived nitrogen oxides. Science 255: 721–723

Schmidt HHHW, Warner TD, Nakane M, Förstermann U, Murad F (1992d) Regulation and subcellular location of nitrogen oxide synthases in RAW264·7 macrophages. Mol Pharmacol 41: 615–624

Schmidt HHHW, Gagne GD, Nakane M ,Pollock JS, Miller MF, Murad F (1992e) Mapping of NO synthase in the rat suggests colocalization with NADPH diaphorase but not soluble guanyl cyclase and novel paraneural functions for nitrinergic signal transduction. J Histochem Cytochem 40: 1439–1456

Schmidt HHHW, Lohmann SL, Walter U (1993) The nitric oxide and cGMP signal-transduction pathway. Biochim Biophys Acta 1178: 153

Sessa WC, Harrison JK, Barber CM, Zeng D, Durieux ME, D'Angelo DD, Lynch KR, Peach MJ (1992) Molecular cloning and expression of a cDNA encoding endothelial cell nitric oxide synthase. J Biol Chem 267: 15274–15276

Snyder SH, Bredt DS, (1991) Nitric oxide as a neuronal messenger. Trends Pharmacol Sci 12: 125-128

Stuehr D, Griffith O (1992) Mammalian nitric oxide synthases. Adv Enzymol Related Areas Mol Biol 65: 287–346

Stuehr DJ, Cho HJ, Kwon NS, Nathan CF (1991) Purification and characterization of the cytokine-induced macrophage nitric oxide synthase: an FAD- and FMN- containing flavoprotein. Proc Natl Acad Sci USA 88: 7773–7777

Tayeh MA, Marletta MA (1989) Macrophage oxidation of L-arginine to nitric oxide, nitrite, and nitrate: tetrahydrobiopterine is required as a cofactor. J Biol Chem 264: 19654–19658

Wilcox CS, Welch WJ, Murad F, Gross SS, Taylor G, Levi R, Schmidt HHHW (1992) Nitric oxide synthase in macula densa regulates glomerular capillary pressure. Proc Natl Acad Sci USA 89: 11993–11997

Xie Q-W, Cho HJ, Calaycay J, Mumford RA, Swiderek KM, Lee TD, Ding A, Troso T, Nathan C (1992) Cloning and characterization of inducible nitric oxide synthase from mouse macrophages. Science 256: 225–228S

Subject Index

2002 C45546

Current Topics in Microbiology and Immunology

Volumes published since 1989 (and still available)

Vol. 152: **Bosma, Melvin J.; Phillips, Robert A.; Schuler, Walter (Ed.):** The Scid Mouse. Characterization and Potential Uses. EMBO Workshop held at the Basel Institute for Immunology, Basel, Switzerland, February 20–22, 1989. 1989. 72 figs. XII, 263 pp. ISBN 3-540-51512-7

Vol. 153: **Lambris, John D. (Ed.):** The Third Component of Complement. Chemistry and Biology. 1989. 38 figs. X, 251 pp. ISBN 3-540-51513-5

Vol. 154: **McDougall, James K. (Ed.):** Cytomegaloviruses. 1990. 58 figs. IX, 286 pp. ISBN 3-540-51514-3

Vol. 155: **Kaufmann, Stefan H. E. (Ed.):** T-Cell Paradigms in Parasitic and Bacterial Infections. 1990. 24 figs. IX, 162 pp. ISBN 3-540-51515-1

Vol. 156: **Dyrberg, Thomas (Ed.):** The Role of Viruses and the Immune System in Diabetes Mellitus. 1990. 15 figs. XI, 142 pp. ISBN 3-540-51918-1

Vol. 157: **Swanstrom, Roland; Vogt, Peter K. (Ed.):** Retroviruses. Strategies of Replication. 1990. 40 figs. XII, 260 pp. ISBN 3-540-51895-9

Vol. 158: **Muzyczka, Nicholas (Ed.):** Viral Expression Vectors. 1992. 20 figs. IX, 176 pp. ISBN 3-540-52431-2

Vol. 159: **Gray, David; Sprent, Jonathan (Ed.):** Immunological Memory. 1990. 38 figs. XII, 156 pp. ISBN 3-540-51921-1

Vol. 160: **Oldstone, Michael B. A.; Koprowski, Hilary (Eds.):** Retrovirus Infections of the Nervous System. 1990. 16 figs. XII, 176 pp. ISBN 3-540-51939-4

Vol. 161: **Racaniello, Vincent R. (Ed.):** Picornaviruses. 1990. 12 figs. X, 194 pp. ISBN 3-540-52429-0

Vol. 162: **Roy, Polly; Gorman, Barry M. (Eds.):** Bluetongue Viruses. 1990. 37 figs. X, 200 pp. ISBN 3-540-51922-X

Vol. 163: **Turner, Peter C.; Moyer, Richard W. (Eds.):** Poxviruses. 1990. 23 figs. X, 210 pp. ISBN 3-540-52430-4

Vol. 164: **Bækkeskov, Steinnun; Hansen, Bruno (Eds.):** Human Diabetes. 1990. 9 figs. X, 198 pp. ISBN 3-540-52652-8

Vol. 165: **Bothwell, Mark (Ed.):** Neuronal Growth Factors. 1991. 14 figs. IX, 173 pp. ISBN 3-540-52654-4

Vol. 166: **Potter, Michael; Melchers, Fritz (Eds.):** Mechanisms in B-Cell Neoplasia. 1990. 143 figs. XIX, 380 pp. ISBN 3-540-52886-5

Vol. 167: **Kaufmann, Stefan H. E. (Ed.):** Heat Shock Proteins and Immune Response. 1991. 18 figs. IX, 214 pp. ISBN 3-540-52857-1

Vol. 168: **Mason, William S.; Seeger, Christoph (Eds.):** Hepadnaviruses. Molecular Biology and Pathogenesis. 1991. 21 figs. X, 206 pp. ISBN 3-540-53060-6

Vol. 169: **Kolakofsky, Daniel (Ed.):** Bunyaviridae. 1991. 34 figs. X, 256 pp. ISBN 3-540-53061-4

Vol. 170: **Compans, Richard W. (Ed.):** Protein Traffic in Eukaryotic Cells. Selected Reviews. 1991. 14 figs. X, 186 pp. ISBN 3-540-53631-0

Vol. 171: **Kung, Hsing-Jien; Vogt, Peter K. (Eds.):** Retroviral Insertion and Oncogene Activation. 1991. 18 figs. X, 179 pp. ISBN 3-540-53857-7

Vol. 172: **Chesebro, Bruce W. (Ed.):** Transmissible Spongiform Encephalopathies. 1991. 48 figs. X, 288 pp. ISBN 3-540-53883-6

Vol. 173: **Pfeffer, Klaus; Heeg, Klaus; Wagner, Hermann; Riethmüller, Gert (Eds.):** Function and Specificity of Á / ‰ TCells. 1991. 41 figs. XII, 296 pp. ISBN 3-540-53781-3

Vol. 174: **Fleischer, Bernhard; Sjögren, Hans Olov (Eds.):** Superantigens. 1991. 13 figs. IX, 137 pp. ISBN 3-540-54205-1

Vol. 175: **Aktories, Klaus (Ed.):** ADP-Ribosylating Toxins. 1992. 23 figs. IX, 148 pp. ISBN 3-540-54598-0

Vol. 176: **Holland, John J. (Ed.):** Genetic Diversity of RNA Viruses. 1992. 34 figs. IX, 226 pp. ISBN 3-540-54652-9

Vol. 177: **Müller-Sieburg, Christa; Torok-Storb, Beverly; Visser, Jan; Storb, Rainer (Eds.):** Hematopoietic Stem Cells. 1992. 18 figs. XIII, 143 pp. ISBN 3-540-54531-X

Vol. 178: **Parker, Charles J. (Ed.):** Membrane Defenses Against Attack by Complement and Perforins. 1992. 26 figs. VIII, 188 pp. ISBN 3-540-54653-7

Vol. 179: **Rouse, Barry T. (Ed.):** Herpes Simplex Virus. 1992. 9 figs. X, 180 pp. ISBN 3-540-55066-6

Vol. 180: **Sansonetti, P. J. (Ed.):** Pathogenesis of Shigellosis. 1992. 15 figs. X, 143 pp. ISBN 3-540-55058-5

Vol. 181: **Russell, Stephen W.; Gordon, Siamon (Eds.):** Macrophage Biology and Activation. 1992. 42 figs. IX, 299 pp. ISBN 3-540-55293-6

Vol. 182: **Potter, Michael; Melchers, Fritz (Eds.):** Mechanisms in B-Cell Neoplasia. 1992. 188 figs. XX, 499 pp. ISBN 3-540-55658-3

Vol. 183: **Dimmock, Nigel J.:** Neutralization of Animal Viruses. 1993. 10 figs. VII, 149 pp. ISBN 3-540-56030-0

Vol. 184: **Dunon, Dominique; Mackay, Charles R.; Imhof, Beat A. (Eds.):** Adhesion in Leukocyte Homing and Differentiation. 1993. 37 figs. IX, 260 pp. ISBN 3-540-56756-9

Vol. 185: **Ramig, Robert F. (Ed.):** Rotaviruses. 1994. 37 figs. X, 380 pp. ISBN 3-540-56761-5

Vol. 186: **zur Hausen, Harald (Ed.):** Human Pathogenic Papillomaviruses. 1994. 37 figs. XIII, 274 pp. ISBN 3-540-57193-0

Vol. 187: **Rupprecht, Charles E.; Dietzschold, Bernhard; Koprowski, Hilary (Eds.):** Lyssaviruses. 1994. 50 figs. IX, 352 pp. ISBN 3-540-57194-9

Vol. 188: **Letvin, Norman L.; Desrosiers, Ronald C. (Eds.):** Simian Immunodeficiency Virus. 1994. 37 figs. X, 240 pp. ISBN 3-540-57274-0

Vol. 189: **Oldstone, Michael B. A. (Ed.):** Cytotoxic T-Lymphocytes in Human Viral and Malaria Infections. 1994. 37 figs. IX, 210 pp. ISBN 3-540-57259-7

Vol. 190: **Koprowski, Hilary; Lipkin, W. Ian (Eds.):** Borna Disease. 1995. 33 figs. IX, 134 pp. ISBN 3-540-57388-7

Vol. 191: **ter Meulen, Volker; Billeter, Martin A. (Eds.):** Measles Virus. 1995. 23 figs. IX, 196 pp. ISBN 3-540-57389-5

Vol. 192: **Dangl, Jeffrey L. (Ed.):** Bacterial Pathogenesis of Plants and Animals. 1994. 41 figs. IX, 343 pp. ISBN 3-540-57391-7

Vol. 193: **Chen, Irvin S. Y.; Koprowski, Hilary; Srinivasan, Alagarsamy; Vogt, Peter K. (Eds.):** Transacting Functions of Human Retroviruses. 1995. 49 figs. Approx. IX, 240 pp. ISBN 3-540-57901-X

Vol. 194: **Potter, Michael; Melchers, Fritz (Eds.):** Mechanisms in B-cell Neoplasia. 1995. Approx. 152 figs. XXV, 458 pp. ISBN 3-540-58447-1

Vol. 195: **Montecucco, Cesare (Ed.):** Clostridial Neurotoxins. 1995. Approx. 28 figs. Approx. 260 pp. ISBN 3-540-58452-8

Printing: Saladruck, Berlin
Binding: Buchbinderei Lüderitz & Bauer, Berlin